The Knowledge Landscapes of Cyberspace

The Knowledge Landscapes of Cyberspace

DAVID HAKKEN

Routledge
New York & London

Published in 2003 by
Routledge
29 West 35th Street
New York, NY 10001
www.routledge-ny.com

Published in Great Britain by
Routledge
11 New Fetter Lane
London EC4P 4EE
www.routledge.co.uk

Routledge is an imprint of the Taylor & Francis Group.
Printed in the United States of America on acid free paper.

Library of Congress Cataloging-in-Publication Data.
Hakken, David.
The knowledge landscapes of cyberspace / David Hakken.
p. cm.
Includes bibliographical references and index.
ISBN 0–415–94508–9 (hbk. : alk. paper) — ISBN 0–415–94509–7 (pbk. : alk. paper)
1. Knowledge, Theory of. 2. Artificaul intelligence. 2. Science—Philosophy. I. Title.
Q175.32.K45H35 2001
121—dc21 2003006068

Contents

Section II. Ethnological Perspectives
Three Sources of Concepts Prerequisite to Answering the Cyberspace Knowledge Question

Section III. Ethnographic Perspectives
The Dynamics of Knowledge in Cyberspace

Section IV. Conclusions
The Analytics of Knowledge in Cyberspace

Acknowledgments

This book advocates changing the ways we approach knowledge. It argues that we focus less on knowledge as a thing and more on the social processes through which something becomes acknowledged as known. One such process is the custom of acknowledging those who, over a sustained period, have stimulated and provoked an author to make her case.

I begin, then, with the State University of New York Institute of Technology's (SUNYIT) Policy Center's Knowledge Networking group, including Barbara Andrews, John Backman, Burt Danovitz, Steve Darman, Phil Endress and Ronnie Tichenor. During a crucial period when the ideas I present here were congealing, this group helped me to create the Knowledge Networking Oneida County project and use it to provoke greater awareness of how social services in the Upper Mohawk Valley get imagined. This is my favorite and I suspect the best way to learn. I also acknowledge Ken Abramzyk and other staff in the Oneida County Department of Mental Health, as well as the volunteer members of its Community Services Board, for engaging our initiatives and ideas.

My first opportunity to think diligently about the knowledge question in cyberspace was during the writing of *Cyborgs@Cyberspace?,* also published by Routledge (1999a). The acknowledgments to this book, and my previous Westview book with Barbara Andrews, *Computing Myths, Class Realities* (1993), include the names of numerous people in Sheffield, England, and in the Nordic countries who have helped me think about technology and social change. My knowledge preoccupations were fed by a number of colleagues in the Community of Anthropologists of Science, Technology, and Computing, especially Joe Dumit, Gary Downey, Diana Forsythe (deceased, sadly), Karl Hakken, David Hess, Pat Sachs and Lucy Suchman. Many of them are also involved in the Society for Social Studies of Science, as are Andrew Feenberg, Penny Harvey, and Langdon Winner. Morgan Tamplin and I have shared an activist as well as intellectual interest in Computer Professionals for Social Responsibility. Carolyn Korsmeyer has graciously tutored me in the way philosophers approach knowledge. Carl Martin Allwood, Gro Bjerknes, Tone Bratteteig, Kari Thoresen, Gudrun Dahl, Ulf Hanerz, Ole Hanseth, Randi Markussen, and Francis Sjersted are among the Scandinavians whose willingness to keep networking knowledge are particularly appreciated, along with their colleagues at the Department of Psychology, Göteborg University; Department of Informatics and Centre for Technology, Inovation and Culture, University of Oslo; The Norwegian Computing Centre, the Department of Information and Media Science, Aarhus; and the Department of Social Anthropology, Stockholm University. I particularly want to acknowledge the continuing strong engagement of Andreas Silow and his colleagues at the Stockholm organizational communications firm Tessla (also, sadly, closed). Dorle Drackle of the University of Bremen and Pille Runell of Turku University in Estonia as well as Randi Markussen encouraged me to think more systematically about the communication dimensions of computing.

Rachelle Hollander ably directs the program in Societal Dimensions of Engineering, Science, and Technology at the U.S. National Science Foundation, which provided a planning grant for a project on Knowledge Networking Knowledge Networking (KN^2). The members of Rachelle's panel with whom I served, and of the two NSF Programs on Knowledge/Distributed Intelligence with whom I also struggled, provided me with an invaluable experience of the palpable reality of scientific knowledge creation networking. The KN^2 project's various life forms drew sustenance from several of those individuals already mentioned. I particularly wish to thank those who attended the Hartford workshop to think up new ways to network knowledge, including Karl Hakken, Tom Hassler, and Jaime Roberts.

Through various summer grants, release time, teaching opportunities, and sabbaticals, the SUNY Institute of Technology has supported work on knowledge and this book. (Grant help has also come from the joint SUNY/United University Professions professional development programs.) For twenty-four years, SUNYIT supported my efforts to discover creative ways to make anthropology accessible to technology students and gave me the honor of being nominated for a distinguished professorship. Unfortunately, this institution's administrators were recently unable to prevent the sociologists from purging anthropological perspectives from the department with which I had been affiliated for all that time. This experience brought home to me in a very palpable way the real costs of trying to think hard about knowledge.

In contrast, my profession and its organization, the American Anthropological Association, have generously supported my efforts to make technology safe for anthropology, including naming me as the first recipient of the Textor Prize for Anticipatory Anthropology. I want to express particular appreciation to Jon Anderson, Jeanette Blomberg, and the other members of the Advisory Group on Electronic Communication of the AAA. More recently, Nuri Soeharto and Eric Thompson have kindled my efforts to become more of a "real" anthropologist by thinking more systematically about technology beyond the North Atlantic.

Individual works of scholarship that I find important to acknowledge are cited in the usual way in the text that follows. I owe a particular debt of gratitude to my family: Nathan Hakken-Andrews for sharing both his Apple expertise and his passion for legal rights in cyberspace; Luke Andrews-Hakken for sharing his experiences and interests, as well as skills as a network administrator in publishing; Karl Hakken for being my closest and most stimulating colleague for the last several years; and Barbara Andrews for being, in more ways that I would ever have imagined, my comrade.

Section 1
Introductory Perspectives

Chapter 1

Introduction: The Knowledge Question in Cyberspace

What the Knowledge Question in Cyberspace Is, and Why It Needs to be Answered

How is knowledge affected by computing? When people talk of "knowledge societies," they presume significant technology-caused knowledge change. Are the knowledge practices common to computer-saturated environments indicative of the profound kinds of change that would justify such labels?

The Knowledge Landscapes of Cyberspace approaches these questions ethnographically, using field research on actual knowledge processes mediated by new, self-controlled information machines as its primary reference points. The results of such research are used to address specific issues—whether knowledge's quantity, character and/or social functions are changing fundamentally, and if these changes are due to AITs. These issues are the chief aspects of the knowledge question in cyberspace, and the book reports the results of some efforts to answer it empirically.

Published in 1999, my previous Rutledge book, *Cyborgs@Cyberspace?: An Ethnographer Looks to the Future*, explored the "computer revolution" hypothesis, the general idea that we are entering a new type of technology-driven social formation. That book justified ethnographic research on the practices surrounding use of automated information technology as an important way to evaluate this hypothesis, and it outlined a broad program of such studies. (*Cyborgs@* . . . was a central part of my receiving the American Anthropological Association's first Textor Prize for Excellence in Anticipatory Anthropology.)

The Knowledge Landscapes . . . stakes a deeper claim to a narrower section of the large territory surveyed in *Cyborgs@Cyberspace?* Changing knowledge certainly *may* be a key part of making cyberspace real. That this might be so is suggested by, for example, the ease with which in the late 1990s the phrase "knowledge revolution" displaced "computer," "information," or even "Internet revolution."

Thus, the knowledge question has some general salience, but why should we strive to answer it empirically—that is, take on the difficult tasks of figuring out how and how much knowledge is actually changing, and how much of the change can be ascribed to the new information technologies? One reason is that while there is widespread belief in knowledge change in cyberspace, its actual existence, as I demonstrate below, has not been documented. As an anthropologist interested in contemporary cultures, I am intrigued whenever there is evidence of belief in a significant process whose existence is open to question. I want to know where such beliefs come from, the nonexperiential grounds that are used to justify them, and the conditions under which they *would really* be justified.

A leap in knowledge's growth rate, its new types (e.g., "machine intelligence"), and a broad expansion of access to its forms are trumpeted most loudly by advocates for rapid and deep integration of automated information technologies (AITs)[1] into the reproduction of current social formations. Fostering change in knowledge has arguably now become the prime justification for continuing to deploy AITs, especially into economies. Recent economic turmoil (e.g., the dot-com meltdown and extended general economic contraction) is itself a justification for problematizing "knowledge society" rhetoric. The high social value associated with *knowledge* makes the term particularly vulnerable to the hyping so characteristic of current technotalk. Further, the Change Knowledge via AIT Now project may also conceal less lugubrious developments. Is the globalizing economy so deeply dependent upon an "IT revolution" that disasters like the creation of dot-com-type bubbles of speculation can no longer be avoided?

Thus, a second reason for seeking answers to the cyberspace knowledge question is that it may be dangerous when the emperor has no clothes. Presumptions about AITed knowledge change are now embodied in important economic policies and social programs. The idea of a revolution in knowledge is at the core of influential programs for changing the central institutions of formal education, for example. If the premises on which these policies are based are false, the institutions based on them will fail, and the policies will need revision in the future.

In a democracy, few could object to the idea that knowledge be more developed, that more use be made of it, that it be "liberated." But what if the doubtlessly greater amount of information in a social formation highly mediated by AITs means more confusion instead of more knowing? Even if there's more knowledge, what if access to it grows even more profoundly unequal? What if new forms of knowledge make it even harder to act wisely? The belief that knowledge changes radically in computered environments has captured the contemporary cultural imaginary. Arguably, inappropriate performances of knowledge revolution rhetoric so often dominate talk about what is good in the human project that they unnecessarily constrain our sense of the possible.

Broader concerns like these are a third reason for seeing the knowledge question in cyberspace as a pressing one for contemporary social thought. In sum, we need to be able to answer it well if we are to cope effectively with the sociocultural implications of the general adoption of new technologies.

Some Problems in Influential Statements of the "Knowledge Society" Idea

Any good anthropology, especially one of the contemporary, pays close attention to popular discourses, practices, policies, and imaginaries. Professor David Schneider told us anthropology graduate students at the University of Chicago that there were really only two basic rules of good ethnography:

1. Listen carefully to "the natives"—your informants, the ones engaged directly in the cultural process in which you are interested; and
2. Remember that the natives might be wrong.

When it comes to AITs and changing knowledge, multiple "natives" are saying a great deal worth listening to carefully, but with Schneiderian skepticism! In this section, I explore some influential performances of the presumption of contemporary, technology-connected knowledge change in cyberspace, in order to communicate more concretely why better answers to the knowledge question in cyberspace are needed.

"Knowledge Revolution" Illustrated: The Geek Ascension

Consider Jon Katz's ideas about "those who constitute the 'high priesthood' of the 'Knowledge Society.'" In *Geeks: How Two Lost Boys Rode the Internet out of Idaho* (2000), Katz presumes to be their herald. The following definition indicates both the high social stakes he sees in the current moment and their connection to knowledge:

> Geek: A member of the new cultural elite, a pop-culture-loving, techno-centered Community of Social Discontents. Most geeks rose above a suffocatingly unimaginative educational system, where they were surrounded by obnoxious social values and hostile peers, to build the freest and most inventive culture on the planet: The Internet and the World Wide Web. Now running the systems that run the world. . . . In this era, the Geek Ascension, ["geek" is] a positive, even envied term. (2000, xi)

For Katz, there are no limits to the Geek Ascension, this emerging "freest and most inventive" new way of life (the one I label "cyberspace"). There is no end to the forms of social life it will colonize utterly:

> Computers and the Net would transform everything; nobody and no institution would remain untouched—not scientists, academics, artists, politicians, journalists, homemakers, doctors, lawyers, or school kids. Computing was no longer the sole province of nerds and engineers but also the new locus of creative people—poets, painters, novelists, critics. These were . . . the geeks." (2000, xxxiv)

As we all become Geeks—or, if not Geeks ourselves, then subjects of the Geek Ascension—old institutions are doomed: "One thing you can be sure of . . . the media . . . were done, over. Newspapers . . . Network TV . . . slick magazines . . . None of them had anything to say to the young, to the future" (2000, xxxv). For geeks, both politics and journalism represent "an insane and useless system that has little bearing on their lives . . . [A geek sees] himself as a citizen of the Net, a separate nation with rules, boundaries, and traditions of its own" (2000, 86).

For Katz, the power base of the new geek ruling class is their knowledge. Their social transformation isn't "about money, but ideas—how they could be manipulated, reproduced, stored, represented, combined, and connected" (2000, xxxiv). Central to the Geek Ascension are profound, even physical transformations in knowledge activities. Of his two once-lost boys, Katz comments,

> [T]he ability of these two to instantly pick up the digital rhythms of their lives and make them focal points of their existence was striking and sometimes dis-

> turbing. Their entrenched inwardness, a profoundly interior consciousness, seemed at times woven into their personalities . . .
>
> Geeks who spend hours of their lives online, programming and gaming, practice what hackers call 'deep magic': They enter a zone unique to the online world, where they are transfixed by the digital environment around them . . .
>
> Inhabitants of a new world with a new culture, geeks often find that the old symbols don't work for them—pep rallies, assemblies, etc. In fact, scholars . . . are beginning to explore the ways in which interactivity and representational writing and thinking are changing the very neural systems of the young." (2000, 85, 122, and 174)

Katz says about Jesse, a "once-lost" geek whom Katz helps enter the University of Chicago, that:

> The net had influenced him enormously, forming his libertarian mish-mash of political views, his suspicion of institutions and corporations, his eclectic interests in a thousand different subjects, his curious notions of intellectual and material property, and his penchant for developing ideas and theories in isolation. Like most geeks, Jesse had turned his computer into a cultural entity. He collected and listened to music with it, used search engines to do everything from ordering an evening meal to resolving a technical dispute . . . turning [his]computers into sophisticated cultural centers . . .
>
> [T]he Net was his life and he wasn't about to let go if it. (2000, 102 and 92)

Along with personalities, basic social institutions like work are also being transformed. Nor is the pace of social change likely to slacken, with the imminent arrival of other transformative technologies based on AITs, for example, "nanotechnology," which will "permit an exponential growth of productive and personal wealth" (2000, 129).

Perhaps most relevant to the argument of *The Knowledge Landscapes of Cyberspace* are the changes Katz claims to see in cyberspace knowledge networking. For example, there are new social analogues of individual-level intellectual functioning:

> On the net, ideas don't need to be pushed. They find their own audience and stand or fall of their own weight . . .
>
> [In their interventions after the Columbine High School events, t]he geek kids had done something remarkable . . . and quite unprecedented: They had

> used their medium, the Net, to join the national debate and had altered the direction of a major story—one that was, thanks to knee-jerk journalists and opportunistic politicians, threatening to become a witch-hunt." (2000, 163 and 159)

This particular manifestation of growing geek social power, Katz believes, could be traced to a new social movement,

> Free software[,] . . . a global geek political movement committed to building good software and distributing it for free, so that a handful of corporations won't dominate the Web the way they do the rest of the world . . .
>
> The open source movement has grown to more than eight million users, perhaps the geekiest social movement on the Net, since its members and passionate adherents are programmers and designers." (2000, 94 and 151)

A Knowledge Road to Cyberspace

Katz's claims truly monumental social implications for his Geek Ascension. Were these implications fully realized, "knowledge society" rhetoric would be quite justified.

I share Katz's enthusiasm for goals like more flexible work and "open" computing. However, there is little in his ethnography that justifies such social conclusions. It is indicative that the victory he claims for his heroic intervention is very conventional—helping one of the Geeks get into an elite college. Similarly, while on the one hand the Geek Ascension is believed to follow from new forms of knowledge networking (e.g., ideas on the Net finding on their own the appropriate, intrinsic level of worth), the knowledge he features, like nanotechnology, is produced in the "traditional" way.

In the coexistence of apocalyptic rhetoric and conventional payoffs, of alleged new forms based on unacknowledged old ones—and in the large gap between pairs like these—Katz's perspective is typical of knowledge society talk. Yet talk like this has a broad hold on popular consciousness. Characterizations of computing-dependent social formations as "knowledge societies" are now common, as in, for example, the tendency of "information society" to be displaced by "knowledge society" figures (e.g., Mankin, Cohen, and Bikson 1996). "Knowledge society" is now a central figure in an ever-broadening lexical field. How has this come about?

One possibility is that its central contentions are valid. Like the claims of many others, Katz's notion of a Geek Ascension depends on the idea that a new social formation is driven by a new knowledge regime. The new knowledge regime is understood to follow inevitably from the use of automated information technology (AIT), from integrating programmable machines for data and information storage and generation—computers—into various social processes. Must computing have such vast consequences?

In her recent (1999) *Epistemic Cultures: How the Sciences Make Knowledge*, Karin Knorr-Certina performs knowledge tropes similar to Katz's, but in a more academic register. She begins, like me, by noting that "There is a widespread consensus today that contemporary Western societies are in one sense or another ruled by knowledge and expertise. A proliferation of concepts such as that of . . . a 'knowledge society' . . . embodies this understanding . . . " Of course, all societies are ruled by knowledge and expertise, "in one sense or another." For her claim to have substance and not amount to some ethnocentric presumption of the transcendent superiority of Western knowledge, Knorr-Certina must demonstrate there is something profoundly emergent in the social function of knowledge. She goes on to claim that:

> The most consistent and powerful line of reasoning . . . is that . . . knowledge . . . has become a productive force replacing capital, labor, and natural resources as the central value- and wealth creating factor . . . [I]t is knowledge that makes our society . . . "post-capitalist," fundamentally changing the nature of production systems . . . (1999, 5)

Thus the sociologist claims that it is an economic phenomenon that transforms society. One might expect Knorr-Certina to go on to offer evidence supporting this political economic account for the transformative impact of the changed character of knowledge, but she merely repeats it, in a rather tautological definitional exercise:

> A knowledge society is . . . permeated with knowledge cultures, the whole set of structures and mechanisms that serve knowledge and unfold with its articulation . . . knowledge as practiced—within structures, processes, and the environments that make up specific epistemic settings . . . . (1999, 6)

Again, to an anthropologist, all social formations are "knowledge societies" in this sense: each culture has its own conceptions of what it means for something to be known, and these conceptions have various institutional manifestations, which we could refer to as "epistemic cultures." Knorr-Certina, however, would have us apply this term to only one particular set of knowledge practices, those characteristic of science; her book title is *Epistemic Cultures: How the Sciences Make Knowledge*. These practices, in addition to being for her at the root of the contemporary transformation of social type, become identified with society: "[W]hat we call 'society' will to a significant degree be constituted by such settings . . . the type of society that runs on knowledge and expertise . . . Epistemic cultures . . . appear to be a structural feature of knowledge societies . . . ." (1999, 7 and 8).

While Knorr-Certina's ethnographic examples of "epistemic cultures"—high energy physics and molecular biology—are rather "cutting edge" in science terms, there is nothing about her argument to limit it to present forms of technoscience alone. Whereas Katz places at least rhetorical emphasis on *new ways* of knowing as the source of the transformation to a "knowledge society," Knorr-Certina's notion of the cause of social transformation rests on ways of knowing that have been around for a long time, although she does not specify when technoscience become structural. Her terminology does privilege knowledge explicitly, while Katz's does it only implicitly.

The purpose of these examples is to illustrate the diversity underlying "knowledge society" rhetoric, but they also highlight an important commonality among those who perform it: Rather than carefully articulating their view of the proper way to conceptualize the knowledge revolution and then going on to make their case for it, most performers merely jump on a generally conceded "knowledge society" bandwagon. By the late 1990s, even the U.S. National Science Foundation joined the enthusiasm, funding an initiative on Knowledge/Distributed Intelligence that was then the largest multidisciplinary initiative in its history. K/DI was framed as a response to cutting edge developments internal to computing and information science (NSF 1998), but it was also explicitly presented as a response to transformed societal concerns, like those surrounding "intellectual property" (see, e.g., Coombe 1998). Conspicuous by its absence in the request for proposals, however, was any attention to whether these technologies were in actuality socially transformative.

Performances of "knowledge society" also regularly frame "intellectual capital" as a significant, often the key, organizational resource and thus, as in Knorr-Certina's comments, the primary economic resource. Knowledge, as "the chief factor of production" (also Davidson, Handler, and Harris 1994), apparently obtains this status because of being (newly?) crucial for the for-profit corporations (*Information Strategy* 1998) that are taken for granted as the key element of collectivizing the social surplus in employment-based market social formations.

Anyone even moderately skeptical, however, does not get far into the rhetoric currently surrounding knowledge without encountering substantial divergence. Consider the implications of the new knowledge regime for the role of different groups in social reproduction. For Davidson, Handler, and Harris (1994), "knowledge workers," because of the potential power they possess due to their centrality to production, constitute a new Marxian revolutionary class, one likely to use its strategic control of knowledge actively to transform society.

The Organization for Economic Cooperation and Development (OECD), both an action and a research group for the primary industrial nations, takes a very different view. In its 1994 *Jobs Study*, knowledge production is also presented as the key factor, but in this case as the link in the technology/job creation chain. For the OECD, the knowledge leading to growth is produced by "Research & Development." It is hardly influenced by workers, who are rather a medium for applying knowledge generated elsewhere. Thus, all workers need is "basic education," which "should be practical and universal; the education-to-workplace transition should be streamlined through cooperation with industry; and workers should be given incentives to constantly hone their skills" (U.S. Chamber of Commerce 1996). While the *Jobs Study* itself makes passing reference to how education should aim to achieve "other fundamental social and cultural objectives . . . as well as . . . extending and upgrading workers' competencies . . . " (1994, 47), only initial education, school-to-work transition, and work-related skills are discussed further.

This position contrasts with that of Davidson, Handler, and Harris. For them, new worker power comes from the scarcity of the skills developed only after necessary and expensive technoscientific training. Thus, education must necessarily push far beyond mere basics. If education *were* to become deeper and longer, a new social revolution might ensure, but if not, it won't.

The Logics of the "Knowledge Society" Hypothesis

If such divergent implications—both more education, and less—can be drawn from commitment to a "knowledge society" position, one would expect sustained interest regarding its various articulations before its scientific merit is presumed. Instead, one repeatedly encounters conundrums like this—over whether workers are the transformative *actors* in the knowledge revolution or are only *acted upon* by it, a socially inert mass serving as a medium for the agency of other social entities—whenever the mechanisms beneath the surface of "knowledge society" talk are examined more closely. The loud buzz about knowledge and technology obscures significant silences as well. These also are in part responsible for "knowledge society" talkers' difficulties with articulating clear mechanisms for their transformation, ones that demonstrate *actual* rather than merely name potential impacts.

This situation justifies a certain skepticism toward the whole idea of a "knowledge society." Indeed, if the knowledge-changing computer revolution is a myth, there is reason to be concerned about the ability of contemporary social formations to reproduce themselves. As *Cyborgs@* . . . argued, if the changes in knowledge and other aspects of social life aren't really all that important, the whole argument that cyberspace is the emerging new social formation collapses.

Taking the cyberspace knowledge question seriously means making explicit, breaking down, and examining carefully the admittedly broadly held social presumptions underlying narratives like Katz's, Knorr-Certina's, Davidson's, and the OECD's. Whether the knowledge rubber really meets the social dynamics road turns on the details of how one conceptualizes the dynamics of the "knowledge society." Yet while Marxists like Davidson and capitalists like the OECD staffers differ over the details of how the change in knowledge changes social functioning, both take the significance of the change as given. In celebrating a presumed knowledge revolution rather than examining whether it actually exists, they follow the pattern of other cyberites, the compputropians and computopians, the Luddites and the cyber-enthusiasts, analyzed at length in *Cyborgs@* . . . . While the regularity and force of their narratives legitimately command the attention of those who would understand contemporary society, following Schneider, a recognition that high frequency and volume do not necessarily indicate accuracy should also be kept in mind.

From what points of view does it make sense to think of knowledge as "growing?" Even if greatly "grown," is the new quantity of knowledge so vast as automatically to entail broad qualitative social change? Are the allegedly new forms of knowledge really all that new? Even if they are, does the character of their "newness" compel social transformation? In the dynamic process known as social formation reproduction, existing social modalities often take on new or changed social functions. Are the social functions of the new knowledge different enough to be referred to legitimately as epochal? Even if they are, what is it precisely about the new functions that is so broadly transformative? How can we be sure that change in knowledge is the cause rather than the effect, or if both social and knowledge change aren't due to some third social modality?

Objectives of the Book

It is the general aim of *The Knowledge Landscapes . . .* to explicate what to expect from knowledges in the future (and thus from knowledge *of* the future). I intend to show that by breaking down "knowledge revolution" talk into a set of discrete, intelligible claims, and then comparing these with what is actually happening in the world, or what is reliably predictable, such claims can be evaluated empirically.

The book's three key objectives follow from this agenda. The first is to explain why, if coming to terms with the cyberspace knowledge question is so important, it is so seldom done well. Stated succinctly, my answer is, first, because the question is seldom asked—that cyberspace is a knowledge society is simply assumed. Second, when the question is asked, it is not answered well because of basic flaws in the analytic tools used in its framing and in the answers to it. These flaws include inadequate understanding of what it would mean to answer the question well, as well as subsequent difficulty relating current ideas about knowledge to what is actually going on in the world. Identifying, analyzing, and illustrating the actual consequences of these flaws—for example, that of the rise and fall of "knowledge management" in organizations—are the tasks, respectively, of chapters 1 to 3 in Section I.

The second objective of the book is to develop alternative, more useful perspectives for thinking about knowledge and its future. One such perspective acknowledges that knowledge is not a single thing but that multiple forms of knowledge exist. A second is to switch from approaching knowledge as content, as "things known," to thinking about *knowledging* as a process; that is, to try to focus not on knowledge itself but instead on "knowledge networking." These two together facilitate a third perspective, replacing individualistic conceptions of knowledge with more social ones. More appropriate perspectives like these are necessary if a better job of studying empirically what actually happens to knowledge in computered environments is to be done. The second section of *The Knowledge Landscapes . . .*, chapters 4 to 6, develops these alternative stances, grounding them in philosophy, anthropology, and informatics (computer science).

Better empirical knowledge about actual knowledges, in turn, opens space for the design of kinds of knowledge infrastructures that really do have transformative implications. The third objective of the book is to show how to use the alternative perspectives. *The Knowledge Landscapes . . .* lays out conceptual groundwork for a generation of technological knowledge infrastructures that *in actuality* facilitate really new forms of knowledge creation, reproduction, and sharing in cyberspace. Section III (chapters 7 to 9) illustrates the perspective's use in research—primarily in holistic, ethnographic studies of knowledge phenomena in protocyberspaces. Section IV addresses the consequences of the new understandings of knowledges developed in the preceding sections for how we should think of culture in cyberspace. Implications for thinking of and intervening in the political economic relationship of technology to social change are addressed in chapters 10 and 11, while ethical, aesthetic, and political implications for both field studies of cyberspace and the design activity I call "cyberscaping" are dealt with in chapter 12.

The Distinctive Feature of the Book: Its Argument for Ethnographic Knowledgescaping

Exploring Cyberspace as Landscaping

Above I began my critique of currently popular performances of the "knowledge society" figure with a survey of some of its elusive manifestations. In speaking of "surveying ," just as in using a phrase like "knowledge landscape" in the book's title, I deploy a metaphor, that of landscape design, of particular importance to my perspective. Landscape design is a professional practice that strives to embody a more or less totalizing aesthetic in large environments, like parks or even cities. Frederick Law Olmstead is often credited (e.g., by Rybczynski 2000) with creating this practice.

I choose to place this extended metaphor, a "knowledge landscape" conceit, at the core of *The Knowledge Landscapes . . .* because aspects of Olmstead's practice suggest important elements of the way I want knowledge to be thought of. To design a landscape, Olmstead had both to create the integrated vision and develop the technologies to make it real. As Rybczynski points out, his arguably most significant notion was that landscaping was not simply a matter of building; rather, it was an orchestrating of conditions that led environments to evolve in desired directions, over long periods of time. For example, as is known by any visitor to Central Park in New York City—or the Utica, New York, park system near which I am writing—Olmstead's visions often had to accommodate to the developmental vagaries of deciduous trees. Indeed, landscape design is a dialectical practice that involves envisioning a difference, taking steps to impose it, but also accommodating to what develops.

Moreover, through his practice, Olmstead taught Americans to "see" landscapes in John Berger's *Ways of Seeing* (1991) sense: He taught them both to appreciate natural views aesthetically and to incorporate into their vision of surroundings a sense of time—an appreciation of how landscapes change. His practice suggests "-scape" as a metaphor for any large-scale, integrative vision realized in a fulsome evolutionary dialectic with already existing contexts.

The anthropologist Arjun Appadurai deploys "scape" in his survey of the implications of *Modernity at Large: Cultural Dimensions of Globalization* (1996). Appadurai views culture as increasingly disengaging from its previous sites of practice, like nations, and becoming manifest instead in transnational "mediascapes" and "ethnoscapes." These variations on "landscape" name for him key arenas in which the ambit of culture is "unleashed." They are now the central moments in globalization, the sites where practices are being transformed while transforming culture, the key sites of modern societies' reproduction.

Notions like "knowledge society" can similarly be translated as "knowledgescape." However, the great changes in knowledge are popularly articulated as coming *after* rather than *during* modernity; they are envisioned as part of a potential new, postmodern rather than modern, way of life. This new way of life is often referred to as "cyberspace."

Other than focusing more on the postmodern than the modern era, *The Knowledge Landscapes . . .* follows Appadurai's analytic approach, exploring whether knowledge landscapes become comparable to media- and ethnoscapes. "Cyberspace" is the label that I accept for convenience for the notional terrain of practices that needs to be surveyed in a serious examination of notions like "knowledge societies". As illustrated above, the idea that its knowledge is profoundly different is now central to conceptions of cyberspace as not just a potential but a really emerging, while still virtual, nonetheless soon to be dominant "way of life."

In one sense, like Appadurai, *The Knowledge Landscapes of Cyberspace* deploys "-scape" as a relatively passive metaphoric extension of Olmstead's idea of landscape. That is, a "-scape"—whether mediascape, knowledgescape, or landscape—is something that exists and can be described, although not of course in any simple empiricist or "abstractly concrete" (Mills 1959) sense. The book explores, as it were, the notional knowledge terrain of cyberspace.

However, I also want to emphasize the other, more active aspect of the "-scape" metaphor. Olmsteadian landscaping is a more teleological process than gardening. It "evolves" its environments toward deliberate, rather permanent, not necessarily immediately obvious objectives. My book argues for "knowledgescaping" as something that can be described because it might happen, *but also* as something that *should* be practiced deliberately. Knowledgescaping actively imagines and aims to bring about new kinds of intellectual spaces.

Indeed, I will argue that scaping new knowledge terrains turns out to be an as yet largely unrealized but essential prerequisite of that new way of life that most other authors seem to presume is necessarily immanent in the widespread implementation of AITs. A great deal of something like "landshaping" goes on already in cyber-environments, in that, as artificial entities, all AIT artifacts are designed. "Scaping," as opposed to merely shaping, means articulating an integrated vision of the way of life that would be created, conceiving of the artifacts that will serve as tools for its creation, and pursuing the vision's implementation as deliberately as the sale of these artifacts. In shifting from knowledge landshaping to knowledgescaping, those who design and/or use AITs should strive to emulate Olmstead's practice as well as adopting the gaze he pioneered.

Knowledgescaping as an Ethical and an Aesthetic Practice

In developing justifications for particular values like these, *The Knowledge Landscapes . . .* goes beyond the mere description of particular cultural practices. It is an argument for a particular professional ethics as well. Moreover, its concluding chapter makes an additional philosophical claim, that new forms of "aesthetics" can support cyber-workers pursuing the practices advocated. Just as nonmodern ethics abjure the purely rational search for transcendent statements of "the good" (chapter 4), nonmodern aesthetics are no longer focused on efforts to articulate abstract standards of "the beautiful." Instead, aesthetics, like ethics and epistemology, has taken an ethnographic turn, concentrating on accounting for how judgments of form, balance, taste (Bourdieu 1990), and so on are actually made by people in different cultural contexts. By understanding the range of aesthetic practices, their contradictions, and their connections, aestheticians have developed ways of talking that illuminate both choice and experience.

Applied ethics involves helping individuals and groups achieve a more coherent embodiment of their notions about how people like them should live. Similarly, applied aesthetics is about understanding why we as individuals and groups respond to cultural representations and performances as we do, finding coherent ways to evaluate these responses, and identifying those initiatives that foster the responses we wish to encourage.

Applied aesthetic practice is based on the presumption that there is a strong connection between having greater awareness of our reactions, influencing them, and being able to pursue those goals that we feel are really important. A good aesthetic greatly aids effective pursuit of ethical action. As argued above, on a nonmodern reading, aesthetics is not about knowing how to separate the beautiful from the ugly. Rather, it is about understanding the choices made by, for example, an artist or designer, the frameworks in relation to which they have been made, and relating these choices and frameworks to those that we would advocate.

Recognition of the importance of aesthetics to informatics—the study of information and the design of machines to automate its manipulation—has grown recently. It came into its own with recognition of the important visual and graphic design aspects of, first, user interfaces and, more recently, webpages. The largely disastrous attempt during the 1990s to develop the Internet primarily as an aid to capital reproduction did have one salutary effect, that of drawing attention to even more general design questions in cyberspace. Under such scrutiny, the relatively narrow practices surrounding earlier design of information systems are being displaced by ones that place AIT design in broader contexts, to foster more effective channels for communication in general (Ehn 1995). This necessitates extended discussion of the aesthetics of cyberspace.

Engineers, Ethnographers, and "Scaping" Cyberspace

To design a landscape, one needs to understand and work with the existing terrain and life forms. Would-be cyberspace knowledgescapers need to understand and evaluate how differently knowledge is shaped in "really existing" cyberspace. Like Appadurai's of modernity, my knowledge of cyberspace is based on my research and practice as an ethnographer, one who has both observed and participated in shaping cyberspace. This experience has convinced me that cyberscaping cannot be—nor should one strive to make it—an isolated, "professionalized" activity. Rather, all the humans and cyborgs with a stake in cyberspace—which is to say, all humans and cyborgs—should participate in it. Because good ethnography must necessarily be participatory, it makes sense to ground cyberscaping on the practice and gaze of the cyberspace ethnographer.

The Knowledge Landscapes . . . asks those involved in engineering automated information technologies to extend their thinking about them by emulating ethnographic as well as Olmsteadian practice. Generally speaking, engineers are expert at finding optimal solutions within problem spaces whose boundaries are externally defined, not at extending, let alone scaping, those boundaries in time and space (Downey 1998). Because ethnography is generally seen as a descriptive, or descriptive/analytical, activity, it may not initially seem an especially obvious part of engineering AITs. For similar reasons, it is not an obvious element of proactive, interventionist practices like landscaping.

However, this view of ethnography is somewhat misleading. In addition to watching what is going on, an ethnographer regularly tests her knowledge of the culture by embodying it in practice, through *participating* observation, which means much more than merely being a fly on the wall. The standard view of ethnography is even more misleading in regard to its practice in cyberspace. Here, intervention is inevitable, and thus advocacy is central to its ethnography. That is, while in the future some fulsomely new—for example, "cyborgic"—life way may emerge, for the moment we can only experience cyberspace practices partially, largely in at most its "proto" forms. As ethnographers' representations of protocyberspace manifest themselves in its assessment discourse, these have fed back into its design and development (see, e.g., Brown and Duguid 2000). Ethnography is participant observation, and participatory study of *how* a social formation coming into being—especially when done from the perspective of assessing the prospects for the new social formation—means that the ethnographer is inevitably implicated in shaping *the ways* it develops.

Ethnographers can choose to participate in cyberspace's creation either consciously or unconsciously, but we cannot avoid participating. The concluding chapter 12 summarizes the ethical case for choosing conscious intervention for the development of culturally informed notions about what we want cyberspace to be, as well as taking the risks, made necessary by things like the indeterminacies of knowledge, inherent in this position.

Thus, I am asking those inventing the future to acknowledge a fundamental need for their professional practice to embrace design theory, philosophy, and ethnography. They need to be more self-conscious about their role as creators of cyberspace, not just of its individual artifacts.

The Book's Organization

As the preceding, relatively simple exercise of breaking general talk into more specific claims illustrates, knowledge is not chopped liver. Its multiple forms and dynamics demand careful attention to its linguistic fields, the broad terms of reference that allow various forms of things like "knowledge" to be central to the culture history of arguably all social formations. This book not only critiques popular cyberspace knowledge discourses; it also develops alternatives to them, alternatives based on readings of contemporary knowledge accounts in multiple relevant disciplines—philosophy, anthropology, business, and informatics (computer science)—as well as on considerable research experience.

Similarly, it approaches ethnographic research as a serious form of empirical study, not just word mincing. Thus, an equally important part of the book is what it takes from field research on cyberspace knowledge phenomena like knowledge management fatigue syndrome, new modalities of knowledge creation in disciplines, design of knowledge technologies, distance learning, and the dynamics of the so-called new economy. Fortunately, fieldworkers and other researchers have constructed an extensive knowledge of knowledges from which we can draw.

Why I think ethnography has an important place in the study of such questions, why I choose "cyberspace" as a moniker to represent the purported changes and their cumulative effects, and why it is particularly important to try to understand potential futures now are similarly important elements of the way I frame my attempts to answer the cyberspace knowledge question. However, they are all addressed in detail in my previous book, *Cyborgs@* [illegible], so I choose not to address them here. The remainder of this introduction instead sketches out the argument that follows.

Chapter 2 of this book is devoted to clarifying the cyberspace knowledge question. It further deconstructs popular knowledge society talk. Through examining in detail the contradictions and silences about knowledge in discourses like those briefly illustrated above, it constructs a sturdy basis for research relevant to the cyberspace knowledge question. Partly this involves shifts of perspective, such as moving from a concept of knowledge that stresses its content to a concept of knowledge approaching it primarily as a process—knowledge networking—that are essential to answering the knowledge question in cyberspace.

Chapter 3, "'Knowledge Management Fatigue Syndrome', and the Practical Importance of the Cyberspace Knowledge Question," grounds the cyberspace knowledge question practically and empirically. It focuses on the recent rise and subsequent decline of "knowledge management," a loosely connected set of automated information technology initiatives in organizations that aimed to take notions like "knowledge society" and "intellectual capital" seriously. The chapter begins with characterization of a new managerial discourse on advanced information technology and knowledge that became insistent in the mid-1990s. Some of its concepts were borrowed from computer science and reflected in phrases like "knowledge engineering." The new approach was most emphatically articulated in the professional discussion about organizations. New management roles, like "chief knowledge officer," were suggested, but it was most clearly focused on the notion of "knowledge management." This is essentially the idea that organizations can be made more successful by reconceptualizing their assets as knowledge and using AITs to control them. However, by 2000 in many regions of the world, interest in the buzz phrase "knowledge management" (KM) had declined in key organizations and "knowledge management fatigue syndrome" had spread widely.

After describing this KM rise and fall, the chapter outlines several ideas about the causes of these dynamics. In this way, it constructs a sharper sense of the practical implications for the reproductive dynamics of contemporary social formations and what is at stake in the various answers to the cyberspace knowledge question. This sharper sense helps motivate the complex analytic moves and descriptions contained in the rest of the book. The chapter also suggests that some substantial part of KM failure results from presuming that all knowledge is essentially the same.

The task of Section II, "Ethnological Perspectives: Three Sources of Concepts Prerequisite to Answering the Cyberspace Knowledge Question," is to construct carefully building blocks for an alternative approach to knowledge of more service to the intellectual cybernaut. Chapter 4, "Toward a Philosophical Basis for the Study of Knowledge Networking," offers a reading of knowledge studies, the topic of "epistemology" in the academic field of philosophy, helpful to answering the cyberspace knowledge question. I first outline the philosophy of knowledge on which the popular but inadequate conception of knowledge described in Section I is built. I then present several of the multiple, diverse critiques of this conception. I locate in neopragmatism a nonfoundationalist, alternative approach to knowledge that meets these critiques. The neopragmatist approach to knowledge requires comparative analysis of the many existing forms of knowledge networking. On this view, it is important not only to base epistemology on actual rather than hypothetical constructions of knowledge but also to recognize the many forms of knowledge, including important, culturally mediated "knowledges." The chapter concludes with a neopragmatist knowledge networking concept.

Chapter 5, "The New Anthropology of Knowledge Networking," addresses the question of whether there are materials in the field of cultural anthropology, including specific descriptions/accounts of knowledge networkings in different cultures, that provide an adequately diverse empirical basis for the cross-cultural knowledge of understandings at which contemporary epistemology aims. There is indeed an emerging ethnography of knowledge within anthropology, but where does it come from, what are its preoccupations?

The knowledge question in anthropology has emerged from recent questionings of the character of anthropological knowledge, including if anthroknowledge is even possible. Concern with anthropological knowledge has led to a more direct, self-conscious ethnography of multiple forms of knowledge. This new anthropology of knowledge has the potential to provide the essential empirical dimension to the comparative philosophy of knowledge as process. However, in many of the same ways as are outlined in chapter 1, making cultural sense of knowledge dynamics (whether of cyberspace or any other social formation type) is impeded by inadequately articulated, contradictory notions of knowledge and its production within anthropology, too. Thus, to answer the cyberspace knowledge question ethnographically, it is necessary to understand both the roots and the forms of this new anthropology of knowledge. Also, the chapter addresses how this new anthro-knowledge can be clarified and specifies what is needed from it, what it can contribute to formulating the knowledge problem in cyberspace, and what study of cyberspace offers the anthropology of knowledge.

Chapter 6, "A Social Informatics Approach to Knowledge in Cyberspace," enters the conceptual terrain of another discipline. Anthropology is the profession that I have been practicing for some thirty-four years. Computer science is one in which I have also been involved for a long time, but more peripherally. This chapter outlines what I see as the proper way to construct an "informatics" (the term for the profession I prefer to "computer science") of knowledge in cyberspace. This approach begins with acknowledging the centrality of knowledge issues to informatics. It then connects many of the conceptual and practical problems of mediating knowledge via automated information technologies (AITs) to the dominant but misleading way of thinking about knowledge in this field. The chapter concludes by identifying and then clarifying for knowledge purposes the alternative, social informatics approach that already exists there.

As opposed to my discussion of the knowledge problem in anthropology, my (hi)story of the knowledge problem in informatics is "unreliable." Like my reading of philosophy of knowledge, my presentation of the knowledge problem in informatics is incomplete. My task is more limited: to provide justifications for thinking about knowledge in social process rather than content representation terms. This is sufficient to allow me to get on with what I identified in the present chapter as the main task I choose to take up, the ethnographic study of knowledge in cyberspace.

Section III, "Ethnographic Perspectives: The Dynamics of Knowledge in Cyberspace," shows how these ways of thinking about knowledge can be moved into the worlds of research and practice. Its main task is to illustrate how the properly conceived focus on knowledge networking developed in the previous section facilitates study of knowledge in protocyberspaces and the construction of effective automated information technological systems to support it. The section is framed in terms of a central issue, how to design AIT systems that facilitate rather than undermine knowledge networking. In particular, to what extent are forms of knowledge networking similar enough to warrant construction of a generic infrastructure, or are they so different as to necessitate distinct systems? If distinct systems are needed, how many? The more generic the system, the greater the range of potential applications (and therefore markets), but there is a trade-off, a concomitant decline in "user friendliness." More specialized systems require less modification to make them useful in the situations they were designed for, but they also are less likely to be useful in other situations.

The main problem in making design decisions in relation to this trade-off is our insufficient knowledge about the differences among the kinds of knowledge networking for which one can design. The better we understand the differences in the social dynamics through which knowledges are constructed, the better we can design systems to support the process. The more diverse the forms under consideration, the less likely we are to create a comparative understanding that inadvertently privileges one sort of knowing over others. For this reason, I have directed my ethnographic research and interventions into three diverse moments of cyberspace knowledge networking. In each of the three arenas—intellectual, organizational, and educational—I have been involved in efforts both to study AITed knowledge networking and to incorporate automated information technology (AIT) as a core element in strategic practices. The chapters in this section share a common format: an initial discussion of the transformationalist case in each arena, a description of my relevant research, and use of research experiences to evaluate the case.

Chapter 7, "Creating Knowledge with Automated Information Technologies, or Knowledge Networking Knowledge Networking," first presents the argument that automated information technology fundamentally transforms the nature of knowledge creation. It then describes my current research project. Through several changes in its relatively brief life, KN^2 has provided useful knowledge creation–relevant experience. The chapter concludes by using this and other research to evaluate critically the transformationist position on knowledge creation networking.

Chapter 8, "Knowledge in Organizations and Knowledge Networking Oneida County," is intended to apply the ethnological concept of knowledge to knowledge networking in AITed organizations. While Chapter 3 is critical of knowledge management approaches to knowledge networking, it recognizes the importance of the knowledge challenges facing the contemporary organization. Chapter 8 thus outlines a more effective alternative. It conceptualizes and focuses on adaptation, the second aspect or submoment, of knowledge networking, specifying its distinctive features. The chapter begins with a discussion of what organizations are and why they are good sites for studying the reproduction aspect of knowledge networking. In an important sense, the fatigue with knowledge management described in chapter 3 results from the failure of KM systems to produce the transformative change promised. In the future, will whatever revolutionary potential there is in AIT continue to be subverted by forces conducive to the reproduction of the organizational status quo? Or is there an alternative, potentially more transformative approach to AITed knowledge networking in organizations? How much should application of this alternative, one based in a better theory of knowledge (e.g., one that recognized the general priority of the reproductive submoment in organizations), increase transformative potential? Would it be enough to justify trying to engineer a real knowledge revolution? How practical would it be to implement such an approach? This constitutes a good census of the issues with which an effective ethnography of knowledge networking in AITed organizations would concern itself.

As a way to focus my theoretical concerns, the chapter's argument is framed in terms of a case for transformative change that I find quite suggestive, what I call the "resocialing work" hypothesis. This is the idea that AIT makes work substantially more likely to produce new and/or stronger social relationships. To test the idea, I describe my experiences trying to promote fulsome knowledge networking among public and not-for-profit organizations in Oneida County, part of the Upper Mohawk Valley in upstate New York. While not yet completely successful, my efforts to "knowledge network Oneida County" have identified some of the social forces with which a truly coherent effort to change the dynamics of organizational knowledge reproduction will have to contend.

Using this experience, I critically integrate ethnological knowledge networking perspectives into recent work in organization theory that aims to come to terms with knowledge. In approaching such issues, the chapter tests as well as applies the ethnological theory of knowledge. The aim is to create an approach to knowledge in organizations as supportive of transformation as possible, especially through "engineering" the resocialing that is the main potential support for real change.

Chapter 9, "The Cyberspace Knowledge Question in Education," has as its purpose to apply the knowledge networking approach to education, mostly its formal forms. I attempt to develop a knowledge networking alternative to the dominant "mechanical brain" approach to the place of AITs in education. As in the previous two chapters, I want to balance my critique of standard approaches to what is called educational technology with an appreciation of their real transformative potential. Also, even more than chapter 7, this chapter draws on my experience deliberately intervening into use of AITs, more practice than basic research. This includes experience as an organizer, as teacher and scholar, and in my professional organization.

Section IV, "Conclusions: The Analytics of Knowledge in Cyberspace," draws out the implications of the two main conclusions of the preceding ethnography of knowledge in cyberspace. First, I remain skeptical that either the character or the social functions of knowledge have changed profoundly in existing protocyberspace social formations. Second, I still see great potential for AITs to contribute positively to knowledge networking, especially if the more social perspectives illustrated in the foregoing chapters can be incorporated into the design of new technologies to support knowledge networking.

Chapter 10, "A Critique of Popular Political Economies of Cyberspace Knowledge," initiates discussion of the implications of the ethnography of knowledge in cyberspace for the larger, more structural aspects of social formation reproduction. Its preoccupation is a critique of several new knowledge political economies, like the idea that a "new" economy has resulted from the shift to knowledge as the chief factor of production. These political economies are critiqued both empirically and in terms of their inattention to changes in the reproduction of capital. Chapter 11, "An Alternative Political Economy of Knowledge in Cyberspace," presents a structural theory of the dynamics of contemporary social reproduction that accounts for its actual, as opposed to only its potential, dynamics.

Chapter 12 on "An Ethico-aesthetics and Politics for Changing Knowledge in Cyberspace," comes to terms with the likelihood of continued efforts to mediate knowledge networking with AITs. What aspects of the answers offered should guide those who maintain either theoretical or practical interests in the question of knowledge in cyberspace? Outlined first are the special ethical responsibilities of the intellectual cybernaut. The ethical responsibilities of the student of any culture begin with understanding its dynamics as accurately as possible and representing them fairly. However, cyberspace culture is only in the process of becoming. Thus, even to intervene by studying it implicates one in its creation. While the cybernaut cannot choose not to be implicated, she can be as explicit as possible about what she would like it to become, and design her interventions, scholarly and practical, to support these directions of development.

Ethics, the practice of considered attention to the likely consequences of actions, is thus an inherently central part of the study of cyberspace. Further, since we want not only to understand likely implications but also to design our actions to increase the chances that our desires will be met, cyberspace ethnography and practice carries concomitant aesthetic implications as well. This is because aesthetics is about the criteria we use when we make design choices, about how not only to be conscious of why we make the choices we do but also to bring these choices into harmony with our core values.

Finally, cybernauts need to work with others if their actions, ethically and aesthetically, are to be effective. The chapter concludes with a political articulation of the preceding perspectives on knowledge in cyberspace.

Conclusion

About twenty years ago, I was asked to teach "Computers and Society," then a regular course in the computer science curriculum at my technical branch of the State University of New York. One day, perhaps annoyed by my complaints about the shoddily argued prognostications of a "computer revolution" contained in the required text, students challenged me to show how to do a better job. I've spent much of my research time since then responding to their challenge, using my ethnographic skills to investigate a series of "technology and futures" questions, leading ultimately to the preoccupations of this book.

My approach to doing anthropology in such an anticipatory mode has been straightforward: I do fieldwork in protocyberspace. My epistemo- and methodologic is as follows: The ultimate goal of general anthropology is to account for the major transformations in human sociocultural practices. The chief methodology of the social or cultural anthropologist is ethnography, systematic and in-depth observation of and participation in ongoing social processes. We live at a time that is, according to a broad social consensus, one of profound social transformation. Thus, we ethnographers have an unprecedented opportunity, at least for practitioners of our discipline, to study fundamental social change firsthand, via fieldwork. In the process, we can also assess many of the presumptions that underlie general anthropology, both our cultural anthropological endeavors and those of our archaeological, physical anthropological, and linguistic cocommunicants.

Moreover, field study of those practices widely held to be prefigurative of the future should allow us a good perspective from which to assess directly the popular wisdom about technology and social change. For example, an important part of the mythos of cyberspace is its liberatory, democratizing potential.

I spend much of my time among systems developers, technology policy planners, listserv acolytes, and distance educators. They actively construct, and are constructed by, the forms of a potential new social formation, which I choose to call "cyberspace." Presuming their experiences to be at least somewhat prefigurative, we can ask: To what extent do their experiences in protocyberspace justify popular liberation views? Even if not fully realized ultimately, what if any alternative social dynamics seem to be being "placed on the social agenda"?

My anticipatory futures project resonates well, I believe, with the noted ethnographer Richard Lee's articulation of the special promise of anthropology, a promise based on the discipline's awareness of how culture can be and is done in so many, so different ways. Lee himself has done this through a long career of participant observation studies of South African gatherers and hunters (1979). He evaluated this corpus himself from this perspective in a 1997 review of The Society for the Anthropology of Work's first twenty years (SAW is a Section of the American Anthropological Association). Lee justified his work on work as meeting an obligation as relevant to the future as to the past: the exploration of forms of work that transcend, or at least are not strictly bound by, the parameters of current capitalist political economies (Lee 1998). Awareness of the possibility that work might be done differently is an essential part of making alternatives real, of influencing the future, not merely being pushed along by whatever dynamic (e.g., technological ones) we unleash.

Cyborgs@Cyberspace? was a broad book, my attempt to identify the key questions around which to organize a study of the culture of the future, as well as an effort to justify use of ethnography as a way to study these questions. Others can judge if the attempt was worth the effort. I was uneasy about the broad focus, and I decided to focus my subsequent work on something more tractable, on just one of the issues raised.

For various reasons, I decided on the knowledge issue. Partly this was due to the practicalities involved in coming to terms, like academic colleagues all over, with distance learning. Partly it was opportunistic, a way to combine my intellectual and practice interests, and partly it was a response to the rise of knowledge management as a phenomenon in organizations. This decision was also theoretical, a realization that whether knowledge was really changing was not only an important issue but also one potentially tractable empirically. Moreover, while my experience on the NSF K/DI panel was in many ways discouraging with regard to voluntary interdisciplinarity, it did at least suggest the necessity of dealing more effectively with knowledge as a potential ground for substantive collaboration with computer scientists. Finally, as I hope was demonstrated in chapter 8 of *Cyborgs@Cyberspace?,* it was relatively easy to develop some conceptual approaches with real potential for facilitating this collaboration. That is, an ethnographically serviceable knowledge construct would also have substantial potential value for the development of better technologies.

As a result, I have spent the better part of the last five years trying to answer the knowledge question in cyberspace. While it does not contain simple conclusions, *The Knowledge Landscapes of Cyberspace* aims to show that being serious about knowledge and technology is both possible and a worthwhile endeavor. It is for those who share this conviction that the book has been written.

Perhaps the main thing I have learned is the necessity in cyberspace studies to balance acknowledgment of the potential for transformation with the reality of continuity. From arenas as diverse as education to voting to publishing, for war and peace, from individual identity construction to group emancipation, for fourth world revolutionaries as well as geeks, cyberspace has come to be the repository of a substantial portion of humanity's dreams. With new technology viewed as a key means of liberation, cyberspace exercises hegemony over large portions of the cultural imaginary throughout the world. Any effort to reform the condition of individuals, groups, or humans as a whole must now incorporate, or at least relate to, some version of cyberspace/AIT in its program.

Nonetheless, cybertalk is still marked by a rhetorical slippage in which potential is transformed into reality. This slippage can be seen in the sustained unwillingness to come to terms with the failed projects of the past, manifest in what Ole Hanseth (2000) calls the "fashion" economy. It appears in a willingness to technologize the study of virtually everything. The creeping tendency to convert knowledge into intellectual property, or medicine into "medical informatics," should be resisted, not because of some mindless commitment to tradition, but because such technologizings depend on conceptual approaches demonstrably febrile. While "social informatics" is a viable, even good "stalking horse" label for a necessary complement to what might be called "technical informatics," it would be unwise to allow the new label to mask a surreptitious colonization of the problem space of social science by computer science. On occasion, I have been driven to arguing that computer science should be seen as more a branch of the social rather than the technical sciences, as a way to overcome the kinds of dogged misframings so present in recent "knowledge technologies."

In carrying out the efforts reported here, I have been guided by a dual practice: on the one hand, attempting to convert the potential of AIT-related transformation of knowledge into the actual, while, on the other, continuing to practice a critical stance, one insisting that knowledge about knowing strive to be "really useful." The book does the former by drawing attention to practices in which the potential takes palpable form, the latter by underlining the many ways in which the dynamics of practice still manifest familiar qualities. In addition, I have tried to act on my own advice about the special ethical onus that falls on those who would study in an anticipatory mode. The reader will judge how well I have succeeded.

Chapter 2

Clarifying the Cyberspace Knowledge Question

The first step in evaluating the various perspectives on knowledge in cyberspace presented in chapter 1 is to make clearer the meanings of concepts that the various perspectives employ and their implications. This is what social scientists call research design, the task of this chapter.

For example, foundational change in knowledge is presupposed in most talk about future work, science, education—indeed, in virtually all the domains in which future knowledge dynamics are described. Such change is assumed whatever is stressed as the key factors linking social change, knowledge, and AITs. This assumption needs to be examined thoroughly, by establishing empirically either its veracity or revealing it to be myth. This chapter presents approaches that enable such empirical work. It uses these approaches to outline the full range of answers to the cyberspace knowledge question, to all of which good research needs to be sensitive.

Preparatory Approaches

Avoid "Optative" Cyber-rhetoric

This section concentrates on what not to do when trying to answer the knowledge question in cyberspace. While perhaps not familiar with the details of the things he describes, I doubt many readers of Jon Katz's book on "the Geek Ascension" found his tone extraordinary. All over in protocyberspace one encounters profligate use, as by Katz, of a speech mode that mixes statements about what is with ones about what one hopes to be. The desired effect would appear to be that readers come to expect what the writer hopes for and thus, in the process, become complicitous in its creation. I call this speech form—manifest elsewhere but especially in cyber-rhetoric—the "optative" form.

Characteristic of the optative form is uninhibited movement across linguistic registers, especially those of verb tense and mode. Katz's comments blur the distinction between present and future, declarative and subjunctive, with abandon. In the optative, one creates an atmosphere rather than making an argument. Perhaps the clearest signal or index of the optative form is the nonsentence declarative (e.g., Katz's "Now running the systems that run the world").

Optative figures of speech are so distinctive of contemporary writing on the social implications of automated information technology as to be almost obligatory, but in *The Knowledge Landscapes of Cyberspace*, I abjure them. This is necessary if I am to be able to compare and contrast what natives say with what they do, and with what is knowable about their social formation's dynamics. It is precisely claims like Katz's that I need to be able to examine: What systems "run the world"? Who "runs" them? What does it mean to "run" the world, or even just run the systems? What is the evidence for the contentions made in these regards? Are there better ways to account for this evidence? Do claims like "systems run the world" make sense? Optative speech impedes our ability to bring such questions into focus.

Avoid Taking Plausibility for Actuality

Figuring out the sense in such performances requires more than avoiding exaggerated styles. Their logic must be analyzable, which means that their presumptions must be made explicit and opened to empirical study as well.

In *Cyborgs@ . . .* , I laid out at length the case for careful analysis of the technologics commonly applied to contemporary society. There I argued for an empirical skepticism as a starting point for cyber-ethnography. This skepticism is limited; it is cognizant of the significant potential for technology-related social transformation. I doubt neither that contemporary knowledge is changing nor that computer-based technologies are spreading. I also accept that society-transforming change via change in knowledge is certainly a *plausible* consequence of automated information technologies. (The centrality of change in individual and collective mental processes to the "cyberspace as social transformation argument" is doubtless related to this plausibility, as is its value as advertising copy!). Further, there are "really existing" examples of distinctively new, achievable forms of knowledge networking in cyberspace, the focus of several chapters to follow.

However, my skepticism means I refuse to take evident potential for proof of significance. The actual significance and breadth of the changes must be demonstrated; examination of a case or two where this appears to be happening is not enough.

More generally, the history of social science justifies skepticism with regard to simplistic causalities, like those that would account for change in knowledge and society in terms only of new technology. Such arguments deserve special scrutiny, as by showing either that technology already *has* transformed society or that such change *is necessarily compelled.*

As described in chapter 10, there are systemic, political economic forces that foster optative performances, of which the confusion of plausibility for actuality is a special form. Whatever the short-run role of such narratives in social formation reproduction, their current performance makes neither changes nor connections necessary in the long run. Yet almost every day during the 1990s, some august body or other publicly performed an "Internet-induced social change" mantra, which was worshipfully reproduced by both mass and so-called many-to-many media, like sites on the World Wide Web. Such performances are important data, but the frequency of their manic performance is itself an indication of their dubious status.

The Knowledge Landscapes of Cyberspace presents a stance from which such ways of speaking can be analyzed, not just performed. In contrast to taking potential for actuality, I assert that under any of the following conditions—

1. If knowledge isn't changing, or isn't changing all that much, in quantitative or qualitative characteristics or social functions;
2. If it is changing in one regard but not the others;
3. If it is changing profoundly in all regards, but these changes have relatively little to do with AIT;
4. If change in knowledge and change in AIT are connected, but change in knowledge is as responsible for change in AITs as visa versa; or
5. If the causal connection is more reasonably seen as a consequence of some other mediating force or forces more central to the reproduction of our social formations—skepticism would be validated.[1]

Avoid Confusing Divergent Change Logics

Breaking apart the claims that logic requires of a change answer to the cyberspace knowledge question allows one to specify the range of alternative arguments made for the change position. One can identify three ways of conceptualizing the purported change in knowledge attributed to technology, as a consequence of either

1. Quantitative growth in knowledge,
2. Change in its quality/character, or
3. Change in its social functions, perhaps involving quantity, quality, or both.

Quantitative Growth Accounts

The quantity connection is the simplest and most popular trope. President Clinton, in his 1998 State of the Union address, performed this cliché when he described profound social implications as following directly from a substantial increase (e.g., "doubling in the past year") in the total amount of knowledge.[2] In other, similar phrasings—those performed in, for example, year 2000 Microsoft commercials—social transformation follows from quantitative factors like alleged wider access to an increasing number of information sources ("Where do you want to go today?").

At a 1996 OECD international conference in Oslo, Norway, in which I participated, an academic variant of this quantitative trope was performed. The conference *Call* defined the "innovation" central to economic progress as "the creative process through which additional economic value is extracted from the *stock* of knowledge" (emphasis added). It went on to argue that as the stock grows, so does innovation.

In the Clinton, Microsoft, and OECD performances, any differences between mere information and real knowledge are ignored. In social evolutionary terms, a punctuated equilibrium is presumed (Gould 2002): Through sheer greater volume, a quantitative change in information has qualitative general social consequences. One suspects that the popularity of the quantitative change trope follows from the fact that once quantified, knowledge is easier to treat as a commodity, a thing of which the more the better.

Qualitative Accounts

Performed somewhat less frequently but more easily mobilized for transformative rhetorics are cyberspace knowledge tropes that stress *qualitative* changes. Social changes "caused" by the "computer revolution" are linked to transformations in the *character* of knowledge (e.g., from formal to situated). The philosopher Charles Ess, for example, refers to "[W]hat many see as a technologically mediated revolution . . . that promises to reshape radically our fundamental conceptions of . . . knowledge . . . " (1996, 2). The Norwegian cyber-ethnographer Henrik Sinding-Larsen (1991) offers a striking qualitative change analogy. He argues that computerization formalizes thought. The impact on thinking in general will parallel the impact of systems of formal musical notation on musical composition and performance. After thorough computerization, the difference in thinking will be on the scale of the difference between a Beethoven symphony and Gregorian chant.

Along similar lines, Kolb (in Ess's volume) argues that the greater orality of computer-mediated communication leads to new forms of discourse. Shank and Cunningham, also in Ess, argue that the Internet fosters semiotically or symbolically unique communication. Because culture is semiotic, Internet use leads to drastic culture change, with communication becoming more nonlinear and less hierarchical. Shank and Cunningham expect the emergence of "multilogues," less restrictive, expanding rather than specializing discussion forms. For them, greater ease of storage and connection leads to more abductive (the invention of new beliefs or revisions of old) than deductive or inductive thought, to nontraditional syntheses of patterns of meaning. These follow from the broader "community" involved in discourse. All of this leads to an "Age of Meaning," as opposed to the recent "Age of Science" (1996, 36–40). Others like Nardi (1993) similarly see new computing modalities like "groupware" fostering palpably new forms of knowledge.

Changed Social Function Accounts

There is, of course, no reason why one can't argue for a knowledge society by combining quantitative and qualitative arguments. More complex transformative discourses typically incorporate both into arguments that AITs—quantitatively, qualitatively, or both—transform knowledge's social function as well as knowledge itself. The economic ideas of Knorr-Certina and Davidson, Handler, and Harris alluded to in chapter 1—for example, that knowledge has replaced land, raw materials, labor, and capital as the chief factor of production—are good examples of this third trope.

Such arguments are often a restatement, in cyberspace register, of an idea at least as old as the 1960s. According to Peter Drucker then, there was already under way movement to a new society, where:

> The basic capital resource, the fundamental investment, but also the cost center of a developed economy, is the knowledge worker who puts to work what he has learned in systematic education, that is, concepts, ideas, and theories, rather than the man who puts to work manual skill or muscle. (1970, 37)

Another version of this "shift from brawn to brain" idea appeared more or less simultaneously in the work of the sociologist Daniel Bell, who saw "knowledge work" as replacing the assembly line (1973). In this view, while the valuable industrial skill *was* proficiency at highly repetitive activity, the new most valuable (soon to be called "cyber"-)skill is the ability to apply scientifically based knowledge.[3]

Drucker's conviction—that knowledge workers being differently as well as more skilled than industrial workers has broad social import—has led him to speak more recently of a "postcapitalist" society. This new political economy of knowledge tends to be presented currently as arising from a profound change in the nature of its creation. In the older mode (interestingly, the one still stressed explicitly by Knorr-Certina and implicitly by Katz), knowledge creation was scientific, independent, and discipline-based. In today's mode, often called "mode 2" (see chapter 6), disciplines and universities are integrated with each other, more closely connected in a "triple helix" (Leydesdorff and Etzkowitz 1998) with states and corporations (Werle 1998).

Driving and driven by such convictions about these changes in the social function of knowledge are energetic efforts at AIT-based educational reform. While debates on knowledge in education also grow out of concern over academic fragmentation, the desire for more interdisciplinary collaboration is now more frequently related to the purportedly more complex profile of knowledge in "high tech" production processes. The incursions of AIT-based new educational technologies—for example, synchronous and asynchronous distance learning—are also justified via a need for new ways to share knowledge driven by its new functions (see chapter 9).

Stressing quantity, quality, and social function change are all plausible but distinct ways of making the changing knowledge/social transformation connection. Many knowledge society arguments skip rapidly and frequently among them despite their differences. A serious knowledge society argument would have to be more deliberate in its articulation of the causal chain. In particular, should it wish to combine them, it would need to show how their causal dynamics systematically interarticulate.

Avoid Deliberately Confusing Knowledge Constructs

Also standing in the way of clear thinking about how much difference cyberspace makes is the possibly deliberate obscurantism of conceptions of knowledge in texts that hype the knowledge revolution. A good example is the various imprecisions and exaggerations in James Bailey's comments on computers and the "Knowledge Revolution." This is, for Bailey, "the true electronic computing revolution [, whose] intellectual impact will be greater than anything since the Renaissance, possibly greater than anything since the invention of language." Bailey offers the following explanation for this monumentality, a change in humans' very species being:

> To the extent that the tasks that these computers take on are tasks that require intelligence when *we* do them, we will refer to them as intelligent . . . . However, the capabilities [computers] develop may bear no relationship to the "intelligence" we have developed . . . because the information-processing milieu is sharply different . . . The result will be . . . [that] humans and computers will exhibit different intelligences. And they will be peer intelligences . . . where thought no longer holds the exclusive franchise.

In line with his penchant for optative rhetorics, Bailey titled the 1996 book that contains these lines *After Thought* even though, were we to accept his argument, he really means, "When Thought Is No Longer Alone." There is no reason to accept his argument anyway. It begins with a reasonable remark on the high occurrence likelihood of a common human linguistic act, use of a simile (reference to what computers do as "intelligent" because they are doing something similar to what we do). This is followed by a peculiar, completely unjustifiably radical critique ("no relation") of the simile. Bailey then substitutes a different metaphoric characterization of a computer's activities, one containing a positive evaluation of their properties ("capabilities"). Thus, instead of concluding "this simile is of limited value," he concludes that computer capabilities and human thought are instead hypercomparable ("peer intelligences"). As a consequence of optative sleight of hand, we need no longer pay attention to human intelligent activity but can concentrate on a transcending machine intelligence, one effectively replacing the human form—one "After Thought." Human thought is no longer relevant, replaced by . . . what, exactly?

How much argument like Bailey's is a hall of mirrors becomes clear when it is contrasted to Harry Collins's *Artificial Experts: Social Knowledge and Intelligent Machines* (1990). Collins focuses on the actual use of AIT-aided equipment in scientific research. He points out that frequently in such research, humans deliberately use their intelligence in very mechanical ways, as in manically counting the number of "special events" in a given scientific artifact, like a particle trace or a microscope slide. Of course, researchers are grateful when a machine is developed to relieve humans of such tedious, inhuman activity. Under Bailey's gaze, such machines would count as intelligent, since they do what humans do, and thus justify a subsequent inattention to the human condition. Yet Collins is quite justified in stressing the artificiality, the triviality, and the *un*humanity of such "expertise." (As taken up in chapter 6, obscurantist tropes like Bailey's have far too much influence in computer science.)

The influential perspective on knowledge of Peter Drucker similarly tends to obscure more than enlighten. He begins in the quantitative register described above:

> But from the 1880s, since the telephone and Frederick Wilson [*sic*] Taylor's Principles of Scientific Management, the amounts of information and knowledge needed for each additional unity of output has been going up steadily at a compound rate of 1 percent a year—the rate at which businesses have added educated people to their payrolls . . .

However, Drucker then switches to a different register and deploys optative rhetoric in a combined warning/celebration: "Like a blast furnace that converts iron into steel, the organization concentrates, processes, and reifies the Knowledge Worker" (both quotes in Marquardt and Kearsley 1998). By dematerializing humans, Drucker grants to knowledge a transcendent, ontological agency, like what Bailey gives to machines. While Bailey evokes machines that abrogate thought, Drucker conjures workers who have become purified knowledge.

It is difficult to create shared frameworks without disciplined discourse, and impossible to have disciplined discourse without thought. If we echo Bailey, there is no way we can address the problem of transforming information to knowledge empirically. Similarly, to network knowledge in actual organizations, we need workers capable of creating social relations, not the dehumaned artificial intelligences celebrated by Drucker.

In so much of the talk associated with computers—not least that about knowledge management explored in the following chapter—forms of expression tend to mystify collective consciousness. Before one can determine, for example, if cyberspace is more conducive to the reproduction of labor or of capital, one needs to be able to state explicitly the causal logic on which various positions depend, so that these logics can be compared with what is actually happening. Notions only implicitly articulated have to make explicit sense. To answer the knowledge question in cyberspace, we need concepts designed to be understood.

Avoid Confusing Knowledge and Information

While theoretically holding knowledge to be "more than" data or information, much knowledge talk fails in practice to maintain the distinctions rigorously. Differences between data, information, knowledge, and wisdom are alternatively constructed and then blurred. "Infotalk" transmogrifies into "knowledge talk"; the two are performed in strongly congruent ways, in similar linguistic garb.

For example, the NSF Knowledge/Distributed Intelligence Program home page initially describes knowledge networking as being about "the integration of knowledge from different sources and domains across space and time." However, it includes among the goals of the research program achieving "new levels of . . . information flow" (NSF 1998). This inclusion implies that more information leads automatically to greater knowledge networking, thus blurring the boundary between knowledge and information.

From such perspectives, as in work like that of Bailey and Drucker, knowledge society talk too often violates T. S. Eliot's advice: to avoid confusing information with knowledge. This is clearly what is going on when an increase in knowledge is interpolated directly from an increase in the number of books published. Such rhetorics are, like many Schumpeterian celebrations of economic destruction (see chapter 10), less analysis than legitimating hype for a gigantic, hippielike, conceptual "happening."

Avoid Suppression of "The Dark Side"

To pose the cyberspace knowledge question properly, we need to be aware of the wide range of its possible answers. The range must include more than the "fundamental change" alternatives so often performed but not clarified.

In Sinding-Larsen's musical analogy, the knowledge resulting from computer-mediated thinking *may* be changing as drastically as musicianship was transformed by the use of formal systems of notation. This is indeed a possible answer, but we need to be aware of others.

In 1996, Diana Forsythe encountered a very different "cyberspacing" of knowledge practices (2001). She was studying a computer-based expert system, a self-diagnostic computer-based tool for people who wondered if they had migraines. The system prompted the "user" to describe her symptoms. It was programmed to repeat precisely these symptoms, whatever they were, as the proper indicators of migraine!

Further, while the system was *capable* of recommending a wide range of therapeutic options, it would not work unless programmed to present only a subset of these options, presumably the ones that the physician or clinic implementing the system preferred. The selectivity here was clearly different from the "automatic reflection" of the first response screen, but it was like the first in one important sense: It took place *without the user knowing it.*

In this case, there is indeed a change in the quality of user knowledge, and the change is indeed related to use of AITs, but the result is greater ignorance, not more knowledge. This migraine legerdemain, while redolent of Big Brother, was legitimated by the system's developers as "promoting user confidence in the system," in order to "increase the likelihood of patient compliance." Perhaps most astonishingly, the system was promoted as one that "increases control by the patient"!

"Cyberspace as knowledge society" tropes like Bailey's, Drucker's, and the promoters of the system Forsythe described imply not just that cyberians do knowledge differently—they do it better. On such readings, cybernauts have leapt to some new quantum, from normal to "hyper" knowledge. One critical perspective on such hype is to call it, as Langdon Winner does, "technological somnambulism" (1980). He places "knowledge revolution" talk firmly in the tradition of technological determinism so much a part of American thought, a way of thinking in which worship of machines fosters policy sleepwalking. In a related bon mot, Winner describes ours as an era of "Mythinformation" (1984).

An Empirically Useful Alternative Approach to the Knowledge Question in Cyberspace

The rhetorical forms and ambiguities described in the previous section have much to do with why little effective empirical attention has actually been given to the knowledge question in cyberspace. For such research to prosper, however, it is not enough merely to articulate critical perspectives. In what follows, I present key arguments for an alternative approach to knowledge, knowledge networking, which can support good research on the cyberspace knowledge question. Rather than just listing practices to be avoided, I articulate positive positions that constitute the core of an alternative.

Develop a Range of Diverse Answers

Whether a knowledge revolution or technological somnambulism is more characteristic of cyberspace is of more than passing interest to the student of the contemporary. The stakes of the current debate on knowledge are high, the issues basic. We really do need to be able to find answers to questions like: Do AITs really foster new roles for knowledge? Do these new roles mean knowledge's character is changed fundamentally? Is it consequently appropriate to privilege certain new forms of getting it?

The doubtful ethics of systems like Forsythe's migraine one suggest why some critics feel that far from a "knowledge society," cyberspace may instead be an era of increased ignorance. To make sense of such claims, let alone evaluate them, one needs a standpoint outside of the rhetorical consensus outlined in chapter 1. This can be obtained by conceptualizing a range of possible answers to the question, "hypotheses" that include profound knowledge change but do not presume either that this is necessary or necessarily a good thing.

Information Needed

Once we have articulated a sufficiently wide range of possible answers to the knowledge question in cyberspace, we can start talking about the information that—if we had it—would allow us to pick among them. An answer to the cyberspace knowledge question would be legitimate if it could specify empirically:

- In what ways the new devices and techniques for handling information significantly change knowledge,
- How the different knowledges actually, not just potentially, transform human activity substantially, and
- Why these transformed activities are meaningfully different—for example, are justly described as "better" than those they displace.

The discourse problems described above, however, make it impossible to move immediately to the collection of this information. They manifest a major problem in what I teach students to think of as "operationalization." That is, to make any empirical specification, one would have to have a clearer, more usable concept of "knowledge" than the ones described in chapter 1.

To go beyond critique to answers, we need a knowledge construct usable in research, one that allows us to compare claims like those recounted with what is actually going on in the world. I believe it is possible to articulate such a knowledge construct.

Differentiate Knowledge from Information

Because the ambiguities in the ways *knowledge* and *information* are used are so fundamental, one thing such a construct must include is articulation of a clear distinction between knowledge and information. The difference must be maintained systematically if empirical questions about knowledge are to be addressed successfully.

Normally, speakers of English do differentiate between knowledge and information. The analytic difficulty lies in doing this consistently. The confusing NSF practice cited above derives ultimately from general linguistic problems having to do with the growing lexical field of both terms. Lack of consistency is encouraged by the optative rhetorical practices critiqued above, but it also has roots in the fact that there are two popular but very different ways to make the distinction. These two different strategies are seldom distinguished from each other, but they are actually quite contradictory.

The Dominant Knowledge Discourse: The Modernist Knowledge Progression

The most common way of keeping knowledge and information distinct draws on what I call the "modernist knowledge progression." The more general account of knowledge in Western culture, it is reasonable to think of it as the standard account.

The modernist progression starts with data. Data are to the modernist more or less directly apprehendable, a view reflecting empiricist philosophy (see chapter 4): Data are "raw," separable in nonproblematic ways from their social and political surroundings, easily plucked from their context.

When such data are arranged in relation to each other in standardized ways, abstracted and placed in some common frame (e.g., an electronic database), the result is "information." The modernist account holds that through further manipulation according to procedures referred to with terms like "verification," information becomes "knowledge." Knowledge is what results when already processed data (information) are processed further in special ways, especially those that aim to verify a bit of information's quality from a "truth" perspective.

For the knowledge scholar, the second transition, from information to knowledge, is the important aspect of the modernist account. Perhaps because manipulations are involved in both the transition from data to information and that from information to knowledge, however, there is a tendency to elide their differences and ignore what is distinctive of each. Indeed, the fundamental accomplishment of academic studies in science, technology, and society/technoscience studies has been to open to greater scrutiny the black box of the second, verification process; one of its chief approaches is referred to as the "sociology of scientific knowledge." These studies illuminate the various but especially the social processes that privilege some of the ways in which this information-to-knowledge transformation is accomplished while marginalizing others.[4]

On occasion, a fourth stage, "wisdom," is acknowledged to be part of the modernist progression. Wisdom can be conceptualized modernistically as knowledge that has been lived or existentially verified. However, to differentiate between wisdom and knowledge requires one to acknowledge procedures beyond verification. Modernists tend to be uncomfortable about such claims, and "wisdom" is usually ignored.

In sum, the core of the modernist knowledge progression is a progressive vector, typically a continuum from data through information to knowledge. Its progressive thrust means that differences between particular stages of "progress" are often blurred. In this way also, the modernist account is biased toward quantitative narratives about knowledge.

A Subordinate Knowledge Discourse: The Postmodernist Knowledge Regression

Wisdom is more easily associated with the second, contrasting, less frequently evoked story about knowledge. For convenience I call this the "postmodern knowledge *re*gression." On this account, one begins at the end of the modernist progression with knowledge (or even wisdom), not data; indeed, the directional relations of the modernist progression are all reversed. The knowledge with which one begins is understood to be situated, embodied, tacit, and so on (Polanyi 1983; Ehn 1988; Dahlbom and Mathiessen 1993). This knowledge, unlike data in modernist tellings, is not easily abstracted from its context, however. Rather, it is understood to be deeply contextualized, often "contained in" actual people. The production of "information" means "desituating" knowledge, shearing it of important context. To get information, one must place knowledge shorn of context in new, artificial conjunctions. Hence, information is inherently constructed, not "natural."

"Data" are the result of further processing of information (denatured knowledge). In a postmodern account, data so regressed are never "raw." Because data are understood as always "cooked,"[5] the relevance of STS insights into the role of sociocultural processes in constructing knowledge is much more obvious in a postmodern account.

The postmodernist knowledge regression may strike the reader as odd, somewhat forced. Because the acts of decontextualization on which it focuses are themselves deeply dependent upon the specifics of context, its steps are certainly more difficult to automate than those of the modernist progression. Yet its insights are an important general caution to the simplistics of the typical modernist conception, like that of the OECD. In such accounts, knowledge is too unproblematically quantifiable, packagable, and objectifiable, its creation too simply planned and predicted; it is too easily bought and sold. Far from being reducible to simple common elements like money value, knowledge, from the postmodern perspective, is generally localized and difficult to describe, let alone quantify.

The implications of situating knowledge (Suchman 1987) have been articulated most clearly within feminist standpoint theory (e.g., Harding 1991; Hartsock 1983). A similar stress on how knowledge is situated socially underlies the position known as "constructivism" in education (see chapter 9). According to Jim Greeno of the Institute for Research on Learning:

> We may require a different way of thinking about knowledge. Rather than thinking that knowledge is in the minds of individuals, we could alternatively think of knowledge as the potential for situated activity. On this view, knowledge would be understood as a relation between an individual and a social or physical situation, rather than as a property of an individual (1988).

Such approaches provide better ways to account for the more "quantum" qualities, such as the leaps of insight, often associated with knowledge.

It is only in a modernist account that the idea of a "body of knowledge," something distinct from social relationships or a collectivity of knowers, makes sense. Experience in cyberspace—for example, the purported decoupling of space from place—is itself identified by some as an important reason to reject such positivist modernist knowledge discourses (e.g., Lyotard 1984).

This is only one of the alternative, nonmodernist accounts of knowledge that have recently gained greater currency. In response to the knowledge management fatigue syndrome discussed in the next chapter, and the failure of the knowledge integration program in informatics discussed in chapter 6, my business and computer science colleagues have again drawn attention to the old programmer's excuse "garbage in, garbage out." They emphasize how computing is still an information *as opposed to* a knowledge technology (and thus provide additional justification for distinguishing between them).

Similarly, teachers but especially librarians find themselves pointing out how the Internet is chock full of material claimed to be knowledge but, given the difficulty for individuals to determine its validity and reliability, such material is best thought of as "information." In general, it may be that only groups of concretely situated cybernauts can effectively evaluate knowledge claims.

The Problems of Using the Progression and the Regression Simultaneously but Unconsciously

In the foregoing sections, I have abstractly contrasted two ways of talking about knowledge. While they are analytically separable, most actual knowledge talk fails to take account of the existence of both and the differences in their derivations. For hyperrealist Jean Baudrillard (1995), for example, AITs render commonsensical notions of "reality" obsolete, since both the real and the "unreal" increasingly exist only in constructed representations. In analogy with the early computer game Space Invaders (or, if you prefer, the film *Star Wars*), the hyperreality of cyberspace is to reality as hyperspace (movie space) is to "real" space. It is not that there is no way to know, but rather so much is already known that individuals can only occasionally make reasonable judgments regarding what is significant.

In chapter 4 of *Cyborgs@Cyberspace?* I used the fact that the content of culture is often contested to critique as simplistic Daniel Dennett's "simple reproduction" model of consciousness (1991). The hyperrealist like Baudrillard in effect extends this contestation, suggesting that our representations of what is are now in some profound sense disciplined hardly at all by an empirical "real world." This is the basis of Baudrillard's famous contention that the Gulf War existed primarily on TV. In such readings, which share much rhetorical space with Bailey's notion of the end of thought, cyberspace obviates development of knowledge in any conventional sense. All cyberknowledge is plural, manifesting multiple voices, multiculturality, and so on. Such articulations of the postmodern knowledge regression are more about the end of knowledge than its transformation. As such, of course, they, too, undermine more than encourage empirical research on the knowledge question in cyberspace.

Thus, keeping modernist and nonmodernist accounts of knowledge separate is not easy. The same terms—data, information, and knowledge—are employed by both. However, as illustrated in chapter 3, many practical problems in making knowledge technologies work derive from failing to keep these things separate. Many organizational knowledge networks, for example, appear to have foundered because they tried to use the same channels to mediate knowledges constructed in very different ways. The results of chemical tests and the stories of regional sales managers *can* both be represented in the same kind of document. However, doing so indexes as similar kinds of knowledge that are actually very different. Such systems grossly flatten the differences between two knowledge landscapes, but, unfortunately, few chief knowledge officers notice.

Part of the reason for their simultaneous, unreflective performance is that both modernist and postmodernist accounts have specific AIT currency. On the one hand, those marketing knowledge systems for computers often perform positivist, progressivist, "bankable," and "more bells and whistles" content-type concepts of knowledge. On the other, they also draw on postmodernist, relativist "narrative," user-centered, constructivist talk. The vaguely constructivist Microsoft ads background positivist conceptions to evoke an aura of contemplation. As Benjamin Woolley puts it in his examination of *Virtual Worlds* (1992), "The significance of virtual reality . . . is that it directly confronts the question: what is reality?"

Via performances like that of Jon Katz, the popular discourse over knowledge in cyberspace implicitly assumes *both*

1. A vastly expanded version of previous forms of knowledge, its production being so automated, technologized, and so on, as to drive social transformation, *and*
2. A decentered, near-hallucinatory version that gives pride of place to paradigm shifts in the very character of knowing, critiquing previous ways of knowing, and projecting multiple knowledges rethought in profound, fundamental ways.

Perhaps taking his cue from Bailey, the keynote speaker at an annual State University of New York conference on AITed distance learning performed the following myth before a polite and apparently appreciative audience. He asserted that the pace of learning technology change was now so fast that we were approaching a "technological singularity." Like particles sucked into an astronomic black hole, once knowledge passes through the singularity, it will be so different that "History will cease to exist." Such circumstances, even the disappearance of thought, need not concern us, for apparently things will be so different, knowledgewise, that we shall have no need to think!

In sum, multiple notions of knowledge are used, their coexistence seldom acknowledged, let alone differentiated. Incompatible knowledge discourses are mixed in practice, often unwittingly. Confusion is not recognized because it is masked by an implicit assumption of cyberspace knowledge's inherent transformativity. Thus forms of knowledge are essentially the same because they are all "content" and all being transformed beyond recognition. These are all important parts of why talk of "knowledge society" must have a more effectively operationalized knowledge construct.

Complexify Cyber-knowledge Talk

It is thus unfortunately not difficult to understand why there is so much nonsensical talk about knowledge in cyberspace: We have learned not to expect it to make sense! The recently increased volume of knowledge of talk is not paralleled by an increased sophistication in its understanding: When evoked simultaneously, as in much current science fiction, the effect of two coexisting but contradictory knowledge discourses is schizophrenic.

While pointing out some of the shortcomings of the two views above, I also indicated how both ways of talking have a certain legitimacy. Since, in the proper context, each has value, trying to decide which view is "correct" will not produce a knowledge construct adequate to the cyberspace knowledge question. The main difficulties in thinking clearly about knowledge derive less from application of "incorrect" than of incomplete or muddled constructs. To account for knowledge in cyberspace, we must be able to acknowledge the possible relevance of both but then keep them distinct in our understandings. Instead of reinforcing naive public preferences for simplistic, progressivist celebration of "discoveries," answers to the cyberspace knowledge question need to embrace complexity.

Possible Answers to the Knowledge Question in Cyberspace

A reflexive knowledge construct, one that systematically illuminates all the accounts out of which any particular claim to knowledge grows, can avoid these problems. Such an approach will need to problematize the presumption that knowledge is being transformed and do so in a way which specifies knowledge networking's connection to both the typical modernist and the typical postmodernist discourse.

A final element of the alternative approach is thus to specify the main possible answers to the cyberspace knowledge question that derive from each of the theoretical accounts examined. On the argument thus far, we can specify two popular answers:

- The progressive modernist transformative or "growth" position, for example, that the accumulation of especially scientific knowledge is so rapid as to transform knowledge's character, a sort of punctuated equilibrium model of the evolution of knowledge (Toffler 1983); and
- The regressive, postmodernist or "deconstruction" position; for example, that the nature of knowledge is transformed via one or more (e.g., qualitative, social function, even textualist), technology-facilitated reconstructions of it (Lyotard 1984).

In the modern approach, change follows from knowledge *disembodied* in new entities, while in the postmodern, knowledge is differently *embodied* in the new. Still, these modernist and postmodernist answers differ less than one might think. While stressing different mechanisms, they produce the same result. Nonetheless, because both take the transformed nature of knowledge for granted, both implicitly justify "knowledge society" rhetorics.

However, just as plausible conceptually, and as consistent theoretically, is a complementary pair of "antichange" answers to the knowledge question in cyberspace:

- A nontransformative modernist position, actually more consistent with positivism in general; for example, that while the *rate* of knowledge accumulation may be accelerating (presuming it can be measured), the *kinds* of knowledge being added to our "stock" remain essentially the same, as do basic social processes; and
- A nontransformative postmodern position; for example, because knowledge has *always* been situated; it is only *awareness* of this condition, not knowledge's "character" itself, that has recently changed. Indeed, by allowing us greater control over our cultural constructions of knowledge, less mystification means intellectual constructions can potentially be more in line with actual social phenomena, so knowledge, and social processes in general, will tend to become more, not less, stable.

Thus, nontranformative readings, based on either modernist or postmodernist approaches to knowledge in cyberspace, are plausible. Either of these two additional readings are compatible with the "AIT as just another technology" conception of computing outlined in chapter 2 of *Cyborgs@Cyberspace?*

From a positivist *anti*transformation stance, knowledge (verified information) continues to accumulate, as it has at least since the Renaissance, but with only gradual, not particularly profound, social consequences. Contentions to the contrary are held to be manifestations of the lamentably necessary exaggerations of the marketplace. Similarly, the alternative reading of the postmodern view stresses how it is our *understanding* of knowledge rather than knowledge itself that changes in cyberspace. Knowledge itself has always been situated; what is new is our no longer being blinded by modernism. We now see more clearly the situatedness that was and will always be. The understanding of knowledge developing in the current era is more mature, more accurate, but it is not knowledge about some new *kind of* knowledge. Nor are there really new *roles for* knowledge; our present new understandings of them do not change the nature of knowledges existing previously. So, at least in regard to knowledge, "cyberspace" is not all that different; we just understand better knowledge's place in it. Indeed, because we no longer misrepresent it, contemporary knowledge should if anything be *even more* continuous with past knowledge.

Possible Answers to the Cyberspace Knowledge Question

In short, to account for knowledge in cyberspace empirically, we need a substantial complexified way to think about it. To get it we acknowledge that, to be non-self-contradictory, the answer we give to the knowledge question in cyberspace can be compatible with any, but only one, of four possible general claims:

1. That cyberspace is a different knowledge society, because of, for example, the accumulation of an increasing quantity of knowledge that has led to a qualitative social transformation (modernist transformity);
2. That cyberspace is a different knowledge society, because recognition of the fragmentation of knowledges and related phenomena has fostered a new form of knowledge with transformative implications for the world (postmodernist transformity);
3. That there is, as yet, no distinct cyberspace knowledge society in any important sense, because, while knowledge is accumulating, such accumulation has been characteristic of modern society for some time without leading to fundamental break (modernist continuity); or
4. That there is, as yet, no distinct cyberspace in any important knowledge-related sense, because, while how we apprehend knowledge has changed, the character of knowledge itself, and therefore society, has not changed substantially (postmodernist continuity).

That is, before being able to use the phrase "knowledge society" in a meaningful way, let alone talk about knowledge as, for example, "the prime factor of production" in cyberspace, one would have to be able to specify which *one* of the two transformity contentions was correct. Even then, one's argument would be suspect until possibilities 3 and 4 had been carefully considered and rejected.

To make such analyses, we need constructs which support thicker accounts of knowledge. We need, for example, to be able to separate the change issue from the debate over the extent to which all knowledges, no matter how long they have existed or how much they have changed, are context-independent (Gross and Levitt 1994) or situated (Suchman 1987).

Coherent discussion of the knowledge question in cyberspace should turn on which of these four positions is most accurate: Whether conceived via progression or regression, are contemporary knowledge construction, reappropiation, and distribution indicative of a foundational change in the character of knowledge? As analysts, we must be able to use both positivist and situated accounts of knowledges while keeping them distinct. Rather than prejudicially committing to an analytic position at the onset, we need to be able to ask at any point: With which of these perspectives are our current data most compatible?

Unexamined tranformationalism, collapsing of the distinctions between knowledge, information, and data, and reductionisms that inadvertently substitute data gathering for knowledge construction must all be avoided. My experience applying a more complex conception of knowledge in practice (described in Section III), and my discussions with scholars familiar with these critiques, indicate that this is possible. Failure to do so constitutes a societal barrier to development of science, engineering, and technology in cyberspace.

An Implication: Shifting the Focus from Content to Process, from Knowledge to Knowledge Networking

Another reason why it is difficult to keep modernist and postmodernist discourses over knowledge distinct is that the two accounts share more than terminology. Whether stressing progress from data to knowledge or regress from knowledge to data, they incorporate substantial aspects of social processes into their knowledge trajectory. This is something we can see more clearly from the current, STS-informed perspective. Because the production, adaptation, manipulation, and sharing of knowledge are social in both, they both justify "thick" conception of knowledge-related processes. It is this shared centrality of the social to which I point with the phrase "knowledge networking," and it is accounting for this process rather an inert "thing" (knowledge) to which attention must turn in knowledge studies.

Ultimately, this shared space for social analysis means that, rather than being foundationally distinct, these two accounts can be reconciled. They can be used to draw attention to different moments in a more general knowledge dialectic, one that I develop systematically in chapter 4 and deploy in Section III. While these accounts of knowledge are in some ways logically contradictory, the main problem is failing coming to terms with *both* their legitimacy *and* their contradictory qualities.

Instead of deploying constructs that encompass the complexity of knowledge, popular accounts like that of the ex-president presume knowledge's singularity. Such tropes, where knowledge's amount is described as for example, "doubling in the past year," are performed either with no discussion of the unit of measurement at all or a facile one, such as the number of books published. Such simplistic discourse patterns also fit badly with the controversies so characteristic of scientific knowledge creation. "Whiggish" after-the-fact "just-so" accounts of the trajectory of technoscience used to explain one side's victory simply in terms of its greater conformity to nature. STS scholars like Latour have demonstrated the necessity of an "equivalence" principle in historical explanation. Rejecting the simplicity that too often forces science "news" to ignore the frequent debates over the nature of knowledge regularly a part of science, STS accounts draw attention to the social processes through which technology action networks are built up and reinforced, or eroded and collapse. "Pure" postmodern accounts, like that of Baudrillard, are equally reductionist.

Achieving some clarity about the possible answers to the knowledge question in cyberspace is important. To achieve greater coherence, we need a notion of knowledge philosophically open to both modernist and postmodernist accounts and that facilitates empirical study of what is actually happening to knowledge in the world. Indeed, my comments about the ultimate theoretical compatibility of the modernist and postmodernist accounts of knowledge suggest that in exploring protocyberspace knowledge phenomena, we will often find ourselves drawing on both accounts.

It is the chief goal of Section II to draw upon three disciplinary discourses for the tools adequate in practice to this complex task. Here I sketch out the basic ideas. The key step in developing meaningful empirical study of the knowledge question in cyberspace is to focus study of knowledge not on knowledge as a thing but on its process aspects, on the social networking from which content results. Knowledge networking is one of the general, overlapping moments in the reproduction of all human social formations. It is as basic as other general moments like work (the production of day-to-day physical necessities such as food, clothing, shelter, and care), physical reproduction of human life forms, the formation of social relations and relationships, and mediation/communication. As a life form, humans are unique in the extent to which the reproduction of their social formations depends upon the creation of new cultural constructs and their integration into human practices.

In an important sense, when some humans acknowledge that something is now known, when they network knowledge, they are creating or re-creating just such constructs.

In the tradition of Western philosophy, knowledge has been approached as content, defined, for example, as "justified true beliefs." One of the advantages of a social networking approach to knowledge is that it moves away from trying to draw universal boundaries, say, between knowledge items as discrete things, separated from mere information by correspondence to "reality." On a networking account, attention is directed away from truth content and toward the social process through which contentions like those about "truth" are constructed and justified. Thus, by approaching the study of cyberspace knowledge via networking, we avoid the philosophical controversy that confronts any effort to establish grounds for transcendent truth claims. Instead, we focus on the social processes that result in some propositions being considered true, while belief in others is suspended, or they are ignored. Out of all of the cultural innovations whose creation is so characteristic of humans (manifest in, for example, the incredible inventiveness of human talk), only some innovations become widespread cultural constructs, "true" in the relatively limited sense of becoming integrated into the reproduction of a specific human social formation. If we can explain how this comes about in a specific case, we have grasped the important element of a particular instance of knowledge networking.

Our notion of knowledge must therefore also be sensitive to the entire range of knowledge-related phenomena generated by human cultures and their sociocultural dimensions. Thinking of knowledge in terms of the social act of networking reinforces this sensitivity. Contemporary social science interest in networking and use of the term is doubtless related to the diffusion of computer talk, but it also has important prior roots, both in sociology and in anthropology. A focus on knowledge networking highlights how the quality of the social relations involved in knowledge processes is quite variable, both within and across cultures. Because knowledge involves communication, argumentation, and conscious collective recognition that, for example, "belief in X is justified," social networking is an essential component, perhaps the key element, in the construction of knowledge. It is therefore intellectually dangerous to separate permanently any knowledge from its social context. As described in chapter 3, this has been among the important sources of problems in recent efforts to "manage" knowledge. (How to avoid doing this is explored more fully in chapter 7.)

As highlighted in the postmodernist regression, knowledge is now generally understood as a product of "communities of practice" (Lave and Wenger

1994; some problems with notions like "knowledge community" are also addressed in chapter 7). Establishment of a sense of shared purpose and trust in coproducers' talk and action are essential components in construction of the actor networks characteristic of contemporary technoscience alliances (Latour 1987). Networking of necessity precedes production of the more or less formal representations, the "knowledge statements" which are these alliance's public performance faces. Still, as even in a modernist account, social relationships are essential to continuing acceptance and necessary revisions of the protocols through which information is verified as knowledge.

In general, before one can decide if knowledge is changing significantly in a new type of social formation (like cyberspace), one needs a concept of knowledge applicable across social formation types. As an ethnologically valid knowledge construct, knowledge networking fosters cross-cultural study, leading to an anthropology of knowledge. Good accounts of the dynamics of past and present knowledge networking are logically essential to deciding if the character of knowledge has changed significantly.

Finally, a notion of knowledge must be rich enough to account for the possibilities for change arising around new, automated information technologies. Indeed, a focus on knowledge as networking encourages us to make distinctions among various submoments of the general knowledge networking moment. As illustrated in Section III, these submoments include the creation or construction aspect, the adaptation or reproduction aspect, and the sharing or education aspect. Each of these aspects is certainly part of the broader knowledge networking process, and they are often deeply implicated in each other. Thus, if we want to understand a particular knowledge creation networking process, we also have to understand the sharing on which it depends. In terms of answering the knowledge question in cyberspace, however, there are also benefits to distinguishing analytically among them. Thus, the "re-creation" of knowledge by the individual learner that constructivists legitimately argue is essential to learning is not the same as, nor should it be confused with, the creation or the development of new knowledge.

At the same time, distinguishing among these submoments leads to an equally important observation: that changes in the dynamics of any one is as important to changes in the dynamics of the others in answering the cyberspace knowledge question. In other words, if our goal is understanding knowledge change in cyberspace, study of distance learning is just as important as study of the Internet in scholarly communication. Moreover, a knowledge construct useful to understanding the diversity of cyberspace knowledge correlates can also greatly facilitate the design of information infrastructures, including ones mediated by automated information technologies. Better infrastructures can encourage, rather than block, the creation, reproduction, and sharing of knowledge.

Theoretical Aspects of the Cyberspace Knowledge Question

Because it inevitably raises such thorny conceptual issues, answering the knowledge question in cyberspace is a complicated activity. Approaches to answering it are easily lost in metaphorical thickets or conceptual conundrums. One way social scientists keep a sense of direction is by locating their specific question (e.g., the character of knowledge in cyberspace) in a broader problematic or problem field. A useful framing for the knowledge question in cyberspace is suggested by Annemarie Mol and John Law's analysis of anemia (1994). In it, they sketch a general problematic for investigations of the interrelational dynamics of science, technology, and society.

Such relationships, as in my deployment of a "landscape" metaphor for thinking about cyberspace knowledge, are often talked about in topographical terms. They are framed—as in the notion of "knowledgescapes"—as "spaces." Such framings invite emulation of geography, a practice that Mol and Law want to problematize. To avoid being imprisoned by this profession's practices, a more meta-analytic perspective is needed. For this, Mol and Law summon topology, a more encompassing practice than topography:

> As a branch of mathematics, *topology* deals with spatial types . . . [but it] . . . doesn't localize objects in terms of a given set of coordinates. Instead, it articulates *different rules for localizing in a variety of coordinate systems*. Thus, it doesn't limit itself to the three standard axes, X, Y, and Z, but invents alternative systems of axes. In each of these, another set of mathematical operations is permitted which generates its own "Points" and "lines." These do not necessarily map on to those generated in an alternative axial system. Even the activity of "mapping" itself differs between one space and another. Topology, in short, extends the possibilities of mathematics far beyond its original *Euclidean* restrictions by articulating others spaces. (1994, 642–643; emphases in original)

Mol and Law argue for studying social "spaces" topologically rather than merely topographically: "The social doesn't exist as a single spatial type. Rather, it performs several kinds of space in which different 'operations' take place." They go on to theorize three types of spaces. The first is the familiar one of "regions," "in which objects are clustered together and boundaries are drawn around each cluster." Regions contain the kinds of topographic, geographylike relationship that are the obvious referent of spatial discourses.

Their second type is "networks." These figure centrally in technoscience studies approaches like actor network theory (Callon 1986), where the connections between the objects in the space or "nodes" are not so much about place but rather "a function of the [quality of the] relations between the elements," "a matter of relational variety." About networked spaces, we want to know things like how many other nodes mediate the connection between given objects, and what kind of connections they have; we are less concerned about how close they are physically. The Internet is now perhaps the social space of the network of greatest interest.

Mol and Law posit "other kinds of space too," and go on to argue for a third kind that they call "fluid." This is of less familiarity to social theory, but they think this typing best captures the quality of anemia:

> Sometimes . . . neither boundaries nor relations mark the difference between one place and another. Instead, sometimes boundaries come and go, allow leakage or disappear altogether, while relations transform themselves without fracture. Sometimes, then, social space behaves like a fluid. (1994, 643)

I believe their terminology allows us to articulate more precisely what kind of question the knowledge question in cyberspace is. First, it highlights the circumscribed value of topographic metaphors when describing intellectual activity; the similarity between knowledgescapes and landscapes is limited. They both are "scaped," in that their "dimensions" result from a dialectic between preexisting context and deliberate intervention, and in that they can be said to "evolve." However, chapter 4 establishes how knowledge spaces are networks, more similar to the Internet than a nation. Thus follows the necessity of conceiving of knowledge in terms of a process of knowledge networking rather than as an inert phenomenon or thing or place.

Further, we can articulate much of the purported difference in knowledge in terms of the kind a shift in modality implicit in Mol and Law's topological typology. That is, the purported changes in cyberspace knowledge processes can be conceptualized as a shift from a place to a fluid. The metaphors evoked in notions like "information flow" suggest just such a possibility: that knowledge ceases to be thought of as a thing and instead becomes a process, the river itself rather than the water in it.

While a movement toward a topology of flows from a topography of stable things would be radical and potentially transformative, this would not be so if what we thought was a stable topography turned out itself to be much more fluid. Through such phrasings, the notion of change in the very character of knowledge begins to take concrete and recognizable form beyond that of cheery rhetoric. It makes intellectual sense to try to answer it, part of the task of this initial section of *The Knowledge Landscapes . . .* being to shift attention away from the question's typical terms of reference. The practical implications of such a shift are addressed in chapter 3.

Broader Benefits of an Empirically Useful Knowledge Construct

As explored primarily in Section IV, there are additional benefits to conceptually sound research on the knowledge question in cyberspace. The highly distributed organizations recently brought into being face real dilemmas, as evidenced by their at least rhetorical stress on teams, virtual organization, and other social forms to replace those previously based on face-to-face interaction. The decline of the dot.coms raise broader questions; for example, can capital successfully reproduce itself in the future?

This is only one aspect of the still important computerization or "computer revolution" issue. Knowledge management and knowledge integration are only more recent examples of the continuing failure of "computer as brain" approaches (see also, e.g., artificial intelligence and expert systems). What is it realistic to expect from computers?

What is the future of the many other aspects of social life? Are work institutions entering a new stage, one of greater sociality, or will computers allow accomplishment of even greater control of workers? How can we relate ethically to the vast contradictions in the dynamics of contemporary social reproduction engendered by computing: that between, on the one hand, the potential for renewal of sociality in organizations and, on the other, how the profound flexibility of the computer as a medium increases the danger of stricter surveillance and hyperabstraction; or between the hopes computers engender for controlling market-fostered creative destruction and the dangers of loss of social infrastructure?

In sum, I share Steve Fuller's (2000) vision of a social science of knowledge that is truly liberatory, one that both recognizes how science and society are connected and fosters a program for how we should strive to reconnect them more effectively. Such a social science needs to be as clear about what knowledge is for as what it is, recognizing that knowledge is inextricably linked to key social movements and thus a central part of the cultural imaginary.

Chapter 3

"Knowledge Management Fatigue Syndrome" and the Practical Importance of the Cyberspace Knowledge Question

For many in turn-of-the-twenty-first-century North Atlantic business circles, "knowledge management" (KM) was the "killer application" that would justify the massive organizational investment in automated information technologies (AITs). KM enthusiasm paralleled the popularity of the "knowledge integration" program in computer science (see chapter 5), and it fed (and fed off) media hype about the "knowledge society." However, in early 2000, well before the dot-com bubble had burst, interest in the buzz phrase "knowledge management" had declined. Many KM projects were being shut down and talk about it discouraged. I found myself talking about "knowledge management fatigue syndrome," an ironic parallel with post-Vietnam "chronic fatigue syndrome."

The fate of KM has suggestive implications for the broader transformative social project referred to with terms like "cyberspace." The primary intent of this chapter, however, is to show how the KM experience illustrates the practical importance of figuring out what about knowledge, if anything, is different in cyberspace. The difficulties of managing knowledge illustrate why better concepts and accounts of knowledge are necessary, and why we need a sharper sense of the social reproduction dynamics that are practically implicated in the cyberspace knowledge question. This sharper sense helps motivate the complex analytic moves and descriptions contained in the rest of the book. In the following Section III, anthropological, philosophical, and informatics accounts of knowledge enlarge upon the alternative framework for answering the cyberspace knowledge question laid out in the previous chapter.

The current chapter begins with characterization of this new managerial discourse on advanced information technology and knowledge, including my own efforts to relate to it. After describing knowledge management's rise and fall, the chapter outlines several ideas about the causes of these dynamics and shows why assessing their relative importance depends upon an alternative, more developed knowledge construct.

Knowledge Management, the Turn-of-the-Century Knowledge and Organizations Craze

In the mid-1990s, central organs in the business media became preoccupied with a new topic: the importance of knowledge to organizations (Nonaka and Takeuchi 1995). The expanding fascination with knowledge emanating especially from the Harvard Business School's *Review* and its press, being performed most forcefully by several leading business consultants associated with it (e.g., Davenport and Prusak 1997). Some of the concepts used to address the topic were borrowed from computer science discussion of notions like "knowledge integration." The new approach was most fully manifest, however, in the professional discussion about organizations. New management roles, like "chief knowledge officer," were suggested, but the concept used most often was "knowledge management."

Knowledge management is a loosely connected set of AIT-enabled initiatives in organizations. KM is essentially the idea that organizations can be more successful by reconceptualizing themselves from a knowledge perspective—in particular, by rethinking certain internal apprehensions and practices as "knowledge assets" and then using AITs to control and "husband" them as if they were capital. KM was held to be compelled by social developments like globalization and changes in the character of knowledge, both driven by AITs.

The interest in knowledge and organizations surely reflected the increasingly overt role of formal science and engineering in production of commodities, but this new talk was more general. Part of the more general talk came to be related to how "intellectual capital" became an organization's "most significant resource." As such, "knowledge" was now the primary productive force (Davidson, Handler, and Harris, 1994), especially among the large, for-profit corporations that now socially collectivize surplus on a global scale (*Information Strategy* 1998). Similarly, for the international Organization Economic Cooperation and Development (OECD 1994), knowledge production was the "key link" in the technology/job creation chain.

The rise of KM talk paralleled the emergence of the new "knowledge society" discourses discussed in chapters 1 and 2 and tended to displace that on "information management." The new talk borrowed transformative tropes from those preceding it; for example, that new forms of knowledge production, based on ever more sophisticated versions of automated information technology (AIT) were seen as "revolutionary." The KM "org buzz" was reflected in the request for proposals of the U.S. National Science Foundation initiative on Knowledge/Distributed Intelligence: "The recent explosive growth in computer power and connectivity is reshaping relationships among people and *organizations*, and transforming the processes of discovery, learning, and communication" (italics added). This initiative, which included knowledge networking as one of three chief foci, was presented as "bleeding edge" computing and information science (NSF 1998).

The "New" in Knowledge Management

What was new in knowledge management tropes was how the new dynamics of knowledge could not be avoided and had to be controlled. They were the first to locate change in knowledge as the chief locus of change and thus to see "knowledge resources" as the most important competitive advantage. It is easy to see how, in a marketplace where previously effective protections against rapacious predators seemed to vanish with frightening speed, such resources were quickly redefined as "intellectual capital" (Coombe 1998). The strategic importance of organizational knowledge networks was quickly presumed.

Talk about "knowledge resources" had other sources, too. It reflected and was further refined in academic business or organization studies. Here it built easily on the previous work of Bell (1973) and Drucker (1970). Knowledge change is central to other "schools," too; for example, the more postmodern argument of Clegg, Hardy, and Nord (1996). Both traditions posited a fundamental, knowledge-related change in organizational dynamics, for example, from brawn as the primary factor of production to generating and/or accessing knowledge and knowing how to put it to use.

A seldom acknowledged but still important background to the new organization talk was concern about the "productivity paradox" (Attewell 1994, see chapters 10 and 11). Emerging in the late 1980s, this paradox was the name applied to a recognition that organizational investment in automated information technology had not in general led to the expected growth in output per worker. Solutions to the productivity paradox tended to converge on the idea that organizations needed to change fundamentally their structure and ways of operating—for example, through total quality management or business process reengineering (Hammer and Champy 1994)—before they could reap the benefits from AIT use.

The underlying rhetorical thrust of KM had already been worked out in, for example, the comments of the management guru Drucker. They were quite compatible with Katz's "Geek Ascension." This new ruling class ruled because work, a basic social institution, is being transformed:

> Geeks keep their computers on round-the-clock and often work irregular hours, in part because Web-surfing, software trading and collecting and ICQ and Hotline messaging are factored in. Geek-friendly companies tolerate, even encourage this: The more wired and current geek employees are, the quicker they are to spot bugs and glitches, the better and more efficiently they can set up operating and security systems, run up the best spreadsheets and database programs.
>
> At companies run by suits, such activities definitely don't qualify as work. (2000, 84 and 85).

Another manifestation of the organized effort to treat knowledge as the chief factor of production was the creation of (or at least talk of creating) a new executive position to foster knowledge generation, sharing, and reproduction. The main task of the "CKOs" (chief knowledge officer—*Information Strategy* 1998)) was to be management of organizational knowledge networking (Quinn 1992). Academic programs in business responded, perhaps most strongly in the United Kingdom, where centers on knowledge and its management were set up quickly. Publications, both online and in print, emerged as well. In sum, the workplace quickly became talked about as the prime site at which change in the character and role of knowledge was impacting social formation reproduction.

Studying Knowledge Management in the Field

Scholars from various fields who have adopted ethnographic methods quickly began studying the knowledge management phenomenon. One appropriation that was significant in fostering an ethnographic interest in organizational knowledge was the adoption of ethnographic methods by some scholars involved in information systems development, a key moment in the emergence of social informatics (Bowker et al. 1997); see chapter 6. A post-Braverman tradition of workspace ethnographies (e.g., the anthropologist Lamphere 1979) highlighted the role of workers' knowledge in successful labor processes, especially their "know-how" (Kusterer 1978) and its situated quality (see also Sachs 1994). Among other things, this research documented the substantial working knowledge of employees and this knowledge's central place in real production, in contrast to its absence from many formal representations of skill (e.g., in educational credentials). It also explained why problems in making production more flexible were less likely to be solved by increasing the degree of formal control over workers bodies (Taylorization), but more likely by increasing the sharing of knowledge among all involved in production.

In such ways, work ethnography actually paved the way for KM. Work ethnographers were a major presence in the late '90s study of knowledge in organizations. For example, Ann Jordan (1998) and Marietta Baba (1999) participated in NSF-funded research on transformation to quality and global organizations, as well as to explore the effectiveness of teamwork in Fortune 500 companies). Kenneth David and colleagues studied KM in a transnational electronics firm (1998). Julia Gluesing (1998, 2000) compared the knowledge networking practices of transnational Japanese and American firms. Christina Garsten (1994) studied the efforts of Apple Computer Corporation to create a unified transnational organizational culture at sites in Sweden, France, and the United States. Jeanette Blomberg (1998) studied the KM marketing of Xerox, part of a long tradition of the study of organizational knowledge-related activities by the ethnography group at Xerox PARC and the Institute for Research on Learning in Palo Alto. Ken Erickson (1998) studied the work of a chief knowledge officer at a multinational greeting card company, Birgitte Jordan (1998) how certain KM accounts become treated as authoritative in organizations, and Charles Darrah (1998) the "leakage" of organizational KM talk into families, and vice versa.

Unfortunately but predictably, results from the ethnographic study of actual KM were seldom integrated into mainstream KM or development of AIT systems to support it. (see chapter 7) Still, the popular concern about organizational knowledge management, with its actual discourses about and related practices that create, spread, and reappropriate knowledge in organizations, made it a highly desirable subject for empirical study. Because the relevant ethnographers shared certain general interests, I thought it would be possible both to learn about and contribute to AITed knowledge networking by establishing a knowledge network to assemble what they were learning about KM. I planned a large "Knowledge Networking Knowledge Networking" project for the NSF K/DI program and obtained planning funds from the NSF Societal Dimensions of Engineering, Science and Technology program (see chapter 7).

The idea was to create an AIT network that would allow those studying KM to pool knowledge of their individual cases to answer questions like: How successful, in general, were these "bleeding edge" organizations' attempts at "knowledging?" How useful were the new AIT products at supporting this? Could electronic tools "keep the conversation going" when:

1. Essential people were leaving their organization at a moment's notice;
2. Former competitors found themselves suddenly working on "teams" in the same organization;
3. Workers who had built careers on careful husbandry of unique wisdom had suddenly to turn 180 degrees and share it; or
4. Engineers with vastly different educational, language, and/or cultural backgrounds were supposed to construct parts of the "same" standards?

The Beginning of the End

By late 1999, in the face of the cancellation or transformation of several high-profile knowledge management projects, several "expert" observers of the knowledge management scene were answering such questions with "no." In an address to a largely corporate audience, Donald Norman (1999), doyen of more humane computer systems, was quite pessimistic about the general ability of current technologies to support real knowledge networking in organizations. He was particularly sharp regarding the adaptation of electronic systems developed for formal learning contexts (e.g., schools and colleges) to organizational contexts. Similarly, Peter Senge expressed little reason to think that current technologies had added much to the capacity of contemporary organizations to learn.

While knowledge remains a popular topic in easy reading "airport" business literature, it is now a common trope in such literature to attack KM. Thomas Stewart's *The Wealth of Knowledge: Intellectual Capital and the 21st Century* (2002), for example, contains a chapter on "The Case against Knowledge Management." This echoes the theme of John Seeley Brown (lately of the Xerox Palo Alto Research Center) and Paul Duguid's *The Social Life of Information* (2000) and Davenport and Prusak's *Working Knowledge* (1997). On U.S. television, the dense Xerox advertising campaign on "sharing the knowledge" (Blomberg 1998) suddenly disappeared. This abandonment occurred about the time Xerox eliminated the Work Practices and Technology Group that had specialized in work ethnography relevant to knowledge. Cancellation of a similar group at McKinsey Consulting was another indication that performing KM rhetoric had lost its appeal.

Indeed, just after I got my grant, I learned that several potential participants had had their KM projects shelved and some of them had subsequently lost their jobs. This created some obvious difficulty for the ethnographer knowledge network. Having difficulty launching the network in the way I had envisioned, I traveled to Europe during the summer of 2000 hoping to find a new approach in the field. I took a monthlong trip to the United Kingdom, Norway, Denmark, Sweden (including a brief residence at the Swedish Center for Organization Research in Stockholm), and Estonia. I visited researchers and consultants involved in a number of knowledge-related projects. Through in situ discussion, I learned much about knowledge in European organizations to complement my own work in the United States: "knowledge management fatigue syndrome" was already setting in. There was considerable embarrassment about knowledge management rhetorics in the United Kingdom and the Netherlands, where high-profile KM projects had also been terminated or transformed.

The experience of one European organization is particularly instructive. This transnational consumer electronics company faced a severe problem. Having decided to take advantage of cheaper labor in non-Western nations to produce components, it discovered a high rate of incompatibility among parts during final assembly near primary markets. The firm established that the problem was not that parts weren't being produced to specification but that what qualified as the "same spec" varied with the different national training experiences of engineers.

The company could have tried to seize control of the engineering institutions of several nations, or it could have tried to graft its own preferred standards from the top. Instead, in line with the late 1990s buzz, it chose to pursue a knowledge management strategy. This involved implementation of a complex internal "communications ecology," much of which was web-based, in order to foster a detailed conversation on standards problems among its geographically dispersed engineering staff. As engineers developed their own conversation in the free space of an intranet, the company hoped that solution of its business problem would follow, especially if channeled deftly on occasion.

The experiment lasted about six months. Then a vigorous discussion was short circuited from the top by production managers anxious to meet immediate targets. Uniform procedures for measuring were imposed from the top and short-term production targets were met. The deeper, more epistemological issues were left unaddressed, possibly to emerge in the future.

In conversation after conversation, I heard stories like this. The organizational focus on KM, and thus the opportunity to use ethnography to study it, had been lost.

Other manifestations of KM fatigue are worth pointing out. Online knowledge discussion groups, like David Skyme's "Entovation" list, started to refer to knowledge management as a "passing fad." Similarly, at a 2001 conference session on chief privacy officers, a speaker wondered if CPO positions "would disappear as quickly as CKOs had."

It is important to keep in mind that "KM fatigue" was unevenly spread. Interestingly, it was less present in the Nordic countries, where enthusiasm for knowledge projects, and "high end" computing in general, remains high. When asked for their thoughts on this difference, my contacts offered explanations that included the higher level of social awareness, better and more rational social infrastructures, and the relative advantages of Nordic nations regarding current "wireless" Internet devices. The remaining enthusiasm, in the face of the evident failure even then of systems like WAP (wireless access protocol), follows from the political economics of knowledge critiqued in chapter 10.

Indeed, in all regions, participants in my preliminary network of research contacts continue to study knowledge dynamics in organizations. Some sought out opportunities to shift organizations' efforts toward enhancing the social capabilities of AITs. Others worked with organizations to base strategic thinking on information via participatory action research. Alternatively, they helped develop more socially informed standards, researched cross-cultural outsourcing of software development, expanded internal organizational media, and "localized" web-based marketing strategies. Several of them used various forms of gaming, many of which involved IT-based simulation.

One can still find KM advertising in airplane magazines, directed especially at economic sectors less known for innovation. In Canada and the Nordic countries—places whose public sphere is relatively less hegemonized by market liberalization rhetorics than the United States, Britain, or the Netherlands—KM is still treated as an important problem, although I think less so than in the immediate past. Thus, in a study of knowledge management initiatives in the Canadian Federal Public Service, Shields and colleagues (2000) present a critical perspective on implementation efforts to date. Still, they presume the continuing relevance of the knowledge management project.

Many countries in what we used to call the 3rd world (India, Malaysia, Barbados) continue to justify the diversion of a considerable proportion of limited resources in "knowledge society" terms. In contrast, in more "advanced" business circles in the United States, Britain, and the Netherlands, KM has become another of the large, stinking dead elephants in the middle of the conference table that most try, unsuccessfully, to ignore.

Understanding the Rise and Fall of Knowledge Management

Roots of the Craze

Was knowledge management another of those AIT-related advertising slogans whose rapid passing is to be cheered and then ignored as we search for "the next big thing"? Is my insistence that we try to understand the rapid rise and fall of KM just "sour grapes" because my nice project got blitzed, making me look rather stupid? I believe the KM debacle has broader significance, perceptible when it is placed in the context of the underlying structural knowledge-related problems that faced 1990s contemporary organizations. One was the practical problem of coping with work processes requiring knowledgeable, skilled workers when, as in the late 1990s, labor markets are tight. This problem is highlighted in the then oft-told urban legend about "calling in rich." The partner of the start-up dot-com firm, upon finding out just how much wealth she now has as a result of a successful initial public offering (IPO), decides she wants "a life" and quits work. Having constructed a highly fluid labor market—for example, one short term–oriented because of frequent downsizing—organizations increasingly encountered the problem of finding enough workers who knew what they were doing.

Concern with knowledge also made sense because merging records from different firms played a strategic conceptual role in the ideological underpinnings of the way "e-commerce" was then being conceived. Knowledge played a similar role in evocations of the broader new social formation type based on information technology pointed at by terms like "cyberspace." Information is of little value if it isn't used, and, among other things, knowledge is used information. We have created technology capable of producing vast quantities of information, but do we make sufficient use of it? Concern that this was too often not the case, and therefore the whole cyberspace project was at risk, informed interest in highlighting the contribution of social science at the National Science Foundation. Director Rita Caldwell expressed the idea that the NSF needed to spend at least a year focusing on what social science could tell us about why we don't use available information and what to do about it.

In such discourses, knowledge is information validated or in verified form. Knowledge is thus the goal of AIT, for which it can only be a means. It is this aspect of the knowledge problem that is an extension of the "productivity paradox" referred to above. Along the same lines, further doubt about the effectiveness of "high tech" commodities and innovations was causing concern about an underlying contradiction in the structure of new organizational forms. This contradiction is between, on the one hand, the increasing scale and greater complexity of production—often referred to as "globalization"—and, on the other hand, efforts to rearrange production in teams and other less rigid organizational forms.

The scale of activity (e.g., production) in contemporary organizations has grown considerably—from small sites to large ones, from one physical site to many, from local to transnational and transcultural, from mass production of single items to flexible production of many, from in real life (IRL) to "virtual" production systems, and so on. Scaling up labor processes, as by expanding operations to include multiple sites, often in different nations, substantially multiplies coordination problems for organizations.

The difficulty of the traditional bureaucratic structure's relative inflexibility regarding scaling problems was one source of a search for new structural alternatives. One way to confront coordination problems is to organize work in teams. Taking advantage of humans' "natural" tendency to form social relations and share perspectives, teams coordinate more "naturally" and "organically" than bureaucracies. Teams, virtual work, telecommuting, matrix structures, and interorganizational networks—all became popular forms. These approaches received much support from the working out, in fields like Scandinavian information systems development (Allwood and Hakken 2001), of notions like "participatory" or "participative" design. Their value was reinforced by contemporary organization theories stressing the limits, especially in a highly volatile economy, of the strongly hierarchical organization.

However, the social relations that are an essential prerequisite to "coordination by teaming" emerge primarily through face-to-face interaction (Baba 1999). Past "natural" coordination depended highly upon face-to-face social interactions, whose quality humans capture in terms like "trust," "commitment," and "community." Such qualities are most characteristic of non-AIT-mediated, more "organic" knowledge-related activities of "unscaled" organizations. Since teams, like knowledge networking, are "thick," their prerequisite social relationships are made even more difficult to achieve by greater scale and broader distribution in space.

Thus emerges the contradiction: Creating and maintaining such social relations is made much more difficult by the very "globalization" that demands it. The contradiction between upscaling and teaming constitutes the material basis for the interest in knowledge in organizations. Practically, something was needed to bridge the gap between the increasing scale of production, which inhibits face-to-face interaction, and contemporary strategies for organizational development, which appear to demand even more of it (see, e.g., Argyris and Schön's notion of the "double loop" learning organization 1978).

It is in this context that mid 1990s efforts to develop and implement AIT-based organizational knowledge commodities emerged. Software engineers in particular seized upon earlier "soft" developments in informatics—for example, Computer Supported Collaborative (or Cooperative) Work (CSCW) systems, Group Support Systems, Joint Application Design Methodologies (described at length in chapter 6)—and attached them to the "off the shelf" products that were a chief characteristic of the preceding "personal computer" era.

In sum, it was not difficult to convince the typical manager that highly touted information technology, as it got more complex, would provide an infrastructure for "sharing the knowledge" among distributed, if not "deplaced" (Giddens 1991), staff. Thus, "Keeping the conversation going" through technology had to be more than just an advertising slogan; it had to solve the central problem in the contemporary reproduction of capital.

KM Fatigue: Technical Explanations

Participants in my KN^2 network and their contacts described the late 1990s knowledge management technologies as not up to this monumental task. They offered several technical reasons for this.

One was related to characteristics held to be typical of IT commodities: their short "shelf life," the (common) "oversell" of early KM products, continuing technical difficulties in using web-based interfaces to merge complex information bases, and the inappropriateness of products designed for education in the organizational context. The short "shelf life" of AIT slogans, and therefore of products tied to them, is a common explanation for the decline of interest in particular systems. The foreshortened product development cycle in high tech itself feeds into this "flavor of the month" syndrome. The problem of knowledge is still salient, but knowledge management lost its appeal as a slogan. This is the view expressed by Kevin Skyme on his knowledge listserv.

Another related explanation, also common with regard to AIT systems, is that failure to live up to their promise is a consequence of overselling. In a crowded marketplace, each product must be promoted as the "greatest thing since sliced bread." A consequence is that any problem in implementation is taken as evidence that the product is "not as good as it was promoted as being," leading to equally exaggerated negativity. Given that many KM technologies were merely renamed data and information networking products, oversell was inevitable. They were simply not designed to address the problems of creating the trust, commitment, and community-life feel of teams/thick knowledge networking, on greater, more complex scales.

Optimists tended to argue that the next generation would be more effective, albeit under a different slogan. However, actual technical failures, not just marketing/profiling problems, were also significant. A common feature promised by many KM platforms was the capacity to merge multiple data sets into a common so-called knowledge base. The increased ubiquity of web browsers and the rising popularity of web-based internal organizational intranets meant that the World Wide Web became the obvious interface mechanism for such systems. However, it has proved very difficult to use the web in this way. Moreover, at least at the time of writing, tools like a form of Java Script able to address these problems were not yet available. Such problems may be a major source of Xerox's cut backs of staff handling knowledge projects and developing related products.

Conceptual Explanations

More relevant to the argument of this book are the deeper, more conceptual problems that plagued the allegedly more sophisticated AIT KNing products. Underlying and tying together all these factors, and more important than any one of them for my informants, was the failure of many KM projects to take sufficient account of the social.

For some time now, I have been directing my anthropological fieldwork toward organizations struggling with knowledge problems. What I have learned provides insight into the possible reasons for the gap between the need for more effective networking of knowledge and the ability of organizations to use systems oriented to knowledge management to do this.

What I learned in several years of reviewing knowledge-related IT proposals for the National Science Foundation offers an initial point of departure. While proposers grasped the rhetorical advantage of distinguishing knowledge from information technologies, they generally failed to carry through the distinction in their projects.[1] Thus it was common for researchers to ask for money to "knowledge network" but then describe a project in which the act of merging existing "information bases" magically transformed them into "knowledge networks." The problems of coordinating an organization dispersed in space go beyond merely distributing information. For example, they involve being able at least to predict how people in diverse sites will use it. In Section II, I will develop further the argument that it is difficult if not impossible to do real knowledge networking if the task is equated with mere information networking.

A related social/conceptual reason for the decline of KM is the difficulties that follow from constructing knowledge as property. Such a conceptualization of knowledge makes sense from a management perspective: If its knowledge is an organization's chief resource, and the job of managers is to control and make best use of resources, then knowledge needs to be made controllable. It needs to be given more concrete form, enough "thingness" so that it can be handled in a manner comparable to the organization's other tangibles, like raw materials, plant, machinery, capital, and labor. In short, knowledge needs to become a commodity.

While convenient for managers, however, notions like "intellectual property" are on their face oxymorons. The value of an idea only exists when understood, when "alienated," not held. The anthropologist Ann Jordan has done an empirical analysis of the low use levels of a common KM system feature, a "knowledge bank." This is an electronic archive into which organizational members are to deposit their knowledge for others to use. Jordan identified a direct relationship between members' willingness to contribute and their expectation of benefit. However, both of these were inversely correlated to members' sense of how much relevant knowledge they already possessed. As a result, those with most to contribute were least willing to do so, and vice versa.

Finally, not the least of the many problems associated with the notion of knowledge as a strategic resource is that each step taken to protect knowledge as property tends to get in the way of its being shared, let alone verifying its knowledge value or especially "leveraging" it to create new knowledge. While perhaps most evident in efforts to market higher education via the Internet course (see chapter 8), KM regimes encouraged many other organizations to try to sell their knowledge. Like thinking of it as a strategic resource, conceptualizing knowledge as proprietary, commodifying it by turning it into something that can be sold for a profit, powerfully inhibits sharing it within the organization, let alone with a broader, interorganizational network. (How knowledge management got knowledge so wrong conceptually is illuminated further by placing it in the long line of machine-based models of thinking-related claims in computer science; see chapter 6.)

The Broader Relevance of KM Fatigue to the Cyberspace Knowledge Question

James Bailey would have us acknowledge that machines abrogate thought, and Peter Drucker would have us imagine workers as purified knowledge. Instead, to network knowledge, we need workers capable of creating social relations. The unfortunate confusions in general knowledge talk discussed in chapters 1 and 2 were manifest in KM discourses and the software sold to support it. As each "intelligent machine" approach has failed to fulfill its promise, the response has not been to inquire why, or perhaps to consider that there is some deeper incompatibility between the sociality of human knowledge and the formality of management and computing systems. Like its mythic predecessors, knowledge management is itself fading without being subjected to critique.

The explanations of the two previous sections suggest that it would be a mistake to ignore the disenchantment of certain segments of especially the Anglo-American-Dutch world with knowledge management. The organizational knowledge problem in cyberspace is real, not a fad whose passing no one should mourn, nor is the failure of knowledge management to provide adequate answers an unfortunate "blind alley"; the knowledge problem is more intractable and continuing. Certainly, knowledge management did not fail because the problem of knowledge at work turned out to be illusory. While for a brief time toward the end of the 1990s, productivity statistics started to go up, by 2001, they had resumed their downward plunge. While a strong case can be made that the "productivity paradox" is an artifact of an hypostatized productivity concept, it has certainly not been "solved." (see chapter 10.)

Indeed, social formations suffering from a new economy hangover (one feels much sympathy for Proudhon's view that property *is* theft!) are now doubly dependent upon solving the cyber-workplace knowledge problem. First, despite commitment to teaming, it is difficult to imagine organizations abandoning their increased scale of operation and, accepting that there is no real alternative to face-to-face interaction, going back to single locations/cultures/nations. In a "globalized" world, coordination remains the dominant economic problem, and some dispersed activity must be found to solve it. Second, the world's economies have by and large taken a deliberate Schumpeterian leap of "destructive destruction." That is, they have hitched the future capacity of capital to reproduce itself to the gossamer but disconcertingly diaphanous fabric of a "high-tech" economy. As manifest in stock markets, AIT commodities—computers and their peripherals of all sorts, software, even "vaporware," as well as hardware—have become the primary site for the realization of profit. As a consequence, use of AIT has been inveigled into most general solutions to our daunting coordination problem. (See chapter 10.)

In sum, the organizational knowledge problem in cyberspace remains to be solved, and AITs are likely to be important parts of the solution. How, then, is this to be done? Can we design something better?

Better designs depend upon placing KM in the context of a broader understanding of the more general process of knowledge networking. This must include clarification of the character of contemporary knowledge and the extent to which it is being transformed. Such broader issues must also be dealt with before an effective approach to the scale/team contradiction in organizations can be developed. (I develop such a program in chapter 7.)

Despite the decline of knowledge management in some places, knowledge issues remain at the center of sophisticated IT practices. Before we can use AIT to solve the knowledge problem in cyberspace, both more generally and its specific organization manifestations, we have to have a better understanding of knowledge, especially its cultural dimensions. Even assessing the relative importance of the various explanations for knowledge management fatigue proffered in the preceding section also depends on this broader analysis.

The rise and fall of knowledge management raises a series of significant and interesting issues for empirical examination. As described in Section III, ethnography of AIT can contribute to explaining why this is so. At the same time, ethnography's contribution is also limited by inadequacies in its own account of knowledge. The goal of Section II is to provide the kind of theoretical account of knowledge that will allow both ethnographic and IT scholars and practitioners to carry out a metadiscourse about knowledge as sophisticated as the knowledges that they would construct and share. As demonstrated in Section III, this approach to knowledge also allows us to conceive more appropriate roles for AIT in addressing our collective knowledge problems as we move toward cyberspace and all that it entails. With such an understanding, it may be possible finally to "get right" the role of AIT in organizations.

Section II

Ethnological Perspectives

Three Sources of Concepts Prerequisite to Answering the Cyberspace Knowledge Question

Chapter 4

Toward a Philosophical Basis for the Study of Knowledge Networking

The Cyberspace Knowledge Problem in Philosophical Perspective

The Necessity for a Properly Philosophical Answer to the Knowledge Question in Cyberspace

Much of Section I of *The Knowledge Landscapes of Cyberspace* focused on how problematic notions of knowledge impede serious examination of "knowledge society," "cyberspace," or other notional evocations of profoundly different social dynamics associated with computers. For example, chapter 2 analyzed the logical conundrums in the popular idea of a computer-based "knowledge revolution," tying them to the multiple, often contradictory implications related to unrecognized difference in constructions of knowledge. Chapter 3 documented how conceptual confusions had practical manifestations in "knowledge management fatigue syndrome."

Besides critique, Section I also introduced some elements of a potentially more fruitful approach to knowledge. These included:

1. Seeing knowledge as a process of networking
2. Conceiving of "knowledges" in the plural
3. Cultivating social conceptualizations of knowledge

However, the standard knowledge construct to which these conceptions stand opposed–that knowledge is a thing characterized by its content, that it is unified, that its prime locus is in individual minds—is taken to be "common sense." Moreover, the standard conceptions are embodied in institutions, like educational systems, that are central to social reproduction as well as Western cultural themes like technoscience.

To rethink knowledge in the ways suggested, to displace the current frame for considering the relationship between technology and social change, let alone to displace a broadly popular notion, is thus a formidable task. Such an alternative approach to knowledge needs to be developed carefully and fully.

Providing such a rigorous underpinning for this alternative conception of knowledge, the central task of Section II, is done by grounding the alternative systematically in three intellectual traditions. This current, first chapter of the section begins by placing the cyberspace knowledge problem and my alternative approach in the context of philosophical approaches to the general problem of knowledge, or epistemology. The goal is to establish the intellectual legitimacy of a view of knowledge that stresses networking, multiplexity, and the social.

The following chapter, on the anthropology of knowledge, concentrates on empirical study via ethnography, the standard mode of cultural anthropological study, of knowledge so conceived. It identifies intellectually and connects to an emerging new approach to studying knowledge in a coherently cross-cultural or ethnological manner. The chapter demonstrates both the general promise of a vigorous ethnology of knowledge and the specific benefit of it to answering the knowledge question in cyberspace.

The agenda of chapter 6, to legitimate the alternative approach to knowledge within informatics or computer science, parallels that of the previous two. By highlighting the centrality of knowledge questions to this field's history—indeed, knowledge is argued to be the underlying preoccupation of the field—the chapter addresses theoretically the problems in mediating knowledge via automated information technologies (AITs) that have so far been approached practically in *The Knowledge Landscapes* . . . . Knowledge problems in informatics are connected to misleading ways of treating knowledge afforded by the scientistic modernist epistemology discussed in the current chapter. Also, like the other chapters in Section II, chapter 6 identifies and builds on an approach that already exists in the field, social informatics.

All three chapters highlight how the alternative approach of knowledge networking offers a better way to answer the knowledge question in cyberspace. Providing intellectual legitimations is a difficult but still limited task. In the process, the chapters also ground certain conceptual tools for the book's main task, using ethnographic research actually to answer the knowledge question in cyberspace.

The reader will notice a difference in rhetorical tone between chapters 4 and 6, on the one hand, and 5, on the other. Anthropology is the profession that I have been practicing for some thirty-five years, so the argument I make for the new approach to the anthropology of knowledge is based on considerable "craft" experience. Like the current one on philosophy, chapter 6 operates on conceptual terrains less clearly my own. (While I have been teaching and researching informatics for some time, I still don't think of it as an intellectual or professional home.) Thus, chapters 4 and 6 contain (hi)stories of the knowledge problem in philosophy and informatics, respectively, that are the reading of a relative outsider. Real histories being themselves major research topics, the readings of the knowledge problem in these disciplines that I offer doubtless contain inadequacies. I present them not because I lack respect for the intellectual work of careful scholars in these fields, but because I cannot conceive of a good answer to the cyberspace knowledge question that does not take philosophy and informatics seriously. However misguided they may be, I hope my comments at least stimulate better efforts to address the cyberspace knowledge question from within these fields.

Section III goes on to use perspectives and concepts developed through the readings of Section II, both professional and "unreliable"/partial, to analyze actual ethnography. The results described there should justify pragmatically the discipline boundary transgressions of this section.

Knowledge Questions in Philosophy

The problematic aspects of knowledge have long been of concern to philosophy, a field whose name means "love of knowledge." It is thus reasonable to begin my alternative account of knowledge in dialogue with philosophical perspectives. Starting here is made easier because, unlike so many of the cybernauts referenced in Section I and described in *Cyborgs@Cyberspace?,* philosophers generally abjure the optative mode of speech. Moreover, in stressing the multiplicity of knowledges, the need to base understanding of them on study of their "really existing" forms, and their appreciation of the processual aspects of constructing them, several current philosophies are quite compatible with the alternative knowledge construct I advocate.

As knowledge problems have long been its preoccupation, however, a thorough discussion of knowledge in philosophy would constitute a general history of the field. This chapter is necessarily more schematic. Initial comments on thinking about knowledge philosophically open conceptual space for the deeper probe of a philosophical version, here labeled scientistic modernism, of the popular but inadequate conception of knowledge critiqued in Section I. Outlined next are several of the multiple, diverse critiques of this conception. While more illustrations than thorough philosophical arguments, they serve to introduce some of the alternative epistemologies at play in contemporary philosophy. One of these alternatives, contemporary or "neo-" pragmatism, provides a particularly sound epistemological basis for answering the knowledge question in cyberspace and is developed at greater length. The chapter concludes by focusing on how a pragmatic alternative approach helps frame more appropriate research on knowledges in cyberspace.

Philosophical Formulation of the Knowledge Question in General

To discuss a topic reasonably, one needs at least a provisional notion of what the topic is. As provisional this notion is, of course, subject to later modification. In this spirit, I start with what *The Cambridge Dictionary of Philosophy* (Audi 1999) refers to as the "standard analysis" of knowledge's typical apprehension, that it is "justified, true belief" (1999, 274).

At the most general level, a "belief" is a describable orientation. Standard ways of thinking of belief include seeing it in terms of inclinations readable from behavior, or as a "strongish" tendency to assent to actual propositions. This latter in particular distinguishes a belief from an idea or a mood. Humans, at least, seem to believe things, and collectively what they believe seems important. For example, at least since Aristotle, Westerners have made a connection between human beliefs and the capacity for deliberate human action, or agency. When we act as agents, we act in ways for which we can be held responsible. Being able to say what we believe, part of accounting for our actions by explaining why we did what we did, is taken legally as indicative of agency. In this way, a connection exists between beliefs and ethics, civility, sanctions, and many other significant aspects of human life. (How agency may be different in cyberspace is an issue to which we turn in chapter 12.)

At the same time, most would acknowledge that what humans believe comes in many varieties and has diverse qualities. Among the important qualitative distinctions often made with regard to a belief is its "truth condition." That is, we distinguish between beliefs that we feel are more or less aligned with the world, to be "true," and others felt not to be so aligned. Further, at least some of the beliefs taken to be "true" are "justified" or "justifiable." For philosophers, for a "justification condition" to exist often means that the belief is warranted or true, but it can also be taken to mean merely that a notion is actually believed—demonstrably taken to be true by the person holding it. That is, the holder has explained, or is apparently at least capable of explaining, why she holds the "truths" she does; for example, that they are "self-evident," proved to be true through experimentation, justified by faith, and so on.

It is this latter, discourse dimension of justification, rather than the "warranted" part, which is of most relevance to the alternative knowledge discourse that this book wishes to foster. Before we can focus on it, however, we must deal with philosophy of knowledge's tendency toward metadiscourse, its being inclined more toward "commenting on"—what is often called "metaphysics"—than "doing" or what Aristotle termed "practical philosophy." Commenting on beliefs has typically involved constructing ostensibly more general accounts for the more specific knowledge justifications offered by individuals within cultures. Especially in the Western tradition, forms of philosophy of knowledge have tended to address questions like the character of the beliefs themselves, what it means for their holders to think of them as true, and the nature of the accounts offered in justification of these claims for truth. "Knowledge" can be distinguished from "mere information" on this line of thought, because the former unlike the latter is composed of the body of convincingly explained and therefore true beliefs. By implication, "knowledge" is a form of "conscious" belief, in more than one strong sense: Not only can its possessor state what the belief is, but she can also provide an accounting for why she holds it to be true, an accounting which an audience is likely to acknowledge as convincing, or at least highly plausible.

Also in this tradition, such an account or an epistemology is distinguished from "ontology," that aspect of a philosophy that addresses what exists, or "being" (or "Being" in several articulations of the tradition). A basic issue in any philosophy of this sort is the extent to which the things that exist are knowable as well as the kinds of things that exist to be known about them. A philosophy's epistemology, in contrast, is its notions about how knowledge is to be accounted for. Given the Western preoccupation with justification, it makes sense that "epistemology" has often been used to label the entire field of the philosophy of knowledge, including ontology, major aspects of which are also about knowledge (a point of periodically particular salience to the knowledge informatics explored in chapter 6).

Thus, to argue in this philosophical tradition for the cyberspace knowledge hypothesis—that there is a fundamental change in the character and/or social functions of knowledge in computer-saturated circumstances—one might make one of three cases. One is that what is and/or its characteristics, including its "knowability," changes (an ontological case.) A second is that how we know what is changes (an epistemological case). Ken Goldberg (2001) coins the term "telepistemology" in raising such an alternative, discussed below. The third is that both the "what is knowable" and the "how we can know it" change. This third case is in some sense the most typical, in that there is almost necessarily a connection between how we can come to know something and what exists to be known about.

Scientistic Modernism, the Popularly Dominant Approach to the Knowledge Question in General and Its Connection to Western Philosophy

Arguments in epistemology often follow a common pattern. Following an articulation of other knowledge accounts, one critiques these, offers one's own account, and then proceeds to compare their quality. In the tradition of Western philosophy, many professional philosophical performances turn on differentiating "better" (one's own) from "worse" (others') strategies for properly justifying beliefs.

For a rather long time, epistemology has privileged science as the exemplar of knowledge creation. For example, Aristotle equates *episteme* with scientific knowledge (Audi 1999, 46), while Quine begins "Epistemology Naturalized" with "Epistemology is concerned with the foundations of science" (1994, 15). Particularly from at least the seventeenth through much of the twentieth Centuries, science, philosophically speaking, was seen to provide better general accounts of how knowledge is achieved than, for example, ones based on faith. Such arguments (e.g., Bacon 1944) tended to stress, for example, that the explicit and formal procedures of science are more accessible than are the indescribable ones of religion.

Philosophical accounts of knowledge that presumed science to be the prototype of knowledge production came to be labeled "epistemology." Despite their declining stock in professional twentieth-century philosophy (see below), they continue to provide the conceptual basis for the more general conceptions of knowledge, including their problems, outlined in Section I. Epistemology tended, for example, to extend itself, claiming knowledge accounting procedures purported to produce *universal* knowledge. Such accounts of knowledge are now commonly characterized as being "foundationalist" or "essentialist." "[F]*oundationalism* . . . [is]...the belief that we possess a privileged basis for cognitive certainty and *essentialism* . . . the belief that, conceding whatever difficulty and self-correction may obtain, the structure of the actual world *is* cognitively transparent to, or representable by, us . . . "

(Margolis 1993, 38). Capital "E" Epistemologists strive after standards of knowledge applicable in any circumstance. Foundationalist accounts aim to be all-encompassing and to account for a presumed essence of knowledge. Emphasis on the universal quality of truths justified in this manner fostered a conception of knowledge not tied to any particular context. "True, justified beliefs" were held to exist more or less on their own and have everywhere the same essential qualities.

In an important sense, foundationalism is an understandable philosophical tendency. After all, what is the point of talk about talk—metadiscourse—if it isn't to find a more generally satisfying or encompassing point of view? It also makes sense historically that forms of philosophy emerging from the Middle Ages, having to create distance between themselves and a foundational Christian theology, would do so as alternative foundational projects, not in opposition to the very idea of identifying universal truths.

As scholars gained a right to inquire independently into the character of the world, Epistemology gained substantial currency. Everyday language increasingly identified knowledge with science. Popular conceptions of knowledge today are colored by this foundational reading, including the idea that "real" knowledge grasps essences not bound by context. As illustrated in chapter 1's discussion of Jon Katz's account of the "Geek Ascension," the fact that real knowledge is foundational is taken for granted even in stories that on the surface seem to be arguments for a new knowledge account.

The deep foundationalist commitments of this science account of knowledge are famously embodied in Newton's mechanics. In it, science is at base a formal epistemology, a set of explicit procedures to produce universal knowledge. Scientific epistemology, if followed correctly, leads to truths that transcend the particular conditions of their creation, including general or foundational truths about the things that exist, an explicit ontology.

Newtonian science is thus scientistic, in that it presumes a correspondence between ontology and epistemology and that there are universal laws that govern physical events, laws that, once discovered via science, can be seen to govern all such events. This presumption that scientific knowledge is composed of universal truths (is knowledge) is arguably the characteristic attitude of modernism. On a modernist account, for example, the point of design is to give knowledge (scientific truth) palpable form. Hence, I will use "scientistic modernism" to refer to the popular epistemology critiqued in Section I that reflects much from the tradition of Epistemology in Western philosophy.

An important aspect of the scientistic modernist view, carried over from Epistemology (and in Western philosophy, at least initially, a legacy of its roots in theology) is the notion that there is a difference between the laws that govern the physical (matter) and the psychical (soul) arenas of universality. Coping with this difference, often referred to as a mind/body dichotomy, is in an important sense the key preoccupation of Epistemology. As mind activity, science came to be viewed as the prototypical use of rational argument, ultimately identified with logic. Identified with values like elegance and parsimony, rationalism fosters preferences for the simple over the complex, the reductionist over the elaborated (Toulmin 2001).

Developed initially in opposition to pure rationalism, as a better way to overcome the mind/body dichotomy, was experimentalism. This alternative was held to be needed because logic could not alone be a sufficient guide to true justified beliefs, perhaps because our mental processes are subject to certain difficult-to-specify distortions, such as those that plague our sense impressions. Perceptual processes must be disciplined if minds are to recognize the laws that govern the physical world. Experimentalism holds that the mind can be disciplined by testing rationalistic notions about laws in controlled conditions. In experimentalism, a laboratory, a physical space for abstracting thought-to-be-basic processes from the deep contextuality in which they are normally encountered, provides an ideal space in which to confirm or disconfirm rationalistically derived ideas of the true.

Experimentalism was eventually incorporated into scientistic modernism. While one can certainly differentiate more rationalistic from more experimentalistic moments in scientistic modernism, they share an underlying commitment to foundationalism: that the ideal forms of knowledge are formal statements about essences, statements decontextualized, statable in terms of formal propositions about relationships among abstract entities. Science is the reliable means to "justified true beliefs."

In the nineteenth-century West, these ideas were combined in terms of a planned, orderly search for knowledge of the social that came to be called positivism. Central to the positivist program was the presumption that knowledge of human affairs could not only ultimately be made unitary, but also was cumulative and therefore subject to planned efforts to get at it. Under positivist science, the dominant moment in gaining knowledge possesses the goal-oriented or teleological quality associated with empiricism, not the negativism associated with criticism. Positivist scientific knowledge is held to be the inevitable result of enough of the proper ratiocination cum experimentation, that these will lead ultimately to the proper understanding of a phenomenon. In the phrasing popularized by Frederick Winslow Taylor and Frank Gilbreath, science promises the ONE BEST WAY, an articulation of what we know that is recognizably superior to all others. Discovery of this one best way provides a positive basis for, in the Taylor/Gibreath case, the scientific form of management. The standpoint from which this knowledge is identifiable would be objective, superior to all other more subjective, less general points of view.

In the discipline of philosophy, despite criticism, ideas like these were long cultivated within the philosophy of science, and they continue to be (e.g., Ackerman 1972). This was especially true for those philosophers of science whose program was designed to be compatible with that articulated for them by empiricist natural and positivist social scientists (see, e.g., Labinger 1995).

In this view, because they are humans with minds, and minds and bodies do not follow the same laws, scientists sometimes get confused, especially with regard to concepts. The job of philosophy of science thus came to be that of concept policing, contributing to pursuit of scientific truth by pointing out the occasional conceptual lapses in scientists' existing programs to coordinate ratiocination and experimentation. Epistemology, the general project of which this kind of philosophy of science was a part, was held to ground knowledge on propositions that had universal scope. It was pursued in some other humanities and social as well as physical sciences, reaching an apogee in the effort of Alfred North Whitehead and Bertrand Russell to articulate *Principia Mathematica* (1927), principles which in their generality were to model how Science would cover all events. In American social science, Epistemology is evident in the work of Talcott Parsons, especially his privileging of statistical hypothesis testing of, ideally, experimental results. If these were unavailable, one could resort to other types of findings as long as they were empirical and especially if these could be trusted to be "objective," because of the "scientific" methodology through which they were "discovered" (not invented!).

I believe that this scientistic modernist road to justified, true beliefs, one mapped on an idealized form of laboratory science, has explicitly or implicitly informed the majority of twentieth-century efforts to create knowledge outside of philosophy, including Knowledge Management. This road encourages reductionism, explaining phenomena at one level of complexity in terms of phenomena at a lower level. Throughout much of the twentieth century, this modernist account of knowledge, including its reductionist tendencies, was as influential among Marxists, like Bernal and Cornforth, as among other social theorists. It is still evident, for example, in the late-twentieth-century effort to create a sociobiology or evolutionary psychology. It is the view of Paul Gross, a self-appointed "science warrior," for whom there is only one Epistemology, not several epistemologies.

Content, Ontological Readings of Knowledge

Corollary to this scientistic modernist approach is a conception of knowledge as an achieved, decontextualized state rather than a process, a "content" reading of knowledge as a "thing." In computer and information science, this content reading of knowledge reaches something of an apotheosis: "The image of knowledge, and of our access to the world, that is projected by the Internet can thus be viewed as a continuation, perhaps even an exemplification, of a set of ideas that lie at the heart of modern epistemological thinking" (Malpas 2001). Considerable ontological concreteness came to be applied to the presumed progression from data to information to knowledge discussed in chapter 2. While in many speech acts, these three terms continued to be used more or less synonymously, these new fields, through much debate, repositioned the middle one as most appropriately the object of their sciences. On the new telling, described in detail in chapter 6, a professional in these fields begins with data, preferably from experimentation. Whatever their source, these data are taken to be "raw"—that is, as things existing without context. When such data are deliberately placed in a context, say, that of a set but in any case in some form of orderly relationship to each other, and then processed in a specifiable way, the result is taken to be information. These bits of information could then be further compared rationalistically, via formal, known scientific procedures (e.g., algorithms) and with existing scientific knowledge. Information further processed in this way and/or confirmed via the comparison is held to be new knowledge. (While popular discourses allowed for a further progressive stage, to wisdom, such moves imply contextualizings different from "informating" ones, and are in any case not held to be necessary to knowledge creation.)

At least in part via such theoretical work in computer science, and based on its dazzling embodiment in computing machines, knowledge lost much of its ideational character in popular consciousness. Instead, it became constructed as a tangible thing, ensconced in artifacts and measurable (see, for example, the work of de Solla Price critiqued in chapter 7). One might say it passed from having a connection closer to the "mind" side of existence to being part of the "body" side, from an act of communication to being a thing. Identification of this new kind of knowledge, and the central role of computers in creating it, led Herbert Simon to declare the existence of a whole new class of sciences, those of the artificial sort (1996). Such sciences don't discover new ideas about things that already exist; they convert ideas into new things, declaring what they see when they study the new artifacts to be science—computer science, not computer engineering.

This shift to more purely ontological readings of knowledge has many analogs, including the current economic idea that knowledge is an (the most?)

important form of special thing, capital. (The knowledge capital theory of value is examined in detail in chapter 10.) That is, whereas knowledge might previously have been considered to be an abstraction, in a knowledge society it has as much "thingness" (is as much a *Ding an Sich* or "thing in itself"?) as any physical artifact.

Ontological Epistemology, Postmodernism, and the Continuing Influence of Scientistic Modernism

I believe ontological readings of epistemology remain an important part of the postmodern sensibility, which in this regard indexes an ongoing influence of scientistic modernism. Extant discourses, like those labeled "hyperreal" by Jean Baudrillard, are sometimes taken to indicate the pending demise, the "lateness" of current modernism. The hyperreal is generally treated as presaging the emergence of a contrasting, postmodern way of life. However, they also indicate beliefs about "hyperthings" in the world, things constructed by—as well as, in some sense, demonstrated to be true by—artificial means and which today constitute knowledge par excellence. In my view, students investigating physical laws by manipulating computer simulations of "the" universe are hyperreal in essentially the same sense as the performance art of someone like Christo. Such reading influences, for example, contemporary studies of material culture that focus less on physical things and more on popular practices. That the postmodern sensibility continues to share much with the modernist one is an important justification for Latour's (1993) and others' (e.g., Hakken with Andrews 1993) efforts to differentiate between "postmodern" and "nonmodern" accounts.

At the end of the twentieth century, a scientistic modernist conception of knowledge as content apprehensible largely free of context, one based strongly on an epistemology that privileges science, remains deeply embedded in Western culture. This is the case despite the obvious contradictions of assigning "natural" reality to something one has designed, the insoluble problem of appropriately separating conceptually the mind from the body, or the misperception of seeing knowledge as inhering primarily in individuals. The goal of this chapter has so far been to bring out the philosophical roots of the popular conception of knowledge critiqued in Section I by describing its correspondences to epistemological notions drawn from philosophy. While some forms of professional philosophizing, especially in philosophy of science, clearly do bear some responsibility for this popular conception of knowledge, it has not been my goal to address the degree of this responsibility. It is beyond my capability to complete that task well; besides, several scholars have taken it up (e.g., Fuller 2000). Rather, I hope to have described why the validity of a popular, scientistic modernist conception of knowledge, one presumed to have been demonstrated by philosophy, is pervasively taken for granted.

Critiques of Scientistic Modernist Knowledge Accounts

Scientistic modernist (SM) accounts of knowledge as real essential truth, arrived at progressively by traveling a formalist, universalizing road, have long been subject to criticism and the projection of alternative accounts. Indeed, it is my understanding that non-SM alternative readings of epistemology are now the most prominent ones in the philosophical study of knowledge. (Parallel nonfoundationalist readings in ethics and aesthetics are taken up in chapter 12.) In the midst of a epistemological intervention central to the second half of the twentieth century on "Epistemology Naturalized," W. V. O. Quine (1994) traces twentieth-century philosophical epistemology's distancing from foundationalism in the following terms:

> Philosophers have rightly despaired of translating everything into observational and logico-mathematical terms. They have despaired of this even when they have not recognized, as the reason for this irreducibility, that statements largely do not have their private bundles of empirical consequences. And some philosophers have seen in this irreducibility the bankruptcy of epistemology. Carnap and the other logical positivists of the Vienna Circle had already pressed the term "metaphysics" into pejorative use, as connoting meaninglessness, and the term "epistemology" was next. Wittgenstein and his followers, mainly at Oxford, found a residual philosophical vocation in therapy; in curing philosophers of the delusion that there were epistemological problems. (1994, 24–25).

Quine is by no means alone in this assessment of contemporary epistemology. In *Groundless Belief: An Essay on the Possibility of Epistemology* (2. ed.), Michael Williams defines "phenomenalism" to include the notion that things have transcendent, discoverable essences. He goes on to characterize his own intervention as

> an argument that epistemology as traditionally constituted, the search for foundational knowledge of knowledge, is tied to phenomenalism of one sort or another, and that it is therefore flawed. Thus, epistemology is not a coherent discipline.
>
> In so far as concern with the theory of knowledge had been a matter of choosing and defending one of these traditional theories, the theory of knowledge will not be something we need concern ourselves with further. (1999, 24)

What under scientistic modernism was the chief characteristic of real knowledge, foundational certainty, is now seldom sought in professional philosophy. While Williams would abandon Epistemology, even epistemology, for others like Quine, epistemology remains a problem that cannot simply be abandoned. Like me, he searched instead for alternative, non-SM approaches to describing how we know.

Linguistic Critiques

I wish ultimately to connect my alternative conception of knowledge to one such alernative epistemology, that of pragmatism. Before doing this, however, I present a variety of SM critiques that illustrate their breadth.

Chapter 2, for example, alluded to the a rather common critique of the progressivist rhetoric characteristic of SM talk about knowledge, rhetoric whose optative character is often employed by knowledge revolutionaries. An alternative, anti–scientistic modernist knowledge construct could also be justified in the manner of anthropological linguistics, in theorizing grounded in everyday language practice. One could, for example, point to the continuing overlap among the meanings in use of "data," "information," and "knowledge" to contest foundational notions of knowledge, and thus by extension how the knowledge progression associated with SM and outlined in chapter 2 is incompatible with what knowledge means in the actual world.

The History of Language Critique

Even stronger grounds for alternative approaches to *knowledge*, ones that, for example, stress process over content, can be found in the history of the term's use. The *Oxford English Dictionary* (*OED*) begins its entry on it by stressing the extreme breadth of the term's signification or meaning in contemporary usage:

> From the fact that *know* now covers the ground formerly occupied by several verbs, and still answers to two verbs in other Teutonic and Romanic languages, there is much difficulty in arranging its senses and uses satisfactorily. However, as the word is etymologically related to [Greek and Latin roots meaning] to "know by the senses," . . . it appears proper to start with the uses which answer to these words, rather than with those which belonged to the archaic vb., to WIT, Ger. *wissen*, . . . to "know by the mind." This etymological treatment of the word, and the uses to which it has been put, differs essentially from a logical or philosophical analysis of the notion of "knowing" and the verbal forms and phrases by which this is expressed . . . (1968, 1549).

The OED's historians of usage here imply that in English the word *knowledge* came to displace multiple other terms, thereby achieving a kind of hegemony over much cultural terrain. This breadth of coverage certainly helps explain why, as noted in Section I, use of the term is often very confusing.

There is clearly an interesting investigation to be done on the historical process through which this hegemony was achieved. Indeed, it may have had much to do with the growing influence of the scientific modernist knowledge conception itself. Stephen Toulmin's *Cosmopolis* (1992) offers an account for just such a shift. He argues that the modernist view outlined above developed in the seventeenth century, out of an intellectual/ruling class desire for certainty after a long period of social turmoil, and to counter the religion-based conceptions at the root of the Hundred Years' War. Especially desired was a separation of knowledge from any particular confessional process. From about 1650 until well into the twentieth century, the philosophy of knowledge was dominated by this seventeenth-century view, embodied in the Cartesian "research program" of René Descartes.

The *OED* entry is also a caution to those who, like me, would try to develop an analytically tighter knowledge construct. We cannot depend upon the history of its use or its historical roots to sort out usage problems for us. Still, while we must ultimately ground coherence elsewhere, a speculation that appears later in the entry provides a useful clue for where to start:

> The origin of **knowledges** [substantive noun] and [verb], and the question of the original relationship between [them] are a difficult problem . . . [I]t is possible that . . . [in Middle English] there was formed a derivative [of "know" meaning] . . . to "become conscious of, make acknowledgement or confession of" . . . If this was so, the verb *to knowledge* was formed first, and the [substantive noun] was derived from it, [so that the verb] would [have provided] the original sense of the [substantive] . . . (1968, 1550)

In other words, our contemporary word *knowledge* may be a derivative of a verb—possibly "to knowledge." Were I to use "knowledge" as a verb while teaching—"I knowledge your awareness of this point"—my students would, with much justice, think I didn't know how to talk properly. Still, because this ancient verb "to knowledge" had a meaning close to the contemporary "to acknowledge," I believe I could give such usage a sensible explanation. The point would be to demonstrate how the word *knowledge*, as it was transformed into a noun, originally meant something like "that which results from a process of acknowledgment, is acknowledged to be."

If we wish to understand some "bit" of knowledge, we often, reasonably, focus on how it came to be treated as known, the very process of acknowledgment. In this regard, then, my desire to shift our concept of knowledge away from content to process is actually an act of *re*covery, a move back toward a more original "process" orientation as opposed to the later "content" view. If, as the etymological speculation of the OED suggests, conceptualization of the processual act of "knowledging" historically preceded content, "thing"-oriented knowledge constructs—that is, that linguistic marking of process preceded marking of content—then for this time attention to process was greater than attention to content. If this is no longer the case, but we wish to make it so again because we think the shift pernicious, both analytically and practically, understanding that the original shift actually took place helps our case. To the extent that content views of knowledge arise in concert with modernist conceptions of science, my efforts have a certain etymological justification: The recapture or at least understanding of more orginal meanings of a term can be a way to backtrack and thereby avoid subsequent confusions.

Compatibly with these thoughts, the OED acknowledges the existence of contemporary verb forms of the term, including the following definitions:

> **Knowledge**, [verb] . . . 1. [transitive] to own the knowledge of, to confess, to recognize or admit as true: =ACKNOWLEDGE . . . 3. To own as genuine, or of legal force or validity; to own, avow, or assent in legal form to (an act, document, etc.) so as to give it validity . . . (1968, 1551)

Thus, as least sometimes, contemporary people "knowledge," engage actively in acknowledgment, as when a speaker urges an audience to "take ownership" of an idea. Rationally, before there can be a content acknowledged to be "knowledge," there was and has to have been a process through which the content is "confessed" or recognized to be known, acknowledged to have validity. Further, because such a "confession" has to be shared, acknowledging its necessity logically implies also acknowledging the primacy of the *social* process through which such a confession occurs. This is precisely the shift, à la Heidegger, that I wish to accomplish: to direct study of knowledge away from its presumed "thingness" as universal content and toward "knowledge networking" as a process, a sociocultural one through which particular contents are acknowledged as knowledge, one that produces, adapts, and shares.

In sum, I ground my rethinking of "knowledge" on several points whose acknowledgment is attainable via such linguistic/etymological exegesis:

1. On the wide dominance of our current word "knowledge," displacing terms used to distinguish among dimensions of understanding which other times and cultures have kept distinct, a linguistic hegemony that is an important part of the scientistic modernist conception of knowledge;
2. Of the likely prior existence of "knowledge" as a verb ("to knowledge"), not just as a substantive noun; indeed, its action/process use may well have preceded its content use;
3. Of how the term's original sense may well be something like the action of acknowledging or certifying; and
4. Of how legitimate uses of "to knowledge" as a verb continue to exist;

That is, in the scientistic modernist account, knowledge's truth is considered to be a quality inherent in it as an object; it is part of its character to be a piece or "bit" of knowledge. There is, however, another, equally legitimate way to think about knowledge, to get it out of its "black box," by directing attention toward the process by which this "acknowledged to be known"characteristic came to be confessed or justified. Since such a process necessarily includes a group of confessors, this alternative approach to the study of knowledge necessarily shifts attention toward the *social* process through which claims of truth come to be acknowledged. It suggests that perhaps the best way to understand what particular kinds of knowledge there are is through apprehending these processes. Further, we may be able to contribute to the production and reproduction of knowledges by facilitating these processes, and we may be better able to design supports for "knowledging" than to manage knowledge.

In short, there are both good linguistic and etymological reasons for directing attention to knowledge networking, the social relations mobilized and implicated in the process of achieving acknowledgment for knowledge claims. We should consider if "knowledging" takes place differently in cyberspace, rather than if knowledge is there a different thing. Knowledge's networking, rather than its content, is the more legitimate focus of inquiry.

An Early Scientist's Critique of Scientistic Modernist Conceptions of Knowledge

I now offer an additional argument for the alternative: that, in addition to being both possible and justifiable on linguistic grounds, it is also necessary. Although the scientistic modernist conception of knowledge has many roots, those in science are privileged and thus perhaps most important to understand. It is on this intellectual terrain of science that a critique of the "content" conception must be convincing if it is to be persuasive.

I do not have the knowledge or the skills necessary to make this critique in the fullness that it deserves (see, e.g., Fuller 2000). I choose instead to illustrate it through some of the ideas of the early-twentieth-century mathematician (better known in the United States as a popular philosopher) and erstwhile Russell colleague, Alfred North Whitehead. I choose Whitehead, despite his current marginal status in the philosophy of knowledge, because of his historic standing in both fields, which means his comments have echo in both of the "two cultures," as well as his prior close identification with scientistic modernism.

Whitehead offered early a valuable theory of the interactions of scientific and philosophical knowledgescapes. Moreover, his views on the proper conception of scientific knowledge were very influential among the first wave of pragmatists; John Dewey called Whitehead's *Science and the Modern World* "The most significant restatement for the general reader of the present relations of science, philosophy, and the issues of life which has yet appeared" (quoted in Whitehead 1925). The early date of his interventions also indicates the time depth of the scientific critique of scientistic modernism.

Because of my focus on popular notions of knowledge, I choose to present Whitehead's arguments as articulated for a broad popular audience from an essay on "Science and Philosophy" in the collection *Adventures of Ideas* (1933). For Whitehead, science and philosophy are necessarily connected but distinct practices. The difference is in the goal or emphasis of their two discourses. Science is based

> upon observation of *particular* occurrences, and upon inductive generalization, issuing in wide classifications of things according to their mode of functioning, in other words according to the laws of nature which they illustrate. The emphasis of philosophy is upon *generalizations which almost fail to classify* by reason of their universal application. (1933, 147; emphasis added)

In drawing attention to foundationalist impulses, Whitehead interestingly associates them more closely with philosophical than scientific practice.

While they have different objectives, for Whitehead there is nonetheless an important connection between science and philosophy, elucidation of which is the task of philosophy of science. Their connection is a conceptual one, but it is not the one articulatedby authors like Ackerman and Labriger, taken up by "journeyman" twentieth-century philosophy of science, where science's speech problems provides the philosophy of science with its program. Nor does philosophy provide science directly with its concepts; neither should dominate the other. Rather the philosophy of science emerges necessarily from scientific practice:

> Each . . . science in tracing its ideas backward to their basic notions stops at a halfway house. It finds a resting place amid notions which for its immediate purposes and for its immediate methods it need not analyse any further. These basic notions are a specialization from the philosophical intuitions which form the background of the civilized thought of the epoch in question. They are intuitions which, apart from their use in science, ordinary language rarely expresses in any defined accuracy, but habitually presupposes in its current words and expressions. (1933, 148)

Whitehead offers here an important application of an "ordinary language" perspective to sciences, that their basic conceptual building blocks may have certain philosophical characteristics ("intuitions") but they generally follow common talk. For Whitehead, in current parlance, science is culture. In this way, he identifies the important but implicit connection between science's intellectual traditions and practices identified in Section I as more general sources of difficulty when trying to understand knowledge in cyberspace.

Whitehead argues that indeed the inductive, generalizing, scientific mind-set cultivated first by the Greeks "has penetrated into habits of thought" and its "knowledge has re-conditioned human life" (146–147). But this has come at great cost:

> European thought, even to the present day, has been tainted by a fatal misunderstanding. It may be termed The Dogmatic Fallacy. The error consists in the persuasion that we are capable of producing notions which are adequately defined in respect to the complexity of relationship required for their illustration in the real world . . . [In contrast, e]xcept perhaps for the simpler notions of arithmetic, even our more familiar ideas, seemingly obvious, are infected with . . . incurable vagueness . . .
>
> During the Medieval epoch in Europe, the theologians were the chief sinners in respect to dogmatic finality. During the last three centuries, their bad preeminence in this habit passed to the men of science. (148–149).

Scientism, like medieval theology, is science being dogmatic when it purports to have overcome the problem of the "incurable vagueness" of its concepts; that is, when it asserts that it has found "truth." Generally ignorant of the philosophically contested knowledgescapes from which their ideas are drawn, including their dependence on them, scientists misguidedly pretend to knowledge that transcends its context of creation.

As himself a distinguished mathematician, Whitehead is not against the practice of science. He is indeed deeply committed to scientific understanding, devoting much of his lecture to a nuanced view of "the dependence of the qualitative elements in the world upon mathematical relations" (153).

What he opposes is dogmatic science-scientism. First, this is because, on philosophical grounds, no truly general conceptual system is possible: "We cannot produce that final adjustment of well-defined generalities which constitute a complete metaphysics." While more limited sciences are of value—

> [W]e can produce a variety of partial systems of limited generality . . . The concordance of ideas within any one such system shows the scope and virility of the basic notions of that scheme of thought. Also the discordance of system with system, and success of each system as a partial mode of illumination, warns us of the limitations within which our intuitions are hedged. These undiscovered limitations are the topics for philosophical research. (149)

The proper task of the philosophy of science is not to do the conceptual scut work of already existing scientific programs, but to focus on the boundary areas (Star 1995) between scientific discourses. Through identifying their partiality, tracing their incompatibilities to only implicitly understood constructs, and suggesting alternatives, philosophy of science constructs its own program.

In Whitehead's sense, *The Knowledge Landscapes of Cyberspace* is philosophy of science, an illustration of how science and philosophy must be practiced together, as a science without philosophy is blind. This is a less foundational, more limited program that the Newtonian one. According to Whitehead, the problem of a scientific practice separated from a philosophical one is that "science makes the abstraction, and is content to understand the complete fact in respect to only some of its essential aspects" (150). While ostensibly supreme, such a scientistic worldview is actually a subjugated one, hegemonized by "the notion of mere knowledge, that is to say, of mere understanding," a situation which he satirizes as " . . . The age of professors[!] . . . " (151).

To illustrate what informing practice by what I call "scientistic modernism" has undermined, Whitehead focuses on physics. For him, early-twentieth-century developments, especially quantum mechanics, laid waste to the earlier inductivist, generalizing, dogmatic science program. This latter was described as "In its day . . . extremely useful; and now that it is dead, it is stone-dead, an archaeological curiosity . . . turned into an obstructive agency" (155).

Whitehead goes on to articulate a theory of knowledge different from scientistic modernism, one built on a resuscitated Platonic notion. This is the idea of a general "Receptacle" for all things, something that "imposes a common relationship on all that happens, but does not impose what that relationship shall be." His relational concept of knowledge echoes strongly the relativity being articulated widely at the time he was writing (the 1920s and '30s), physically by Einstein and culturally by Boas, Mead, and Benedict. About this relativity,

> The real point is that the essential connectedness of things can never be safely omitted. This is the doctrine of the thoroughgoing relativity which infects the universe and which makes the totality of things as it were a Receptacle uniting all that happens.

While his Platonism may seem arcane, much of the rest of Whitehead's case is prefigurative of later articulations of the "strong" program in interdisciplinary study of science, technology, and society (e.g., Bloor 1976):

> [I]n the absence of some understanding of the final nature of things, and thus of the sorts of backgrounds presupposed in such abstract statements, all science suffers from the vice that it may be combining various propositions which tacitly presuppose inconsistent backgrounds. No science can be more secure than the unconscious metaphysics which tacitly it presupposes. The individual thing is necessarily a modification of its environment, and cannot be understood in disjunction. All reasoning, apart from some metaphysical reference, is vicious.
>
> Thus, the certainties of Science are a delusion. They are hedged around with unexplored limitations. Our handling of scientific doctrines is controlled by the diffused metaphysical concepts of our epoch. (157–158).

Whitehead further specifies, as an aspect of his critique of what I call "scientistic modernism," what C. Wright Mills was later (1959) to call "methodological individualism" in relation to sociology: "Observational discrimination is not dictated by the impartial facts. It selects and discards, and what it retains is rearranged in a subjective order of prominence. This order of prominence in observation is in fact a distortion of the facts." This is a direct attack on the modernist knowledge progression. The act of arranging data in some frame, of creating information, is a distortion more than a reflection of the world. Whitehead illustrates his argument in reference to the displacement of Newtonian physics:

> We merely require to understand the contrast between the most general notions respectively underlying Newtonian physics and modern [e.g., quantum

> mechanic] physics. Newtonian physics is based upon the independent individuality of each bit of matter. Each stone is conceived as fully describable apart from any reference to any other portion of matter. It might be alone in the Universe, the sole occupant of uniform space. But it would still be that stone which it is. Also, the stone could be adequately described without any reference to past or future. It is to be conceived fully and adequately as wholly constituted within the present moment. (159)

In the Newtonian view, knowledge is concrete in its thingness, just as knowers came to be seen as independent individuals. However, in the contemporary, nonmodern view, data are not ontologically "raw"; they already exist only in relation to each other, like knowers. The Newtonian ontology has been displaced by a view of the world in which the mind/body dichotomy is being overcome:

> This . . . full Newtonian concept . . . bit by bit was given away, or dissolved, by the advance of modern physics. [Under the contemporary view, w]e have [had] to discover a doctrine of nature which expresses the concrete relatedness of physical functionings and mental functionings, of the past with the present, and also expresses the concrete composition of physical realities which are individually diverse.

This new (I would say, more cyborgic) view requires a switch from presumed-to-be-stable content to ever-changing process:

> For [contemporary, non-Newtonian] physics, the thing itself is what it does, and what it does is this divergent stream of influence. Again the focal regions [of Newton] cannot be separated from [his] external stream. It obstinately refuses to be conceived as an instantaneous fact. It is a state of agitation, only differing from the so-called external stream by its superior dominance within the focal region . . .
>
> [A] complete existence is not a composition of mathematical formulae, mere formulae. It is a concrete composition of things illustrating formulae. There is an interweaving of qualitative and quantitative elements.

In summarizing his view of the philosophy of knowledge in an aphorism—"The final problem is to conceive a complete . . . fact"—Whitehead accomplishes a switch in attention back from content to process. He rejects an ontologically oriented, Newtonian conception of knowledge as being about independent things for an epistemologically oriented one. Rather than presuming the existence of isolated, contained-in-themselves "things" in the manner of the empiricist, the contemporary scientist must constantly attend, and attend critically, to the processes by which she creates them, always sensitive to avoiding the dogmatic fallacy. In short, knowledge is multiplex, processual, and social.

Recent Historical Critiques of Scientistic Modernist Epistemology

An important goal of this chapter is to show how, once the misleading knowledge construct of scientistic modernism has been displaced and replacements found, alternative approaches to knowledge like Whitehead's can help us better survey the knowledgescapes of cyberspace. To accomplish this, we need to understand how thoroughgoing the critique of scientistic modernism needs to be.

On Toulmin's reading, one key assumption of the Cartesian research program is that the true locus of knowledge is personal and individual—"*I* think, therefore *I* am." However, contemporary epistemology, following Wittgenstein, substitutes an acknowledgment that, since "all knowledge is socially and culturally situated . . . its primary locus must be collective . . . " (1995, xii). Because knowledge is collective, it cannot be understood in terms of individual minds.[1]

Further, because the character of knowledge changes with social group and culture, there is no way in which its building blocks can be based on another Cartesian assumption, that there are transcendent "certainties," like the "cogito" of an individual man. There is thus no point to the inner search for self-evident truths at the heart of the Cartesian project (or the apodictic ones of Husserl). This Cartesian program is now in its "death throes," according to Toulmin: " . . . [T]he debate about knowledge in the last years of the twentieth century can no longer be committed to—or constrained within—these axiomatic [Cartesian] assumptions" (1995, x).

Alternatives to Scientistic Modernist Epistemology

While I agree with philosophers like Williams that searching for a foundational epistemology no longer make sense, one point of Section I is that we still need some defensible theory of knowledge in order to develop good answers to the knowledge question in cyberspace. Are there nonfoundational studies of knowledge worth pursuing?

Despite Williams's claim, some philosophers continue to pay attention to the way claims to knowledge are justified. Joseph Margolis, for example, argues that the point is to find nonfoundational answers to questions like those asked by Kant; for example, how is knowledge possible? In this section,

I offer some flavor of this alternative philosophy of knowledge. Epistemology is dead; long live epistemology!

Evolutionary and Other Biology-Informed Epistemologies

Much attention has recently been given, for example, to the notion of evolutionary epistemology (e.g., Derkson 1998). At the most elementary level, this is the idea that thinking about how we know evolves, changes over time in ways at least somewhat analogous to those of biological species. Tracing the evolution of these changes is a good strategy for illuminating specific epistemological styles, which then can be compared and contrasted.

This is a consummately empirical orientation to the philosophy of knowledge, one that bases thinking about knowledge on historical and comparative study of the way peoples arrive at and justify what they take as known. As I understand their program, evolutionary epistemologists are disinclined to make general claims, like the idea that epistemologies tend to go through similar developmental progressions (à la Piagetian or Kohlbergian psychology), or that they tend toward some ideal goal. Rather, evolutionary epistemologists adhere firmly to the central tenet of modern evolutionary biology, as articulated by scholars like Stephen Jay Gould. This is the idea that there are no fixed goals or even general direction of evolution, that evolution is nonteleological. Species do adapt via natural and sexual selection, but adaptations to one environment convey no necessary selective advantage to any other. "Fitness" is a concept that makes sense only relativistically, in relation to a specific set of environmental conditions.

By analogy, an epistemology may fit its culture well or badly, but this does not make it either superior or inferior in any general, let alone universal, sense. In rejecting the notion of any "natural" trajectory to the evolution of epistemologies, evolutionary epistemologies share much with STS theorizations of the production of technoscientific knowledge like actor network theory (Hakken with Andrews 1993). That is, an epistemology, like other technoscience notions, develops broader influence through the efforts of its advocates, through recruiting powerful, influential allies to extend its reach. The making of an epistemological system is thus at base a social affair.

Sociologists like Susan Leigh Star use another biological analogy to make a similar point (1995). The proper framing of the sociology of scientific knowledge (SSK), in her view, is to examine what she calls *Ecologies of Knowledge*. Not only are there multiple forms of knowledge, but also these "knowledges" exist in complex interrelationships. These interdependencies are as essential to them as are their internal logics.

Some philosophers have attempted to make even stronger biological connections in thinking about knowledge. In *Darwin, Machines, and the Nature of Knowledge*, Henry Plotkin argues that as basic evolutionary events

> Adaptations are themselves knowledge, themselves forms of "incorporation" of the world into the structure and organization of living things. . . . [A]daptations are biological knowledge, and knowledge as we commonly understand the word is a special case of biological knowledge.
>
> [T]he relationship of fit between parts of the organization of an organism, its limb structure for instance, and some feature or features of the world in which it lives, such as the terrain or medium though which it must move, is one in which the organization is in-formed by the environment.
>
> This is the only way to understand the effectiveness of adaptations [,] . . . this in-forming relationship between parts of organisms and their world is knowledge (biological knowledge). Human knowledge conforms to the relational quality of fit that adaptations have. (1994, xv–xvi)

In a number of books, George Lakoff and colleagues have attempted a similar rapprochement between the philosophy of knowledge and biology. For Lakoff, for example, reasoning is not objective and abstract but embodied and specific. He states the main conflict between his and the standard modernist view of knowledge in the following terms:

> Do meaningful thought and reason concern merely the manipulation of abstract symbols and their correspondence to an objective reality, independent of any embodiment beyond physical limitation? Or do they concern the nature of the organism doing the thinking—including the nature of its body, its interactions in its environment, its social character, and so on? (1990, xv–xvi)

Overcoming the scientistic modernist mind/body dichotomy is obviously of central concern to such strong biologizations of knowledge. Maturana and Varela have developed an influential, similarly biological theory of cognition and knowledge formation (1998; see Helmreich 1999). Their approach has important parallels with the recent attempts of Tim Ingold to relativize our conception of physical nature (1995). For example, he rejects conceptions of things like "genomes" that imagine them as independent units, advocating instead that the meaningful units of life configure genes dialectically with environment, in a much tighter relationship, one more of production than of effect.

Readings like these greatly undermine simplistic notions of genes being the "program" of human life. As such, biological readings of knowledge networking have much to recommend them. In *Cyborgs@Cyberspace?* I used a biological conception of information similar to Plotkin's to critique the idea that information technologies only came into existence in the contemporary era. Rather, all lifeforms obtain and use information from their environment and are thus users of information technologies of some sort—for example, "biological" ones. Moreover, every human social formation has structured its culture to "manage" such information exchange and thus all can be said, in a strong, even deliberate sense, to be a deployer of information technology—for example, "cultural" technologies.[2] It is not information technology that is new, but the deployment of technologies that automate the production, adaptation, and dissemination of information (and knowledge). This is an example of the kind of perspective both necessitated and facilitated by current rethinkings of epistemology.

Cyborgic Informatics Epistemologies

However, I also used such arguments in *Cyborgs@Cyberspace?* to focus equal attention on the *non*biological, to the invented, the "extrasomatic," the technological, in constructing proper understandings of human practices like knowledge networking. There is a similarity between the way a species adapts to environmental change and the way I adapt to colder weather by putting on a jacket, but there are also important differences. All designed adaptations, including epistemologies, are artificial, and they are thus at least potentially amenable to conscious direction. While it is true that our epistemologies become (collectively) embodied and therefore somewhat biological, our bodies in the process also become designed. The cyborg, a unitary biomechanical entity, is ultimately a more accurate synecdoche than a pure bioform for the entities that carry culture. Hence, it is also a more appropriate conceit for overcoming the mind/body dichotomy than is biological adaptation (Hakken 1999a).

Nonmodernist Epistemologies

In chapter 2, I offered a postmodernist knowledge regression as a first alternative to the modernist conception of a knowledge progression. The critique began with "knowledge," not "data." In this reading, one begins with knowledge, not "raw" but highly situated, embodied in particular collective biographies and sculpted in cultural reproduction. In this reading, "information" does not exist on its own; it must be created by being abstracted from knowledge. Similarly, data are information further processed according to consciously articulated procedures. Thus, data are always and highly "cooked."

Those committed to a strong science program raise charges of "relativism" when notions like this regression are articulated. This has been more or less the position of the self-appointed "warriors" for science, like Paul Gross, in the so-called science wars: If all knowledge is deeply situated, there is no basis for comparing qualitatively one form of knowledge with another. This position holds, of course, only if we demand that a proposition be foundational/universal to count as knowledge. I see no reason for acceding to such limitations. Indeed, by shifting attention away from the content of knowledge to the process of knowledge networking, we necessitate comparative study of different knowledges/knowledge processes. Pursuing such comparative study is the core task of the ethnology of knowledge outlined in the following chapter. As we understand better the social and cultural contexts in which particular knowledge networkings evolve, we will be able to make more meaningful comparisons as well as construct more appreciative understandings of each particular knowledge practice.

Naturalizing Epistemology

W. V. O. Quine did not himself identify with those who reject epistemology *tout court*. Rather, he conceptualized its future as a subfield of psychology, as an investigation of "how the human subject of our study posits bodies and projects his physics from his data." What is distinctive about this new philosophy of science is that those who engage in it "appreciate that our position in the world is just like his" (1994, 25).

One of the difficulties of achieving a rapprochement between contemporary philosophical epistemology and other contemporary intellectual trends (e.g., the sociology of scientific knowledge program of STS) can be traced to an apparent contradiction. On the one hand, current epistemology rejects the foundationalism of scientism, but on the other, like Quine, it calls for even greater attention to scientific knowledge, even though much of this remains firmly foundationalist and essentialist! How to free scientific knowledge from the presumptions of scientism remains, of course a serious issue for the philosophy of science.

This is one reason why other philosophers (e.g., Kim 1994), while acknowledging the futility of the search for epistemological essences, have resisted Quine's efforts to reduce epistemology to psychology. They have sought to maintain a distinct role for philosophical, as opposed to scientific, work on the character of knowledge. As Kim reads Hilary Kornblith, their natural epistemology, like the evolutionary epistemology referred to above, begins with recognition of the existence of "natural kinds" in the world. They argue that natural things follow from physical processes that often reach particular states of homeostasis or relative stability; apparently related structures are usefully apprehensible as "real kinds." Further, according to Kornblith, successful human reproduction as a species suggests an evolutionary, adaptational advantage to mental structures closely oriented to the perception of these naturally occurring kinds. Drawing on Chomskian notions of the pronounced human capacity to learn languages with standard human characteristics as an example, Kornblith posits our species as one with mental structures that provide us with pronounced proclivities to act as if there are real natural kinds. Over time, as it were, we "evolve" mental structures that increasingly fit the structure of the world.

In this "natural" way, the mind/body dichotomy is gradually being overcome. Kornbilth is careful to present this as a historic, not foundational, process. He is particularly critical of reductionism in science. The ideas of kinds on which humans come to operate develop over time, through a process like natural selection on random variation, in relation to specific contexts, not because they correspond to some logical way to underlying structures. Human tendencies toward conceptual conservatism—to stick to the known conceptualization rather than to experiment with the new—and systematic ignoring of the laws of probability appear arguably to be efficacious in a world where kinds, and therefore structures, are largely stable.

As an Americanist anthropologist, one committed to biocultural, integrated accounts of human phenomena, I see strong parallels between this philosophy and the work of, for example., Tim Ingold. However, two factors appear to me to limit its value. As argued at length by Marshall Sahlins in his critiques of "practical reasons," whether of mechanistic Marxist (e.g., Marvin Harris 1968; Sahlins 1976a) or sociobiological (Sahlins 1976b), the overwhelming impression one gets from the study of cultural kinds (as opposed to "natural" ones) is of their diversity. To the considerable extent that what we perceive is cultural, it is difficult to imagine that one can explain its diversity in terms of some set of homeostatic "realities." As I argued in *Cyborgs@Cybersapce?* contra Daniel Dennett, the main difficulty in accounting for cultural forms is to explain their semantic diversity rather than their uniformity.

When one tries to construct accounts of human knowledge in cyberspace, a further apparent limitation of Kornblithian naturalistic approaches to epistemology becomes more evident. The main problem for human knowledge in cyberspace to account for is the greatly increased variability in time as well as form in human representations of the world. Perceptual categories appear to be increasing in number and variety, not tending toward some limited set corresponding to "natural kinds." While Kornblith acknowledges the irreducibility of the biological, he does not address the cultural, including the inherent sociality of concept construction/reproduction.[3]

Current Pragmatist Epistemologies

Still, in presenting conceptual innovation as a culturally dependent knowledge "happening" rather than as something driven by underlying forces, Kornblithian natural epistemology indicates the value of comparing collective knowledge construction in different contexts. Such comparative study could, for example, include attention to the dialectic between science and philosophy described by Whitehead. Indeed, such culturally sensitive, nonfoundational approaches to knowledge are the chief characteristic of the trend toward pragmatism, in the direction of which "the entire movement of contemporary philosophy . . . ," according to Joseph Margolis, "has moved increasingly . . . " (1993, 143).

The original pragmatists were named and thereby constituted as a philosophical "school" by the American William James in 1898. While it is doubtful that the "members" of the school were actually committed to a substantial set of philosophical positions, their sudden rise to prominence (and equally rapid decline in the face of later analytic Philosophy) probably had to do with the set of questions with which they were preoccupied. Louis Menands describes these as including

> the emergence of theories of cultural pluralism and political progressivism, the fascination with pure science and the logic of scientific inquiry, the development of probability theory as a means for coping with randomness and uncertainty, the spread of historicist approaches to the study of culture, the rapid assimilation of the Darwinian theory of evolution, and the Emersonian suspicion of institutional authority. (1997, 56 and 58)

As Menands goes on to comment, "None of these developments is 'pragmatist,' but pragmatism was one of the places where they came into focus." If there was a habit of mind that connected the early pragmatists, it is echoed in one dimension of the contemporary uses of the term "pragmatic," the notion that ideas should be evaluated more in terms of their utility than in terms of an abstract ideal of degrees of "truth." Like contemporary pragmatism, this habit of mind is consonant with the contemporary turn in social science to the study of practice, that is, of behavior culturally informed. Here, explanations of behaviors are evaluated in terms of what the behaver gets as a consequence of acting this way, how much "mileage" they offer. In regard to epistemology, for example, in the words of pragmatist John Dewey,

> When a theory of knowledge forgets that its value rests in solving the problem out of which it has arisen, viz., that of securing a method of action; when it forgets that it has to work out the conditions under which the individual may freely direct himself without loss of the historic value of civilization—when it forgets these things it begins to cumber the ground. (quoted in Fenstermacher and Sanger 1997, 5)

In such ways, the pragmatists distanced themselves from epistemology in its Cartesian pursuit of "truth" form. Theirs was a rejection of the foundational preoccupations of Western academic philosophy.

Pragmatist philosophizing should not be confused with another contemporary implication of the term "pragmatic"—when this is equated with the obsequious realism that Wright Mills labeled "crackpot." Neither the first pragmatists nor their contemporary epigone were organizational "yes-men," giving acquiescence to prevailing power or wisdom. Dewey, while perhaps the most visible social reformer through his educational work and activities with Jane Addams of Hull House, was by no means the only one. In an important sense, pragmatism provided philosophical justification for the Progressive political interventions of its era.

As Margolis suggests, a number of contemporary thinkers choose to locate themselves *chez pragmatiste*. Some have an unfortunate tendency to overstate the difference between pragmatism and science. Charles Sanders Peirce, the man whom James credited with inventing pragmatism, was apparently committed to an arguably universalist program for grounding science (Ayer 1968; Margolis 1993, 77 sees a basic contradiction in Peirce's approach to these issues). Peirce is generally credited with founding structuralist semiotics, for example. Dewey, as in his decimal system for library science, seems equally optimistic about the ability of science to produce stably universal knowledge, although his enthusiasm for Whitehead means it is a nuanced optimism.

In an important sense, what distinguishes most contemporary pragmatism from its first period is the former's developed critique of modernist tellings of scientific knowledge. Current pragmatism combines the earlier emphasis on practice and practicality with an appreciation of the limits of scientific knowledge quite similar to that in science, technology, and society. For example, pragmatist scholars like Russell Goodman attribute considerable importance to the work of Thomas Kuhn (1970). A stream of younger STS scholars, including myself, have used Kuhn's *The Structure of Scientific Revolutions* to legitimate conflict theories of "normal" science, concluding that each science's trajectory has as much to do with the dynamics of competing knowledge discourses as any easily referenced external world. Along the lines of actor network theory, we construct accounts or histories of technosciences that stress discourse dynamics and power in the rise to hegemony of one school over others. Such hegemonies are accounted for, not by their embrace of transcendent truths but by their access to valued, mobilizable social resources with which to impose a favored way of doing things, or paradigm.

Steve Fuller (2000) argues that Kuhn was much more committed to the value of a reigning paradigm than people like me have read him to be, that he was perhaps more a Peircean than a Jamesian pragmatist. Goodman finds a similar dialectic separating Hillary Putnam from Richard Rorty among influential contemporary pragmatists.

Thus, as in the nineteenth century, there are considerable differences in contemporary pragmatisms. Whereas in past work I have tended to identify most closely with Marxist realist (1993 with Barbara Andrews) and critical realist (1999a) philosophy, I now think it important to supplement these self-identifications with an alternative conception of knowledge on pragmatist grounds. An extended contemporary presentation of the pragmatic perspective on knowledge that I find particularly useful is that of Joseph Margolis. His *Pragmatism without Foundations: Reconciling Realism and Relativism* (1993) presents as "the characteristic pragmatist doctrines . . . [:]

> [1] . . . the denial of any foundationalist theory of knowledge, either in terms of self-evident percepts or sense or the like, or in terms of self-evident truths on which the body of science depends and must depend;
> [2] . . . the claim that human inquiry is continuous with, and develops out of, the biological and precognitive interaction between organism and environment and that a theory of science must account for that continuity;
> [3] . . . the affirmation of some form of empirical realism, to the extent at least that human organisms are perceptually in contact with the external world, whatever the internal conditions on which their sentient capabilities depend – though of such a sort that contact is inextricably qualified by their conceptual schemes and *vice versa*.

Margolis glosses his pragmatism as follows: "In effect, this suggests how a pragmatic account of knowledge concedes the indissoluble linkage of realist and idealist features and, eschewing every form of foundationalism, legitimates science by construing it as continuous with and emergent from the cognitively tacit practices of actually surviving societies . . . " (1993, 259 and 260).

In Margolis's pragmatism, any knowledge necessarily focuses on praxis or what he also calls "the praxical":

> [T]heorizing even at the level of science and philosophy is primarily praxical, grounded in and pertinent for the smooth and effective functioning of particular societies. Hence, the praxical thesis conceptually or logically precludes (and denies the need for) any independent, ahistorical, cognitively competent, higher-order capacity to review and assess the would-be science and practical knowledge of an actual functioning society. (45)

In this pragmatism, both science and Epistemology clearly lose their privilege. They are displaced by praxical analyses that meet a more general criterion of salience:

> The point about salience is that there can be no privileging of practice without the recognition of salient phenomena within the life of a particular practice; for that signifies the collective recognition of certain detailed features of an experience world that, however affected by our diachronic effort to understand and influence the nature of things, are . . . so robust that theory cannot ignore them and cannot erase them . . . as mere artifacts of itself. . . .
>
> Saliences change – and in many ways; and so salience provides a confirmation of any particular claim. Salience marks only the compelling core of a society's experience . . . Practices, which are the natural home of every promising inquiry, are themselves plural, risked by shifts of contingent fortune, historically open-ended. (6)

As well as being anti-foundationalist and to be grounded in really existing social practices (historicist), Margolis's pragmatism is both relativist *and* realist, materialist *and* idealist: "[W]ithin our [pragmatic] enterprise, we can hardly fail to acknowledge that the problematic of our time concerns at least the reconciliation of realism and relativism within the changing contexts of history . . . [,] . . . the theme of what follows . . . " (6).

The flavor of Margolis's pragmatism is neatly captured in his following comments on the central role of technology to any philosophically sophisticated theory of knowledge:

> [O]ur best clue about the validity of such arguments lies with the stable technological features of social praxis . . . [—] . . . broadly speaking, the consequence of Marx's thesis about the relations of production and consciousness. The technological, therefore, performs a double role. On the one hand, . . . it signifies how reality is "disclosed" to humans – primarily because it is through social production, invention, experiment, intentional action, and attention to the conditions of survival (both precognitively and through explicit inquiry) that our sense of being in touch with reality is vindicated at all; but . . . the correction of all theories of cognition and reality thus informed is itself inevitably historicized and subject to the ideological limits of any successor stage of praxis. There is no escape from the historical condition, but the recognition of that fact itself is the profoundly simple result of transcendental reflection within the very condition of history – which obviates . . . the inescapability of Heidegger's various pessimisms. On the other hand, the technological signifies how the study of the whole of reality – of physical nature, of life, of the social and cultural activities and relations of human existence – is unified in terms of our own investigative interests. Hence . . . not only can the theory of the physical sciences not afford to ignore the systematic role of the actual historical work of particular human investigators (for instance, against the unity-of-science program); but, also, we can neither preclude the scientific study of man nor insure that the human sciences must conform to any canon judged adequate for either the physical or life sciences. The primacy of the technological, therefore, facilitates a fresh grasp of the methodological and explanatory peculiarities that the human studies may require – for example, regarding the analysis of causality in the human sphere, the relation of causality and nomologicality, and the bearing of considerations of rationality, understanding, interpretation, on the explanation of human behavior. (206–207)

In this way, Margolis appreciates the profound role of technology in the production and reproduction of knowledge. His argument is another reason for attending to the knowledge question in cyberspace and for approaching it via empirical study of knowledge in practice.

Conclusions

The "Regressive Knowledge Progression," an Illustration of Pragmatic Knowledge Networking

With regard to epistemology, contemporary pragmatist perspectives like Margolis's support in multiple ways the substitution of a notion of knowledge networking as a time- and culture-based process for a conception of knowledge as stable content. On such pragmatist grounds, one can construct philosophies of knowledge with which to answer the cyberspace knowledge question. In contrast, neither the modernist knowledge progression nor the antimodernist regression presented in chapter 2 provides on their own a sufficiently encompassing theory of knowledge. Both, however, capture important aspects of the way knowledge is networked in Western social formations. Under pragmatism, they can be seen as alternative moments within a broader, on-going, developmental but nonteleological knowledge networking dialectic.

That is, while from one perspective all knowledge is ultimately situated in some cultural frame, there are also occasions when it makes pragmatic sense to treat it as if it were abstract. All the moves in knowledge networking conversations and or discursive acts, whether situating or abstracting, are narratives of justificatory performance. Words and other symbols are used to construct persuasive accounts for particular beliefs. Such symbols can only be persuasive because they draw upon a collective background of often contested but still "understood" cultural constructs. Social networking in general, but especially knowledge networking, can proceed only because of shared willingnesses to act as if individual constructions are in fact shared.

With regard, then, to the frequent occasions on which we in the West act as if data are "raw," the task on pragmatism, as opposed to postmodernism, is not to "deconstruct" such illusions but to understand what "natives" mean when they appear to believe that something worthwhile has been accomplished. Similarly, historicization of statements that appear to be abstract universals and/or are "transcendant" in form helps us avoid "misplaced concreteness" in relation to knowledge. In short, we can combine the progression and the regression evolutionarily, recognizing that both "progressive" and "regressive" moments can be of value. Coconstructions of shared understandings can be based on a variety of forms of knowledge networking. An armamentarium of multiple knowledge resources is necessary for coordinated action in the face of the mazeway and cyborgic character of culture outlined in the next chapter.

Summary

In sum, current philosophy offers a process approach to thinking about knowledge that complementarily foreshadows the anthropological and informatics process approaches to be developed subsequently. The argument for this view began with provisional adoption of the core concept of knowledge in the Western tradition, "justified true belief." This provisional step was taken so that what was useful in this view could later be differentiated from the scientistic modernist and Cartesian presumptions that normally and popularly encumber it: the centrality of introspection, the search for apodictic statements as essential justification strategies, and so on. The subsequent bridge to a useful alternative philosophy of knowledge drew attention to the curious etymology of the word. This bridge justified giving priority to the process of acknowledgment, the overt, even public process by which any purported claim to truth comes to be acknowledged. Additionally, out of consideration of the alternative modernist and postmodernist discourses on the relationships among data, information, and knowledge, a framework was created that incorporates both positivist and deconstructionist moments into a single account. This account was given additional support through connecting it to the contemporary revival of pragmatism in philosophy.

In such ways, turns to the social, the complex, and practice in the philosophy of knowledge were legitimated. In the next chapter, an anthropological project to take advantage of this philosophical grounding is outlined. This project involves empirical study of multiple knowledges focused especially on the social practices central to acknowledgment, within the same and across different social formations. This anthropological knowledge project is intended to become a rich resource for the new philosophies of knowledge outlined here, both evolutionary and pragmatist.

A dialectical, pragmatist approach to knowledge has great promise for answering the cyberspace knowledge question. However, Whitehead cautions us that

> always there remains the same beacons that lure. Systems, scientific and philosophic, come and go. Each method of limited understanding is at length exhausted. In its prime each system is a triumphant success: in its decay it is an obstructive nuisance. In the end—though there is no end—what is being achieved, is width of view, issuing in greater opportunities. (1933, 159–163)

Thus, it is important that the philosophy of knowledge described here be understood as a tool to help us get beyond a specific impasse in trying the answer the cyberspace knowledge question, not as itself some transcendent universal law of knowledge. The following chapter surveys elements in contemporary anthropology which make me ultimately optimistic about how perspectives like those outlined in the current chapter can lead to a fruitful comparative study of knowledge networking.

Chapter 5

The New Anthropology of Knowledge Networking

Introduction: Cyberspace and the New Knowledge Anthropology

My argument thus far has been that a great deal (for example, social policy) rides on the claim that humans are moving toward a new way of life, which for convenience I call "cyberspace." The claim rests centrally on a questionable link, that the new culture is driven by significant change in the character and/or social functions of knowledge. Whether this link is valid is the cyberspace knowledge question. Our ability to answer it is severely restricted by both ambiguity regarding what we mean by "knowledge" and by insufficient understanding of knowledge's actual forms in real cultural contexts. Our limited answering abilities have practical consequences, as demonstrated in organizational knowledge management fatigue syndrome.

Section I introduced several elements of an alternative conceptualization of knowledge, knowledge networking, intended to enable development of sufficient understandings. This current section II is a guide through academic discourses that support and extend the alternative approaches. The previous chapter used perspectives from the philosophy of knowledge to critique scientistic modernism. Instead of searching for transcendent, foundational truths, epistemology should aim to construct a discourse that better reflects the multiple ways humans actually come to know.

Indeed, answering the cyberspace knowledge question involves two kinds of activity:

1. Theoretically reconceptualizing knowledge to stress its social, multiplex, and processual character
2. Fostering empirical study of the various ways that knowledge is actually networked

These two tasks are closely intertwined. More complex understandings depend upon comparative study of how actual knowledges arise, which itself requires better conceptual tools to recognize knowledge networking when we encounter it. The more diverse the types of knowledge networking we understand, the less likely we are to construct comparative understandings that inadvertently privilege one sort of knowing over others. More diverse understandings would also help those concerned about the practical problem identified in chapter 3, including how to design automated information technological (AIT) systems that facilitate *all* the forms of knowledge networking relevant to an organization. The better our knowledge of how understandings are actually constructed, especially their social dynamics, the better we can design systems to support them.

Anthropology is the academic discipline that deals most directly with describing and accounting for empirical diversity in practices and in cultures. Anthropology has been especially concerned with discursive practices like statements of belief. It is thus a logical place to turn for comparative study of knowledge networking.

Indeed, a new appreciation of and approach to knowledge has recently emerged within anthropology. While in part a response to the idea of a knowledge change–induced cyberspace, the new knowledge anthropology has origins and preoccupations largely internal to the discipline, some of which strongly critique the empirical value of anthropological knowledge.

Do anthropological descriptions of and accounts for knowledge networking in different cultures, whether old or new, provide sufficient empirical materials for rethinking knowledge? Is knowledge anthropology both deep (specific) and wide (diverse) enough to provide generalizing understanding of understandings? How useful is knowledge anthropology to answering the cyberspace knowledge question? What does study of cyberspace knowledge networking offer anthropological knowledge, both to conceptualizing knowledge networking as a universal human/cyborg process and to the anthropology of cyberspace?

Knowledge Anthropologies and Their Critique

Knowledge Questions in Anthropology

Philosophers deal with concepts, anthropologists with patterns. Like philosophers, anthropologists often begin by abstracting out for examination a particular pattern of or for culture. On a linguistic analogy, they then try to specify the distinctive features of the pattern in peoples' actual practice.

The last chapter argued that philosophically speaking, what distinguishes knowledge from other forms of belief is that its holders explain, or are apparently at least capable of explaining, why they hold the belief that they hold to be true. By extension, the culturally distinctive feature of a pattern of knowledge networking, as opposed to other forms of networking, would seem to be the (expected) presence of moderately deep metadiscourses on why some things are believed to be true and others are not.

Such metadiscourses are present in every culture of which I am aware. Practices that socially engender knowledge creation, reproduction, and sharing take place sometimes in kinship groups, age groups, bands, and/or local living units, while at other times via "invisible colleges," academic disciplines, or institutions like universities—but they do take place. Whatever form it takes in specific social formations, it would appear that knowledge networking is a very common human practice, even a cultural universal.

Knowledge networking's universality has in effect been a methodological premise of the kind of reflexive, immersive ethnography typically advocated and practiced by anthropologists like me. We actively and often successfully elicit comments by "natives" on their culture. We call those who deliver them "informants" and tend to treat their comments as a prime, increasingly privileged data source. We go on to embody our understandings in action. How natives respond to our highly participative form of participant observation is arguably the most privileged data for our analyses of cultures (our "ethnographies"). All this relies heavily upon the regularity of natives' explications of their own culture.

Discourses about the knowledges of discrete cultures do not automatically lead to effective analysis of knowledge networking in general, however. Among anthropologists, optimism has varied over time regarding how successfully our discipline can go beyond description of patterns in specific cultures (ethnography) to general accounts for differences across cultures (ethnology). Hopeful ethnological moments have nonetheless been frequent. Much current anthropology—for example, on globalization (Appadurai 1996), multisited practices (Marcus 1998), or transnational activities (Hannerz 1996)—features phenomena claimed to be manifest in many, potentially all existing cultures. Their theorizers aim to identify dynamics generative of such processes. Candidate dynamics include the spread of computer-mediated communication like the Internet (e.g., Jacobsen 1999; Ito no date) and other computer-mediated activity, like the cultural presumptions implicit in database construction (Harvey 1999). Also examined are the cultural implications of related technoscience practices (Helmreich 1999; Downey 1998), and the relationship of culture and currently new information technology (Forsythe 2001; Fujimura 1996; Traweek 1988).

These phenomena and dynamics are relevant to the knowledge question in cyberspace. Echoing broader social preoccupations and intellectual trends, at least in the background of such discourses is a classical ethnological question, whether some new type of society or culture, a new social formation or "way of life," is emerging today. In *Cyborgs@Cyberspace?,* I employed the language of one tradition of ethnology to ask if cyberspace will constitute a new "level of sociocultural integration" (Service 1971; Sahlins and Service 1960) I pointed out the many problems that follow for ethnography in failing to address this question directly, and I outlined a program for taking up the issue systematically. I have increasingly sited my personal fieldwork in "proto-cyberspace," among the automated information technologied (AITed) practices popularly conceived as predictive of a new social formation type.

Two decades of actively promoting the study of contemporary technology and social change within anthropology led me to see knowledge as a key cultural arena, central to determining if existing and likely future change is indeed transformative. In anthropology, the positing of a phenomenon as important in one social formation type increases attention to it in other, even all types. Anthropologists like Lucy Suchman, Diana Forsythe, and Bryan Pfaffenberger have also pointed out the centrality of knowledge issues to the question of contemporary sociocultural change.

Cyber-concerns, however, are not the only ones leading to more explicit attention to knowledge anthropology. Current attention also arises reflexively from within the discipline and derives from basic concerns over the very character of anthropological knowledge, including whether anthro-knowledge is even possible. Such concern could prod more effective, direct, and self-conscious ethnography of multiple forms of knowledge as well as contributing the empirical materials essential to a comparative philosophy of knowledge as process. However, knowledge anthropology is impeded by inadequately articulated, contradictory notions of knowledge, some of which are implicated in a radical questioning of the very possibility of anthropological knowledge. Our knowledge talk reflects many of the limits, outlined in chapters 1 and 2 and manifest in business, informatics, and foundationalist forms of epistemology in formal philosophy, but there are also some limits more distinctively our own. The work of Peter Worsley considered below, for example, is overly dependent on binary oppositional framings like those that have so often misled sociologists—albeit different framings, for example, between Western and non-Western, rather than "primitive" and "modern."

For ethnography to make its greatest contribution to answering the cyberspace knowledge question, it needs a fully robust concept of knowledge. While the knowledge understandings that many professional ethnographers have to offer are not yet thick enough, the situation is changing. A truly comparative ethnographic study of knowledge, in both cyberspace and other social formation types, is emerging from the critique.

In presenting my views on such a fully robust anthro-knowledge construct, I begin by locating intellectually the current introspection in anthropological disciplinary history. Anyone who wants to use current anthropological knowledge of knowledge networking, whether philosopher or practitioner of IT systems development, needs to understand the preoccupations which produced these events. What are its roots and how do older forms of anthro-knowledge compare to and contrast with newer ones? What clarifications of the proper relationship of the anthropology of knowledge to anthropological knowledge move us beyond critique to new empirical study?

The "Oldest" Anthropology of Knowledge

There is a long history of anthropological attention to knowledges in other cultures, and to differences among knowledges. Like Marilyn Strathern, I feel that the (problematic) form of this attention has been quite stable:

> It is of course a convention in social anthropology to take knowledge practices in the plural, and the discipline has a long investment in the study of different modes of apprehension . . .
>
> [T]wentieth century anthropology . . . has taken the contextualisation of knowledge as one of its epistemological foundations . . . (1995, 3)

Strathern connects this ethnographic contextualizing of knowledge directly to the field's own theory of knowledge:

> Anthropology has long used for its own heuristic purposes ([and based on] investigating the local distinctiveness of people's conceptions of themselves) those constructs about person or society that people often offer as the most

general or global statement they can make about the human condition. (1993, 5)

As suggested above, the probable ubiquity of such "general or global statements" helps explain disciplinary interest in knowledges, and they are in practice reasonably close analogs of "justifying true beliefs." Explicating their logics became commonplace relatively early in the discipline's nineteenth-century, preacademic history. "Primitive" peoples' alleged preference for "magical" as opposed to "scientific" justifications fit neatly into the discipline's earliest project, explaining why European cultures were superior (Bird-David 1999). Thus, an early disciplinary activity was the construction of compendia of such logics, including Sir James Frazer's *The Golden Bough* (1997). Such compendia were used to test ideas about which practices were unique to certain cultures, which were common in particular kinds of cultures and therefore likely relevant to theories of cultural evolution, and which were the most general.

Thus, it is not the existence of a comparative study of human knowledges within anthropology that needs explaining, but the idea that there needs to be a "new" one. What was wrong with the old?

Early, "Modern" Critique of the Initial Knowledge Anthropology

The problems of fitting actual practices into "procrustean" evolutionary schema led twentieth-century anthropologists to distance themselves from such compendia. A common criticism was that compendia's purported knowledge was inevitably misrepresentative. More nuanced description first supplemented and gradually tended to background explication. The latter could never really disappear, however, because of the need for translation. Besides, good description at a minimum requires cross-culturally applicable, and therefore to some extent general, concepts. Still, the creation of anthropology as an academic discipline was largely a response to criticisms of this early, compendia approach to anthropological knowledge.

What we now think of as ethnography was articulated most famously by Branislaw Malinowski to include long-term, deeply immersive, language-based field study. As a research methodology, ethnography was adopted to guard against the distortion that comes of ripping traits out of cultural context. False understandings of understandings were attributed to reliance on such simplistic analytic moves that distorted what was really happening "on the ground." The term "ethnography" also became used to refer to the genre of writing, monographs integrating dense description and analysis of practices in a single social formation, in which culturally contextualized results were reported. The constitution of American anthropology as a four-field affair, to include linguistics, physical anthropology, and archaeology along with cultural anthropology, was Franz Boas's attempt to guard against decontextualizing analyses.

In sum, the Malinowskian and Boasian traditions together constituted "modern" anthropology in this critique of anthropological knowledge. Especially in the United States, the knowledge produced by these two early-twentieth-century interventions, a kind of knowledge not conducive to the broad theorizations regarding culture in general that were the characteristic product of the earlier comparative method, came to be framed as "historical particularist."

More Recent Critiques

For several reasons, however, the pendulum swung against historical particularism and back toward general theory in the post–World War II years. One reason was the popular Parsonian effort to develop a general justification, and therefore reasonable division of labor, among the social sciences. In order to participate in this new academic dispensation, anthropology needed to articulate a general project to complement its culturally particularist one. A second reason was the complementary swing toward positivistic scientism, the search for "laws" becoming the preoccupation of "real science" in the era of rising technoscience. A third reason was the emergence of an applied anthropology whose practice required a common, cross-culturally applicable disciplinary knowledge base. Finally, the engaged anthropology arising in the 1960s, fostered by anti-imperialist struggles, also required morally defensible "rules for action" in both liberation and postcolonial development programs.

By the early 1980s, it seemed likely that some new knowledge schema would replace historical particularism, although which one—Marxist, materialist, idealist, or "realist" (in Mills's "crackpot" form)—was unclear. What emerged instead of a new theoretical paradigm, however, was the most explicitly reflexive and extensive meta-critique of the character of anthropological knowledge since the rise of modern (Malinowskian/Boasian) anthropology.

Disciplinary Displacement Critique

One root of this new metacritique was certainly material (e.g., placement difficulties). Strathern draws a direct connection between anthropology's new knowledge reflexivity and changes in academic institutions, especially their penetration by a business notion of "organizational culture." This has affected the very character of anthropological knowledge:

> Not all anthropologists work in universities, but they invariably start here, and it would be surprising if the new traffickings in university culture did not affect their production of knowledge . . . [Culture's] promotion is what counts. Universities must promote values . . . that must in turn be made increasingly explicit. It takes time to do so . . . The culture of performance – having to perform culture – is one epitome of instrumental rationality. A specific form of ambition that once moved some colleagues but left others untouched is now institutionalised as management, however empoweringly it is presented. Modernity might call itself postmodern, but that is only to lighten the unbearable weight of paperwork.
>
> I only break the taboo in talking about the conditions of production in order to underline the recontextualising effect of new practices. Academic anthropologists are as much caught up in culture change as anyone. What was once a necessity to analysis (thought, reflection) is now regarded as a luxury; the temporal duration and lapsed moments of an unfolding project must be compressed; the means to synthesis (writing as composition) and to communication (publication) have become ends in themselves." (1995; see also Hess 2001a)

It is particularly ironic that kinds of anthropological knowledge (e.g., the notion "organizational culture" itself) that became valuable coinage in the homeland displaced the importance of understanding "the other." Such uses of our knowledge mean the production of new anthropological knowledge may be deeply compromised.

Internally Contradictory Critique

The changed material conditions noted by Strathern have affected academic knowledge production networking in other disciplines, too, but they have had a particular effect because of anthropology's self-image. The discipline has aspired, consequent to its Malinowskian and Boasian heritage, to be both humanity and science. To the extent that the knowledge practices of humanistic and scientific traditions evolved in contrast to each other, a contradiction was built into any inclusive conception of anthro-knowledge.

On the one hand, most anthropologists would agree with the humanistic idea articulated by Strathern that knowledge is among the most culturally relative phenomena. It is created, communicated, and reproduced in a manner heavily dependent on culturally specific practices. This "cultural" conception of anthropological knowledge had much in common with the humanities as normally conceived. The humanities/study of culture aspect of anthro-knowledge was heavily influenced by the Germanic/Romantic study of, for example, *Volksgeist* and thus had roots in the search for cultural essences. Indeed, it was preceded historically by an explicitly racist project, that of explaining white/European supremacy in biocultural terms (Sahlins 1996). (Thus, notions of "cultural self-determination" could be mobilized by white South Africans in their defense of apartheid.)

While he insisted on method being scientific, Malinowski never critiqued this romantic position on anthro-knowledge. Instead, his "revolution" (Jarvie 1964) grafted a self-identified "scientific" process onto this humanistic analytic tradition. Thus, modern anthropology was equally a child of the Enlightenment and of Romanticism.[1] This meant that, in contrast to the multiplicity of the knowledges we studied, the one we ourselves created was treated as unitary and transcendent, like other "scientific" knowledge. (See, for example, Sahlins's critique of his mentor Leslie White's search for the culturally "superorganic" (2000).)

Under the particularist doctrine of "cultural relativity," it was deemed inappropriate to judge a knowledge claim generated in one culture by justificatory criteria from another. Knowledge claims could be legitimately redeemed only in their own cultural context. (No wonder anthro-knowledge has been repeatedly invoked by critics of positivism—e.g., sociologists of scientific knowledge (like Latour 1999.) At the same time, scientific methodological justifications did enable anthropological participation in the "technoscientification" of the American academy. This led to a certain kind of positivism that deemphasized the humanistic side (although less so than in the other social sciences). Thus, even the '60s critique of anthropology, including the Science for the People program of the Anthropologists for Radical Political Action in which I participated, echoed Engels's called for a "more scientific, less Utopian" anthropology.

In sum, while committed to relativism with regard to *other's* knowledge claims, Mailinowskian ethnography justified *itself* as science. It is ironic that an ethnographic *methodology* originally justified as providing more rather than less science within a discipline was chosen as the methodology of preference in many postmodern critiques (e.g., Rothschild's feminist methodology 1983). Thus, in describing and analyzing their own knowledge of particular "local knowledges," anthropologists generally followed Malinowski and deployed knowledge discourses drawn from scientistic, essentialist epistemologies.

More recent material stresses, exacerbating a tension inherent in this almost "anti"-discipline, have prompted new efforts to square a contradictory circle. For Strathern, as for many of the other anthropologists advocating a "rethink" of anthropological knowledge, the consequence is a (very

Whiteheadian) intellectual crisis in the "middle range" or "half-way houses" of anthropology's disciplinary constructs:

> Traditionally, . . . bringing together separate orders of knowledge had been accomplished through concepts such as "level" and "context," "structure" and "event," or through the conventions of comparative analysis. Such solutions have in turn rested on middle-range constructs of "culture" and "society" which served as reference points for evaluating the significance of diversity and homogeneity. These middle-range constructs no longer seem sufficient in the face of transformations attributed to world movements, cultural creolisation, international consumerism, proliferating nationalisms and so forth.

One construct in particular, " . . . culture[,] ceases to work as a middle-range construct . . . [because] its current ubiquity has at once aggrandised it and trivialised it out of all proportion" (1995, 3).

Attempts by those outside the discipline to borrow ethnography are now seen as running into trouble (beginning, I would argue, with the Lynds' attempt in *Middletown* (1925)), because this contradiction is often not even perceived, let alone dealt with. But if the material stresses and global social movements make the old "their knowledge is relative, our knowledge of them is transcendent" gambit no longer tenable, what is to be done? Are we to "scientify" their knowledges, or relativize our knowledge constructs?

Postmodernist Critiques

In the 1980s and '90s, anthropological knowledge came to be viewed through the lenses of the Foucauldian critique of disciplinary power, and the "relativize our constructs" strategy came to the fore. Critiques of anthropological knowledge like Strathern's are usually associated with either a "Linguistic Turn" (Scholte (1986) or "the rise of Postmodernism" (Geertz 2001)). The most strident current critiques of anthropological knowledge derive from this discourse, among which is that of James Clifford, who also rhetorically maps out a preferred "way forward":

> Intensive fieldwork does not produce privileged or complete understandings. Nor does the cultural knowledge of indigenous authorities, of "insiders." We are differently situated as dwellers and travelers in our cleared "fields" of knowledge. Is this multiplicity of locations merely another symptom of postmodern fragmentation? Can it be collectively fashioned into something more substantial? Can anthropology be reinvented as a forum for variously routed fieldworks—a

> site where different contextual knowledges engage in critical dialogue and respectful polemic? Can anthropology foster a critique of cultural dominance which extends to its own protocols of research? The answer is unclear: powerful, newly flexible, centralizing forces remain. (quoted in Geertz 2000: 115)

This antiethnological criticism of anthropological analysis focuses on the scientistic presumptions of modernist anthropological data gathering. For our previous knowledge gathering practices, as expressed by Clifford Geertz, Clifford would substitute

> an anthropology in which fieldwork plays a much reduced or transformed role . . . [a] rerouting, and "rerooting," of anthropological research: to turn it away from static, high-resolution . . . descriptions of this or that people, in this or that place, living in this or that way . . . toward loose-limbed, "decentered" accounts of peoples, ways of life, and cultural products in motion—traveling, mixing, improvising, colliding, struggling for expression and domination [in] . . . "contact zones." (115).

In short, rejecting a scientistic conception of our method's capabilities, we would relativize our knowledge to be a rough equivalent of that of the "natives." Empirical criteria like "accuracy" are displaced by those of style used more characteristically by the art critic.

Implications of the Current Critiques

To those somewhat outside the discipline like Clifford, or to those exposed to it first through postmodern-era theorizing, the critique of anthro-knowledge on the linguistic turn/posmodernism was quite telling. The basic contradiction regarding anthro-knowledge—externally to relativism, internally to foundationalism—seemed to expose anthropology strongly to Foucauldian challenge. As chairman of the American Anthropological Association's Committee on Scientific Communication in the late 1990s, I saw these dynamics acted out in battles over the editorship and direction of the AAA's journal of record, the *American Anthropologist.*

Outside the discipline, Foucauldians began to frame anthropological knowledge narrowly, restricting it to, in a popular Geertzian phrase, "local knowledge." For such appropriators, ethnography is the stick with which to beat foundationalism out of social science. They see a gradual but marked shift in disciplinary talk away from defending "scientistic" practices, such as the formulation of quantitative hypotheses, formalistic testing of them, and the like.

The contradiction between content relativism and process foundationalism is indeed a problem. The old dispensation cannot hold. Foucault's impact, as with Clifford, is quite noticeable. Appropriations of ethnography that merely jettison its scientistic epistemological pedigree are not viable, however. Some (e.g., Van Maanen 1983) have tried to deal with the problem by purging from ethnography any discussion of field methodology, opting for an ethnography that is merely one writing genre among many.

However, such ethnography, shorn of deliberate fieldwork design and ethnological analysis, tends to spawn the very endless paeans to localism disparaged by Clifford. Strategic adoption of restricted "local knowledge" views, ones that dissolve the ethnological moment in the ethnographic one, is a self-destructive act for a discipline that requires at least quasi-universal categories, some useful descriptive cross-cultural categories remaining a communicative necessity. Ethnographies merely shorn of explicit ethnological awareness are subjectable to more or less constant, apparently telling criticisms of unacknowledged "foundationalism." Without descriptively useful cross-culturally applicable categories, legitimate "alterity" based on critiques of ethnocentrism would increasingly be displaced by a (probably doomed) search for representational media that permanently foreground the "native" and background the ethnographer.

In short, one needs ethnology to do ethnography. As suggested above, ethnographic forms of knowledge gathering probably presume the universality of knowledge networking.

"Foucauldian ethnography" is perhaps a doubly empty category. Yet in Foucault, one cannot divorce construction of knowledge from concern for power, and a search for/presumption of universals is inherently suspect on power grounds. Is the Foucauldian challenge to anthropology fundamental? Does this mean that any substantive anthropological knowledge is merely illusion?

The Anthropology of Networks: An Illustration

The anthro-epistemological developments chronicled above have also fostered something of a backlash: for example, the efforts of those like Marvin Harris, Roy d'Andrade, and Dwight Reid to develop more explicitly formal representations of anthro-knowledge as a way to resolve the Malinowskian contradiction, often with the help of AITs. A focus of many of these attempts has been the representation of sociocultural networks, a commonly recurrent topic of disciplinary interest (Strathern 1996). A focus on networks is of course relevant for a networking approach to knowledge like the one I advocate. Contemporary social science interest in networking, including the frequent use of the term, is doubtless related to the diffusion of computer talk, but in anthropology it also has important roots in a distinct, relevant, prior body of anthropological network study. For these reasons, network study is a good illustration of the many strands of the current knowledge question in anthropology.

Talk of networks emerged among British urban Africanists shortly after the Second World War (e.g., Mitchell 1969). The prewar social anthropology they learned in post graduate study had given analytical precedence to groups based on lineage as the primary organizing principle of a "social structure," the British alternative to the Americans' choice of "culture" as the central disciplinary construct. However, social life in the cities and urban societies in which these new anthropologists did their field studies had a much less bounded, more evanescent quality. They looked for an alternative to "group" for framing analyses of urban societies, and they chose "network." They conceptualized networks as covering a broad middle section of the continuum of types of socialities, distinguishable from quite stable, grouplike highly dense *relationships*, on one side, and very unstable, loose, often individually centered or dyadic *relations*, on the other.

These British urban Africanist studies meant anthropology might have been well poised for comparative study of diverse forms of contemporary social networking, including technological and knowledge as well as other types. Grounding the study of knowledge networking in this anthropology would have "naturally" shifted focus from knowledge per se (e.g., as content) to its creation, reproduction, and sharing via diverse social relations. The recent preoccupation with the standing of its own knowledge makes anthropology open to alternative approaches to study of knowledge.

However, the British tradition of network studies within the discipline also had some negative baggage that stood in the way. Africanist-informed studies of networks quickly adopted a scientistic program (e.g., Mitchell), one aimed at the discovery and articulation of general laws of network types that transcended culture, mathematical statement of networking dynamics abstracted from context.

While this approach has much in common with the scientistic alternative promoted recently in the new Department of Scientific Anthropology at Stanford University, within a generation the Mitchell networking program was largely abandoned within the discipline. Networks have recently been rediscovered, seized upon by some as a basis for a (in my view premature and overly abstract) rapprochement between, say, anthropology and computer science (see Bernard 2001). Social relations are framed as being about information, which can be abstracted and represented via, for example, digraphs. More a study of networks than networking, this topic remains a stronghold of formalism within the discipline. It is yet another new "strong" program in anthropological computing; traditional ethnography continues to provide data, but computers are substituted for the analytic moment of discovering deep structure, for abstracting knowledge from information. (For a fuller critique of this approach, see chapter 6.)

A Knowledge Problem Recently Imposed on Anthropology

Metadiscourses on all sorts of academic disciplines, some prompted by Foucauldian and similar interventions, have occasioned some sharp general confrontations, including the so-called science wars (Ross 1996; Gross and Levitt 1994; Fujimura 1999). More than previous contestations among disciplines, these have tended to congeal around the status of knowledge: Is the knowledge produced by scientific activity to be privileged? If so, privileged absolutely or only relatively? If not, is science only one form of "knowledging" among many?

The position one takes on the status of scientific knowledge profoundly frames whether and how one studies technoscience's societal dimensions. If scientific knowledge is privileged, study of its societal dimensions can be justified only on relatively narrow practical grounds, as in terms of how such study can help science avoid distortions in its search for truth (Labinger 1995). However, if science is only one of many ways of knowing, scientific texts are, like others, legitimately open to broader and/or critical cultural inquiry (Hakken 1995a, 1999a). Less than absolutely privileging positions on scientific knowledge have been manifest particularly in studies of science, technology, and society, while culturally relativist positions were championed by feminists (e.g., Rothschild 1983) and advocates of black studies (Bernal 1989). Often like Rothschild invoking ethnographic knowledge to justify the presumption of multiple meaningful epistemic moments, authors like Leigh Star (1995) praised efforts to foster knowledge diversity as an appropriately ecological approach to scientific knowledge.

Strong attacks on those who come close to this last position were launched very publicly in the 1990s by self-appointed "warriors" in service to science. In one very specific way, the science wars focused additional particular attention on anthropological knowledge. Technoscience anthropologists like David Hess and Linda Layne (Hess 1992), and Chris Furlong (1998) claimed that the knowledge(s) about "natives," now framed to include both indigenous peoples and professional, even scientific "tribes" with distinctive "craft" practices, deserve to be studied seriously but in the comparative, non-privileging epistemological manner outlined in chapter 4. Those expressing such views (e.g., Fujimura 1999) found themselves in the science wars line of fire. People like Paul Gross (in SCI-TECH-STUDIES 1996) took particular aim at such ethnographic claims, arguing in contrast that there is only one epistemology and it is that practiced most fully by science.

The opening of an explicitly anthropological front in the science wars reverberated with the discipline's internal discourse. Attention to the way other sciences are cultured has doubly fostered reflexive concern among anthropologists about our own claims to knowledge. This reflection has also been heightened by the rapid appropriation of anthro-method by others (e.g., "quick and dirty" ethnography in information systems development (Forsythe 2001 and chapter 6).

Summary: The Present Moment in Anthropological Knowledge

Thus, even before but especially since publication of Patrick Tierney's *Darkness in El Dorado* (2000), considerable explicit attention was already being focused on anthro-knowledge and the way it's produced. This attention results from both internal, Foucauldian, and external moments of "knowledge in cyberspace," "science war," and ethical reflection. A crisis of theoretical confidence attendant to the linguistic turn/Foucauldian challenge engendered multiple new positions. In addition to the "mere" localness of anthro-knowledge, they include, for example, the idea that ethnography is not really a research methodology but "mere" writing (Van Maanen 1988); that the alterity inevitably at the center of the anthro-ethnographic enterprise is inherently colonialist and therefore that the only knowledge worth having can only be created by the "natives" themselves, via subaltern studies, and so on. Adoption of such positions entailed further conceptual complications, "knock on" intellectual crises additional to those highlighted by Strathern and Clifford. For example, to embrace subalternism, anthropology would arguably have to abandon the critique of ethnocentrism, including the idea that those most deeply enculturated into a culture will have the most difficulty seeing its deepest dynamics, such as the silences that are likely to be among its distinctive features.

The prominence of such concerns is indicative of the absence of broad consensus on what anthro-knowledge is and how to get it. This lack of consensus is manifest in our own disciplinary "mini" science wars, (e.g., over editorship of the *American Anthropologist).* It is likely connected to other indicators of disciplinary discomfort, such as the decline in applications to the U.S. National Science Foundation's cultural anthropology program. There are currently many forms of anthro-talk about either knowledge or networks, even some on knowledge networking, and thus many ways to get lost in them. Our discourse contains multiple, even contradictory conceptions of knowledge, both in general and in its proper anthropological form, with strong roots in the discipline. Our contradictory knowledge constructs mean our discourses often do not articulate well with each other. We have multiple, even contradictory conceptions of knowledge, both in general and in its proper anthropological form. All these factors limit anthropology's capacity to provide the comparative study of knowledge networking needed to answer the cyberspace knowledge question.

The "Moderate" Knowledge Renewal Project in Anthropology

Like Geertz, many of my contemporaries and I remain skeptical of the need for as radical a reorientation as that envisioned by Clifford. Foucauldian critiques of the power dimensions of knowledge in anthropology strike us as interesting but not particularly devastating. Anthropologists are used to the hurly-burly of more or less continuous "paradigm wars" over the discipline's knowledge discourses. The field had at most only a weak "cannon," honoring "worthy" ancestors more in the breach than in the observance. The commitment to science has never been that thick, more to the ideal of disciplined (i.e., rigorous) inquiry rather than to "tight" laboratory or lablike methods. Similarly, the degree of actual relativism was also overstated. After all, we regularly deployed a conceptual armamentarium, honed over many years, which was presumed to be cross-culturally applicable. Admittedly, we tended to wield this in the blissful ignorance that Whitehead described as typical of science.

One can thus identify at least three basic positions in the current profusion of anthro-talk about knowledge. First, "strong Foucault" à la Clifford would relativize knowledge through relativizing both field and analytic method. Second, what at Stanford is called "scientific anthropology" would essentialize knowledge via "truly" scientific method and a purging of relativism. Third, a moderate, perhaps Geertzian position would (relatively) relativize our knowledge construct via its reconstruction in line with the critique of epistemology offered in the last chapter. Anthro-knowledge would be more comparable to that of the natives, but would still retain some distinctiveness.

Anthropology's capacity to provide a base for comparative study of knowledge networking is currently limited, but, in the Geertzian reading, our current preoccupation, a radical questioning of anthro-knowledge, need not, even in a Foucauldian form, be fatal. Rather, it can prod us to "relatively" revitalize anthropological knowledge. Further, concern about the status of anthropological knowledge should also foster a new ethnography *of* knowledge, prod more effective, direct, and self-conscious ethnography of knowledge's multiple forms. I believe such a development would provide the essential empirical dimension to current developments in the philosophy of knowledge (epistemology) as well as informatics and organization studies.

A Neo-Foucauldian Perspective?

Stung by the '60s critique of its role as a "handmaiden of imperialism" (Gough 1968), anthropology was responsive to Foucault's general critique of the power dimensions of disciplined knowledge. Because, as was the case in many other disciplines, anthropology's "linguistic turn" drew heavily on Foucault, it makes sense to begin articulation of a Geertzian alternative conception of anthro-knowledge by responding to his approach.

Paul Rabinow is the anthropologist most responsible for bringing Foucault to his colleagues' attention. In his introduction to *The Foucault Reader* (1984), Rabinow highlights a similarity between the Foucauldian approach and the resistance to glib generalizing inherent in anthropology. He begins with Foucault's critique of the notion "human nature": "In the history of knowledge the notion of human nature seems to me mainly to have played the role of . . . designat[ing] certain types of discourse in relation to or in opposition to theology or biology or history." Rabinow then goes on to make a carefully balanced comment:

> Foucault is highly suspicious of claims to universal truths. He doesn't refute them; instead, his consistent response is to historicize grand abstractions . . .
>
> This position does not entail any preconceived reduction of knowledge to social conditions. Rather, there is a consistent imperative . . . to discover the relations of specific scientific disciplines and particular social practices. (1984, 4–5).

In Rabinow's reading, Foucault's suspicions about disciplinary knowledge production turn out to be in many ways similar to Whitehead's:

> One of the hallmarks of Western political philosophy, in Foucault's interpretation, has been its devotion to such abstraction, first principles, and utopias—i.e., theory . . . .But, Foucault claims, it is exactly this emphasis, this "will to knowledge," that has left us totally in the dark about the concrete functioning of power in Western societies. (5–6)

For Foucault, universalizing disciplines pursuing "knowledge" are central to the reproduction of power, not a way to avoid implication in it: "[K]nowledge of all sorts is thoroughly enmeshed in the clash of petty dominations, as well as in the larger battles which constitute our world. Knowledge is not external to these fights; it does not constitute a way out of, or above the fray . . . " (6–7).

However, Rabinow stresses that while for Foucault much disciplined knowledge is critiqued as dangerously essentialist, this is not always or necessarily the case. Instead, disciplined knowledge can be revitalized, turned back on itself: "Rather, for Foucault, the 'will to knowledge' in our culture is simultaneously . . . a tool to combat that danger . . . .'[K]nowledge now calls us to the sacrifice of the subject of knowledge.'"

So that we can no longer be "refusing to separate off knowledge from power," "the subject" must be sacrificed. This subject has two important senses, according to Rabinow. One sense is that on which a discipline focuses (and thereby inevitably tends to colonize), its "subject." The second is something much broader, "the modern subject." This latter, similar to what Marxists call "the Bourgeois subject," is "what [Foucault] sees as the greatest threat—that strange, somewhat unlikely, mixing of the social science and social practices develop around subjectivity" (7). The most necessary intellectual project, then, is not to destroy disciplines, but to deconstruct this notional subject, to create "a genealogy of the modern subject, to analyze the peculiar ways in which, by mixing the exercise of power with claims to knowledge, subjects (disciplines) create a peculiar social construct, 'the subject.'" Rabinow restates

> Foucault's most general aim[, which] is to "discover the point at which [disciplines'] practices became coherent reflective techniques with definite goals, the point at which a particular discourse emerged from these techniques and came to be seen as true, the point at which they were linked with the obligation of searching for the truth and telling the truth." (7)

In short, Foucault's project can be seen as drawing attention to the ways in which, via a notion of subjectivity, social science is captured by a Cartesian, scientistic modernist epistemology. I develop a Foucauldian critique of subjectivity in my comments on ethics and agency in chapter 12. My aim here is to focus understanding of his challenge to anthropological knowledge as a way to construct a more serviceable notion. The first task is largely self-evident: Foucault demands the creation of knowledge about knowledge, to put knowledge itself on the scholarly agenda. That this objective has already been achieved in anthropology is manifest in, for example, the strikingly extensive incorporation of philosophical terms like "epistemology" into recent anthropological writing. Clifford Geertz describes, for example, how both philosophy and anthropology "find themselves, these days, repeatedly invaded and imposed upon by interlopers claiming to do their job in a more effective manner than they themselves, trapped in inertial rigidities, are able to do it" (2000, ix). Whether they are overtly perplexed by it, like Geertz (2000), or just getting on with it (Sahlins 2000; Strathern 1995; Hastrup 1994; Worsley 1997; Nader 1996), the (re)emergence of an anthropological knowledge question is indexed by this change in terminology.

A second, stronger element of the Foucauldian challenge is to divert the aim of anthropological knowledge of knowledge away from discovery of universal "laws" of knowledge; indeed "His strategy is to proceed as far as possible in his analyses without recourse to universals" (Rabinow 1984, 4). From this perspective, an important early task for anthropology is to construct a nonessentialist epistemology different from the one Malinowski rhetorically championed. Sahlins recounts his own later intellectual biography as a course back toward strongly particularist, almost Boasian, ideographic anthropology:

> Physical things have cause, but human things reasons—symbolically constructed reasons even when they are physically caused. And this makes anthropology a science of another kind, different from the natural sciences, because its object and its method are the same kind of thing [as reasons . . . What] I came to practice [is the idea that] the distinctive character of anthropological knowledge is that it involves a substantial unity of the knowing subject and that which is known. (2000, 28 and 29)

While anthropologists differ about this eventuality, it is important to keep in mind that for Foucault, the point was not to destroy disciplined knowledge, but to revitalize it as a tool in the battle to alter power relations. While moments of critique of disciplined knowledge are essential, so are those when disciplined knowledge is turned back on itself to be revitalized. In Paul Rabinow's account, "Rather, for Foucault, the 'will to knowledge' in our culture is simultaneously part of the danger *and a tool to combat that danger . . . .*'" Rabinow himself, along with, at least, Downey (1998), Forsythe (2001), Hakken (1999a), Hess (2001a), Heath (1994), Helmreich (1999), and Suchman (1987)—see Hakken (2001), has been a central figure in the creation of a distinct anthropology of technoscientific knowledge, arguably now the most collaboratorily extensive effort at explicit ethnography of knowledge.

New Epistemological Framings

Thus, while I accept that there is a valid critique of some of anthropology's knowledge constructs but also of the ultimate value of some forms of existing anthro-knowledge, I am optimistic about the eventual prospects for moments of real revitalization. There is spreading recognition of the need for new ways to think about knowledge, both in general and in our field, as well as for a more focused ethnography of knowledge in diverse social formations. Enough anthropologists, both those studying culture in non- and protocyberspace social formations, have called for more explicit attention to how we cook knowledge soup to make me optimistic that it will one day be thick enough, that the internally and externally stimulated discourses discussed above can be brought into dialogue with each other.

In particular, we are beginning to construct a shared understanding of at least the criteria by which an alternative vision of anthro-knowledge should be judged, if not yet what that vision *is*. In what follows, I discuss some exemplary current efforts to revitalize thinking about anthropological knowledge by thinking about knowledge more anthropologically, moving at least toward partial visions.

Acknowledging the existence of a knowledge problem in anthropology is the first step. Indeed, Henrietta Moore begins her *Anthropological Theory Today* (1999) with the comment: "It is very tempting to begin a book of this kind with the statement that there is no such thing as anthropological theory." Lamenting the temptation, she identifies an unclear discourse on knowledge as the reason for this temptation, because our

> confusions have only been deepened by debates in the last ten years or so about the purpose and pretexts of anthropological knowledge. The inclusion of the anthropologist and their role in knowledge construction within the parameters of theoretical critique has had the effect, among other things, of linking anthropology as a practice to questions of power, domination and discrimination in ways that have highlighted moral and ethical dilemmas for practitioners individually

> and collectively. The results have been diverse, but in some sense have involved not only a retreat from theory, but even from the project of anthropology itself.

Yet despite such problems, Moore sees the situation as having improved enough to justify putting together an anthology "with several calls for a renewal of theoretical thinking and an emerging note of optimism about the future of anthropology" (1 and 2).

Merely highlighting reluctance to theorize is, of course, not enough. Kirsten Hastrup's argument is a good illustration of optimism about revitalizing theory by making the rethinking of anthro-knowledge a central project. She, like Moore, feels that an effective discipline can emerge from the current epistemological self-examination. Rejecting the older mapping, "view from nowhere," "explorer" orientations of modernist conceptions of knowledge in the discipline, she argues instead for a clearer focus on the actual experience of ethnographic fieldwork. In this way, we can refine constructs and root out scientistic vestiges. In particular, she identifies overt acknowledgment of the bodily experience of the ethnographer as a means to overcoming the modernist mind-body dichotomy. A more nuanced understanding that in fieldwork we embody practice has demonstrated "the need for dissolving the 'Cartesian anxiety,' that is, the fear of not having a fixed and stable foundation for knowledge, a grounding for reference" (1994, 230). Ethnography's potential access to the way "culture becomes incorporated" in the body, plus awareness of the way in which the fieldwork experience reshapes "the body's actual ability," opens up "a centre of knowing which has [so far] remained obscure" (230 and 231):

> The important point is to realize that the field-world is not experienced through the fixed coordinates of a semantic space. The world is always experienced from a particular point in a social space . . . Moreover, the point from which we experience the world is in constant motion.
>
> The shift to an ego-centric approach to referentiality is a shift from a semantic to a pragmatic view of culture, and of science. This does not exclude "semantic" interests of course; it only integrates studies of meaning in studies of practice . . .
>
> The agent of scholarship is a living person, not just a mind. (234 and 245)

Such an ethnography, one manifesting the focus on practice described by Sherry Ortner (1984), need not be ideographic because solipsistic. Hastrup hastens to add that

> fieldwork is quintessentially an intersubjective experience. From there we realize that [to quote Vendler,] "subjective states —sensations, feelings, and emotions —cannot be found, recognized, or discovered in bodies but are attributed to them on the basis of certain observable manifestations that warrant such attribution." Attribution of feeling demands a degree of personal involvement because particular phenomena warrant attribution of feeling, not because there is a scientifically established chain between experience and behavior, but because we have learned what it means to "be in pain . . ."
>
> To reach maturity, anthropology must work by its own lights and bring to methodological effect the fact that there is no disengaged standpoint of knowing. We cannot know except by way of our own presence and questioning. Knowledge is profoundly embodied. (235 and 236)

Importantly, Hastrup connects her reconstruction of anthro-knowledge as "incorporated" not only to romanticism ("to dignify subjective experience, not to deny reality") but also to philosophical neopragmatism: "This implies a view of truth and objectivity as based in rational acceptability (Putnam . . .); and in agreement within a scholarly community of potential dissenters (Rorty . . .)" (237). (This connection is one basis of my view that a revitalized anthropological ethnography of knowledge can also play a role in regrounding philosophical epistemology.)

As evidenced by many of the comments already referenced, a third element essential to revitalizing the anthropology of knowledge is already well on its way to being realized. This is a commitment to explicit epistemological framings of the knowledge question in anthropology. Through their recent, more direct encounter with philosophy, anthropologists are better positioned to understand why they need a more coherent notion of what anthropological knowledge is all about and what benefit they stand to derive from it.

The Advantages to Anthropology of an Explicit Ethnography of Knowledge

As argued in chapter 4, one thing we stand to learn from our engagement with contemporary philosophy is the value of explicit understandings of "actual" knowledge networking, of a renewed ethnography of knowledge. We need field studies of knowledge networking in as wide a variety of cultural contexts as possible, ones that have a comparative moment like the original knowledge program but are informed by both modernist and postmodernist critiques.

These new, explicit field studies provide the content impetus to the emerging discussions referenced above.

The basic ethnographic critique of other social science techniques of knowledge production (opinion surveys, formal interviews, secondary analysis of data gathered for other purposes, such as unemployment compensation) is that these substantially distort through decontextualization. Abstraction from context tends in turn to foster multiple misunderstandings, increasing the chances of implicitly imposing inappropriate, ethnocentric questions and framings—for example, sociobiologists' glib discussion of "incest" and "marriage" among other life forms. (See Helmreich 1999 for an extended discussion of this problem among technoscientific "Artificial Lifeers.") This argument is essentially a restated version of Whitehead's critique of the way in which at most semi-aware borrowings from the culturally taken-for-granted inevitably limit all sciences.

Since Malinowski, ethnography's alternative to such research has been extended field study. Initially, this meant largely empiricist study of "whole cultures." Over the years, ethnographers have shifted away from such simplistic studies toward more problem-oriented ones. This had much to do with recognition of the impossibility of actually "studying everything," but it also had to do with learning ways to avoid the tendency of earlier ethnographers to presume that which needs to be established first. For example, one tendency was to presume that the culture under study could be understood in isolation, on its own terms as a "primitive isolate."

Arjun Appadurai's analytic points about how culture is now more than ever "at large" in the world (1996) are in a long line of anthropological works that question the utility of such assumptions. For him, indubitably—but for others like Ulf Hannerz (1996), not so clearly—a "modern" phenomenon, culture now often functions transnationally, not within the frame of a particular geographic group. According to its critics (e.g., Marcus and Fisher 1986), the conventions of earlier "ethnographic present" writing styles constitute genera counterparts of the old assumptions. As I argued in *Cyborgs@Cyberspace?,* the presumption of uniformity within the boundaries of a culture has been replaced by attention to how it is that cultures maintain a sense of coherence in the face of substantial variation. So what, indeed, *is* a culture?

An explicit ethnography of knowledge can help answer this question. I suggest the following as a useful heuristic: It makes sense to treat a bunch of people who share a distinctive set of ways of knowing, share patterns of knowledge networking, as if they share a culture. Conversely, when in dealing with "bunches" among whom it is difficult to identify a distinctive set or shared patterns of knowledging, we should avoid describing them as "having *a* culture."

One value of such a "knowledge networking" approach to the culture question is that it allows us to mediate what I would call the "preconscious/conscious" divide in the study of culture. That is, anthropologists have long argued that much of culture works "behind our backs." Constituting what Bourdieu called a "habitus" (1978), culture is inveigled into what we do often without our being aware that it is there. Study of "ways of knowing," acts by which "natives" justify what they believe to be true, provides an entrée into this world of the preconscious. This is because such acts of justification often make the "normally only taken for granted" explicit, raising to consciousness the "already presumed to be true." Because such acts need to be convincing to an audience, "natives" themselves explore and often state that which "is (preconsciously) presumed to be given." By the same token, the absence of any situations in which one can try to convince most others about what is is itself prima facie evidence of the absence of a shared culture.

Like Sahlins, I would argue that such shared sets and patterns are not sufficient in themselves to constitute a culture, that they alone are not the distinctive features of each separate cultural unit. Knowledge networking is, however, central enough to cultural distinctiveness to justify, on efficiency grounds, studies of knowledge networking in networks of human relations that other reasons lead us to believe *may* be distinct.

As implied above, historical particularism fostered continuing attention to discrete knowledge systems. Because of their detailed contextualizing and despite occasional scientism, ethnoscience studies (Frake 1980) deserve specific mention. Other loci of the continuing cultivation of concern for anthroknowledge were the overtly reflective (Powdermaker 1966; Stocking 1990) texts on ethnography.

The more explicitly epistemological concerns of recent years have led to a number of overtly knowledge-reflexive ethnographies. They include applied studies of the use of knowledge about cultural difference to the transnational operation of organizations (e.g., Baba 1999; Gluesing 2000), studies of knowledge in work and organizational contexts (Suchman 1987; Jordan 1998; David 1998; Erickson 1998) and of media (Appadurai 1996; Spitulnik 1993). Sahlins's polemic *How Natives Think: About Captain Cook, for Example* (1996), a response to Obeyesekere's critique of Sahlins work on Hawaii, is richly informed by the epistemological critiques discussed above. Sahlins updates Levi-Strauss's *The Savage Mind* (1966) in a way that carefully articulates a Foucault-informed "moderate" epistemological program for the field.

Helen Verran's *Science and an African Logic* (2001) is an interesting complex autocritique of her own former stress on the relative incompatibility of Western and Yoruba number systems. In a manner very reminiscent of the pragmatism of Margolis described in the previous chapter, she locates both universalism and relativism as foundationalist programs. Nurit Bird-David's "'Animism' Revisited: Personhood, Environment, and Relational Epistemology" (1999) is a particularly suggestive effort to find a basis of equivalency between "native" (in this case, gatherer/hunter) and Western epistemological stances. Her demonstration of how a relational epistemology is manifest in practice should be greatly suggestive to ethnographically oriented philosophers.

Beyond Evaluative Criteria to Real Vision

How far has this new, reflexive anthropology of knowledge gone? Peter Worsley's *Knowledges: Culture, Counterculture, Subculture* (1997) is indicative of how far we have come and need still to go. It is an especially good example of how studies of particular knowledge systems can lead to powerful articulations of broader programs, for both the anthropology of knowledge and for anthropological knowledge. Although requiring analytic modification, this book comes closest to exemplifying the detailed kind of explicit ethnology of knowledge that I advocate.

Worsley's introductory account of how his study evolved reflects the changes in ethnography of knowledge. A relatively recent work, his book began "with the very limited idea of writing an account, fairly quickly, of the scientific knowledge of one Australian Aboriginal tribe" (1997, 1; about his use of the term "scientific," much more, reflexively speaking, below). Early in his distinguished ethnographic career, Worsley had been struck by certain patterns of knowledge among his informants:

> [I]t was not simply that they had an awful lot of knowledge about plants and animals—in the aggregate, as it were. They also had systematic ways, first, of distinguishing trees and plantts from animals and, second, of identifying the latter as either land, water, or marine animals, and creatures that lived in the air. They had, that is, a biological, quite non-religious taxonomy—a systematic way of classifying things. (3)

At that time, Worsley wrote several monographs on these knowledges. He later found himself in dialogue with Claude Levi-Strauss's attempt at "a new kind of theory, not just about what [Levi-Strauss] called 'the savage mind,' but about human knowledge in general." Worsley critiqued, "Levi-Strauss . . . [who] was still looking at Aboriginal thought as if it was all of a piece. I argued, conversely, that there were several distinct modes of thought in Aboriginal culture." Worsley goes on to specify four Aboriginal "ways of thought" manifest in relation to bioforms: religious, gastronomic, biological, and linguistic (3), later adding a fifth, mythological form.

Two decades after this encounter with Levi-Strauss, publication of a detailed ethnoscientific study of the groups he had studied earlier reengaged his attention. He moved quickly from an effort to write a nontechnical account of Aboriginal biological knowledge to engagement with a broader range of ethnology of knowledge issues. These included the idea that there were "similar modes of thought in other cultures" as well as development of justifications for serious studies of ethnoscience: "[I]t would have been news to many," he felt, "that hunters and collectors too had their own forms of science" (5).

Yet as he worked further along these lines, Worsley

> became sensitized to two further kinds of questions. The first was a comparative one, about the social distribution of knowledge: how uniform, in fact was the thinking of people, not just in Western cultures but even in cultures most people think of as "simple"? Were there not, rather, *sub*cultures or even *counter*-cultures, in any culture? My second question was a simple [!] matter of identification: when we talk about the ideas of this or that "people," who exactly are we talking about? For thought does not think itself; it is individuals who think. Furthermore, who precisely were the people who created, codified, and transmitted the thinking we describe as the cultural ideas of, say, "the Lilliputians". . . ? (6; emphasis in the original, exclamation added)

Thinking anthropologically about knowledge thus led Worsley to the core theoretical issues of contemporary anthropology addressed above, the kinds of pattern differences that are internal to a culture, those that distinguish one culture from another, and what makes such differences significant as boundary markers.

Additionally, Worsley's macrocomparison of Aboriginal and Western food classifications led him to recognize parallels between the debates of taxonomy biologists and those among social scientists. This in turn led to comparison of Western and non-Western medicine, in which he discovered the utility of distinguishing "wisdom" from knowledge (9). Macrocomparison led back to microcomparison, and a realization of just how rare were anthropology of knowledge "studies which look at different kinds of thinking, side by side . . . *within* the same culture . . . " (8; emphasis in original).

About knowledge in general, Worsley concludes:

> Knowledge, then, is necessarily plural. There are knowledges, not simply Knowledge with a capital K. The social distribution of knowledge is plural, too, for although everybody thinks, not everyone has the same amount or kind of knowledge. It is the distinctive social activity of "ethnointellectuals" to develop, consolidate and transmit these systems of thought . . . (10)

and thus, in an important sense, a key task for ethnographers to seek out these "ethnointellectuals."

The Also Necessary *Ethnology* of Knowledge

Such general conclusions led Worsley straight to a strongly Foucauldian knowledge/power comment: "There is inevitably an ethical dimension to all this . . . Making comparisons between different kinds of knowledge in Western and non-Western cultures raises questions about the status of 'our' knowledge." For example, he suggested at a peace conference that more attention be given to "positive" aspects, like their ethnoscience, of Aboriginal thought. A member of the African National Congress responded, "Yes, that's important, but why, Peter, do you call it *ethno*science?" Worsley is driven to wonder,

> Are the kinds of scientific knowledge which are usually labeled "ethnoscience" simply part of a unitary and universal science which all cultures have developed to different degrees? Or are they different kinds of science? Is Western science always superior, across the board? Or do other knowledges possess special strengths that our science lacks? Is Western science . . . , however distinctive and powerful a mode of thought, nevertheless only a subculture coexisting with other subcultures . . . ? (13)

Almost overcoming his British cultural preference for rhetorical question over direct assertion, Worsley concludes, "If, then, there are many kinds of knowledge in all societies, should we not, instead of contrasting 'their' knowledge and 'ours' across the board, specify *which* of their (and our) kinds of knowledge we are talking about?" (14; emphasis in original).

In arguing this way, Worsley illustrates the inevitability of an ethnological moment, the development of descriptive categories, as a necessary complement to any ethnographic one. While in agreement with the general drift, I find myself somewhat anxious about the ontological finality in Worsley's tone. He seems to merely assume the universality of the four or five forms of knowledge that he has identified and that provide the structure of the following book.

Rather, more rigorous comparative study of knowledge networking is necessary before one concludes that useful descriptive categories like "religious, gastronomic, biological, linguistic, and mythological" are universal ontological forms. After all, these terms are all drawn directly from only one, Western, cultural tradition. One or more of them may not be applicable everywhere, and there are likely other forms which Worsley has not identified.

This illustrates again why the creation of an equally explicit ethnology of knowledge is an essential part of the "symmetrical anthropology," one which brings the knowledges of all cultures into a common frame, for which Latour calls (1993). Such an empirical, *wissenschaftlich* ethnographic knowledge of knowledge activities should greatly expand capacity to place our own disciplinary discussions in the context of both current scholarly movements and broader cultural developments—for example, our possible entry into cyberspace.

Essentially ethnological arguments for greater consistency and rigor in knowledge-related constructs are, I believe, at the center of key contemporary theoretical works in anthropology (Geertz 2000; Clifford and Marcus 1986; Marcus and Fisher 1986; Taussig 1993;). Implicit arguments for an ethnology of knowledge, they build on the already considerable comparative knowledge research base of earlier generations.

In pursuing disciplined, empirical study that is symmetrically generalizing (ethnological) and yet pays attention to the specific context of cultural practice (ethnography), several recent efforts explicitly articulate useful but still partial anthropological theories of knowledge. Michael Taussig's appreciation of mimesis in ethnoknowledge systems (1993) has much in common with Bruno Latour's attempts to articulate a "nonmodern"—neither modern nor postmodern—science (1993). Laura Nader chronicles various efforts to valorize non-Western anthro-knowledges as an essential prerequisite to urging similar moves on other sciences (1996). The geographer David Harvey's efforts (e.g., 1989) to operationalize "postmodern" conception of knowledge materialistically has found echo among many colleagues (Hakken 1999a; see also Barth 1974).

An Anthropology (Both an Ethnography and an Ethnology) of Knowledge in Cyberspace, or Why a Tempered, Empirically Based Analysis of Social Change and AIT is Essential and Focusing on Knowledge Is a Good Strategy for Doing It

Doing Anthropology of Cyber-knowledge: Donna Haraway

It is perhaps Donna Haraway's articulation of the goals of research that most strongly influences the emerging generation of anthropological scholars who focus on technoscience knowledge. As they often have to have feet in both the doing of technoscience and the study of its practices, an explicit metadiscourse on knowledge is part of every effort to "'scape" its cultures. Any serious ethnographic study of technoscience will be deeply situated and polyvocal and involve what Haraway calls "trafficking": "Reading and writing on the razor edge between paranoia and denial, I venture to consider the syntax of intellectual property . . . " This practice is indexed by her 1997 book's "title's Internet address": *Modest_Witness@Second_mllenium.FemaleMan©_Meets_Oncomouse*™. Haraway describes her knowledge as "contaminated" in that she participates in, while critiquing, "sociotechnical production" of knowledge. As a Foucauldian, she acknowledges the inevitability of "knowledge-power processes that inscribe and materialize the world in some forms rather than others" (1997, 7) and thus the limits within which she works, but she throws herself into the knowledge production process anyway.

I share her conviction that though this may be all we can do, it is worth doing. Yet how are we to relate to the knowledge produced by one who is both an insider to the production of technoscience knowledge and an outsider in the sense of being an ethnographer, potentially an ethnographic critic (Forsythe 2001)? Such dual positioning is not just a consequence of Haraway's preferred ethical stance; some substantial professional standing is increasingly a prerequisite of access to technoscience knowledge production.

A Foucauldian-engendered preoccupation with the knowledge question in anthropology intersects directly in Haraway with the broader implications of new, technoscience forms of knowledge. Concern over how to conceptualize knowledge today is given impetus by the potential changes in knowledge networking associated with automated information technologies, of, in a phrase, the cyberspace knowledge question. Good accounts of the dynamics of past and present knowledge networking are of value on their own, but they are essential if we are interested in whether current changes in knowledge's production, reproduction, and/or sharing are of transformational significance.

Henrik Sinding-Larsen's Anthropology of Information Technology

Blocked largely by the same implicit presumption of transformation manifest in the broader cyberculture, however, examination by contemporary anthropologists of cyberspace change in knowledge has emerged only slowly. Why this is so is an issue taken up at length in *Cyborgs@Cyberspace?* As argued in chapter 2, for example, the cyberspace change in knowledge issue has gotten mixed with the more general issue of whether knowledge is situated or independent. We need to extricate the change issue, the idea that knowledge(s) is/are situated substantially differently in cyberspace, from the debate over the extent to which all knowledge(s), whether previously existing or new ones, is/are context-independent (Gross and Levitt 1994) or situated (Suchman 1987). This, too, will only be possible when we have accumulated a sufficiently broad understanding of the very different ways in which diverse forms of knowledge networking are situated.

One of the first anthropologists to articulate clearly the need for a specific study of the knowledge question in cyberspace, and to connect it to the need to revitalize anthropological study of knowledge, was Henrik Sinding-Larsen (1987). Explicating as flaccidly tropic the tiresome question of whether computer thinking will replace human thinking—while still acknowledging that in modified form there is an issue here that needs addressing—Sinding-Larsen switches attention instead to the challenges that use of these machines implies for how knowledge itself is to be approached. Will mediated doctors, like children who use calculators and therefore have difficulty mastering and remembering the procedures of arithmetic, lose the skills of diagnosis?

> The new information technology has altered the conditions for the acquisition of experience and other learning processes. Currently this technology only affects very limited areas of knowledge. However, with the developments we are witnessing in "artificial intelligence," many activities which were previously considered intellectual challenges may soon become boring routine work. This may result in large quantities of knowledge being forgotten and disappearing. Knowledge has to some extent always become obsolete and disappeared. However, this can take place much more rapidly and completely today than ever before, resulting in a vulnerability of considerable proportions. This vulnerability follows partly . . . because increasing standardisation of knowledge leads to rigidity in relation to necessary changes. (1987, 95)

In the next chapter, I take up various overestimations of AI's potential, as I believe Sinding-Larsen's to be. What I wish to focus on here, however, is how his Nordic cyberspace concerns lead him to outline a rigorous, diligent comparative study of existing knowledge as process. He proposes an "anthropology of information technology" that would focus, ethnologically, on the different standards, and thus different practices, by which cultures evaluate knowledge:

> One aim of a comparative study is to clarify the way in which various cultural ways of managing knowledge is [sic] related to various forms of information technology . . .
>
> It is not the first time in our history that new information technology has altered the rules for the management of knowledge. However, it may be the first time that this has happened so abruptly that the process has become apparent to the extent that it has become the object of a comprehensive research program. (97).

Sinding-Larsen's program would focus attention on the increasing extent to which knowledge is "externalized" and is no longer "stored intracognitively." This is for him a question in cultural evolution, suggesting that any ethnography not only requires an ethnology but an ethnology with evolutionary dimensions. He sketches this externalization from early forms, such as the geographic semiotics of baboon water source location, through Mesopotamian writing systems:

> What we regard as knowledge is typically the structures which guide us when we are carrying out specialised tasks. Since structures can be stored in written form, we do not need to remember all of them, we can "look them up" in a book and just follow the procedure given by the text. But we cannot escape the fact that we must understand the structure in order to be able to carry out the operation . . . We must possess knowledge about both the structure and the process . . . This is not the case with knowledge stored in the form of computer programs. In this case both the structure and the process can be stored and we can have operations carried out without us human beings having to understand or be aware of what is happening. The process of externalization is more complete . . . (99)

While phrasing in terms of "an anthropology of information technology," Sinding-Larsen is clearly not talking about a narrow focus on artifacts. His comparative discussion of what is at risk when humans change how they "manage" knowledge is clearly linkable to the implications of the use of AITs in organizations to promote knowledge management and the fatigue this produced. Just as important as the analytic advantages of thinking about knowledge by studying knowledge networking are the empirical and practical ones (the focus of Section III). Our ability to understand what is going on in such situations is nonetheless dependent, Sinding-Larsen argues, on the existence of an independent, less design-driven study of knowledge processes in general. The challenge, then, is not to construct an ethnography of knowledge in cyberspace alone, one that avoids issues of evolution or change in type of social formation, but to construct a more general discourse, and to do so in a way that escapes the Procrustean bed of foundationalist, teleological evolutionisms.

Toward an Ecology of Knowledge: Lucy Suchman

Lucy Suchman outlines another alternative, somewhat more artifact-centered approach to the study of knowledge in computer-rich environments. Her artifacts are nonetheless not simple machines encased in metal or texts embodied in print. Rather, a computer or a document like an environmental impact statement is a point of entry into complex "ecologies of social-material relations" that, if we are to understand them, demand a different approach to knowledge:

> Recent practice-based theorizing includes a reconceptualization of knowledge and action as located in "ecologies" of social-material relations (Fujimura 1996; Star 1995). These relations are not given by nature, but are the product of ongoing practices of what John Law has termed "heterogeneous engineering" . . . drawing attention to the diverse discursive and material, human and artifactual elements that must be assembled together in the construction of stable organizations and artifacts . . . The intellectual traditions that underwrite these reconceptualizations . . . view knowing and acting as always and necessarily embodied, and therefore as located in particular, historically and culturally constituted settings. The generality of knowledges, in this view, comes not from their contextual disembedding but from the extent and stability of relevant social-material relations.
>
> At issue here is not knowledge as a self-standing body of propositions, but identities and modes of action established through ongoing, specifically situated moments of lived work, located in and accountable to particular historical, discursive, and material circumstances. (2000, 312–313)

To restate a Law comment in a form that Suchman encourages,

> Perhaps there is knowing, but there is certainly no knowledge. This is because . . . knowledges are never complete. Instead, they are more or less precarious and partial accomplishments that may be overturned. They are, in short, better seen as verbs rather than nouns. (313)

Cultural Difference and Computer Artifacts: Lorna Heaton

To see the direct relevance of such approaches to the knowledge question in cyberspace, consider the case of those allegedly more sophisticated automated information technology-based knowledge networking products often referred to as Computer Supported Collaborative (or Cooperative) Work (CSCW) systems. According to Lorna Heaton (1999), CSCW systems designed in Japan aim fundamentally to create simulacra of face-to-face interaction. Here, the problem of coordination in deplaced organizations is seen in hierarchical terms. Heaton argues, for example, that these systems are designed primarily to facilitate the sensitivity to the views of high-status individuals characteristic of organizational interactions there, where deference to the highest-status individual in the group means that others attempt to accommodate their reactions to hers. Lower-status members are particularly sensitive to nonverbal cues. Thus, Japanese CSCW systems have been largely oriented to providing visual information, via multiple video images, especially about facial movements. It is very difficult for such systems to work as well as being in the same place; "virtual" copresence is never as good as the real thing.[2]

Heaton contrasts this design orientation with that of Nordic systems also called CSCW. In the Nordic countries, a contrasting ethos of equality, even "sameness," is dominant, in organizations as well as social life in general. Performing higher status is frowned upon, as is overt sensitivity to it. Nordic CSCW systems are oriented toward maximizing the participation of each, stressing the collective rather than the individual, through things like shared editing spaces, text-based, multiple drafting, and support for diverse conversational ecologies, for example, the early PORTACOM system for electronic communication. Nordic CSCW gives priority to promoting the collective participation in work of stakeholder groups, downplaying the contributions of any particular individual, including the boss. As such, they marginalize precisely the kinds of social content that are central to Japanese CSCW.

Both types of systems give thicker attention to the social than do standard products. However, they do so in such substantially different ways that it is difficult to say what CSCW is if they are both CSCW. Heaton argues that ignoring such differences in cultural context of application is a main problem in getting CSCW systems to work and, by extension, of constructing a meaningful "knowledge base" for computer science. There is a similarity between Heaton's observation and the difference between Dutch–Anglo-American fatigue with knowledge management, on the one hand, and the continuing enchantment of Nordic organizations with cyberspace, on the other. This similarity is that, even when we talk about an apparently universal technical process like computing, an appreciation of cultural context is essential if we are to pursue AIT goals successfully.

Summary and Conclusion

In sum, the central claim to originality of this epistemologically informed, symmetrical anthropology of knowledge is its shift of attention away from knowledge as a thing, object, or commodity and toward the social process of its creation. This shift focuses study of knowledge on the discourses associated with actual knowledge networking. In line with the original intent of the Africanist/social networkists like Clyde Mitchell, it can document the variable quality of the relations involved in creating, distributing, and reproducing knowledge. As highlighted in linguistic turn accounts, intense social networking is recognized as an essential component, perhaps the key moment, in the construction of knowledge, something produced in what my colleagues like to call "communities of practice" (Lave and Wenger 1994; but see below).

Work in several other fields is as relevant to analyzing knowledge networking as is that in anthropology. Like many philosophers, I find the pragmatist tradition of Peirce and Dewey of particular interest (especially its newer practitioners like Calhoun 1995 and Frasier 1989), as described in chapter 4. Perhaps no group of scholars have raised more pointed questions about knowledge than those feminists who assert its inherent "situatedness," drawing attention to, for example, how claims for universality can also be claims for intellectual "turf" (Smith 1989).

Establishment of a sense of shared purpose and trust in coproducers' talk and action are essential components in construction of technoscience alliances (Latour 1987) and the more or less formal statements, the "knowledges," which are their public performance faces. Even in many modernist accounts, social relationships are essential to continuing acceptance and necessary revisions of the protocols through which information is verified as knowledge. As the prominence of technoscience ethnographers among the exemplars of this chapter attests, there are several dimensions of the knowledge question especially foregrounded by the development of cyberspace that are worthy of attention at the current moment. One is the commodification of knowledge that would seem to be an almost necessary component of current attempts at "distance learning," an issue pursued at length in chapter 9. This is only one of several practical knowledge questions made pressing by the spread of computing.

An array of empirical studies of, for example, the history of technoscience, as well as phenomena like "invisible colleges," have emerged in STS (science, technology and society, or science and technology studies) and related fields. These of course inform and are informed by studies in the sociology of knowledge (Mannheim) and science (Merton). Along with the anthropological ethnographies of knowledge described in this chapter, these provide important empirical materials for a reconceptualized, more self-conscious, comparative study of knowledge, a study necessary if we are either to take full advantage of AIT's potential or avoid the Whiteheadian discourse failures of scientism.

Simpler Alternatives?

Constructing such an epistemologically reflexive empirical array of anthroknowledges of differing knowledge networkings will be time-consuming. One might be tempted to try to hasten the new analysis of knowledge by imposing a priori one of the more orderly intellectual frames suggested in chapter 2 or some other knowledge paradigm of preference. Intellectuals are seldom successful in such endeavors; others pay little penalty for ignoring us. Moreover, the extensive discourse disparities outlined above suggest that transformations in how knowledge comes about may indeed already be under way. Their outcome would inevitably be prejudged by an imposed standard.

Another way to initiate study of knowledge networking empirically would be to identify a set of practices as "about" knowledge and observe and analyze them. To do this, one could try to operationalize one's study by defining abstractly what it is about a practice that makes it "about" knowledge. However, as suggested by Whitehead, any such a priori approach would inevitably distort the emergent picture of the general knowledgescape. Alternatively, one might operationalize as "about knowledge" all those things that "natives" call knowledge. This approach would likely founder on the many contradictions and silences in knowledge discourses in actual cultures, like those identified in chapters 1 and 2.

Revitalization of the study of knowledge in anthropology must instead proceed empirically, through study of existing knowledge practices. Fortunately, knowledge networking, the actual discourses and other practices that create, reappropriate, and spread knowledge, is highly accessible to empirical study. Moreover, the general problem of how to conceptualize knowledge has today been given further impetus by changes in knowledge networking purportedly associated with AITs. This returns us of course to the cyberspace knowledge question. Good accounts of the dynamics of past and present knowledge networking are not only central to understanding the general dimensions of knowledge(s). They are also essential if we are interested in the cyberspace knowledge question: whether current changes in knowledge's production, reproduction, and/or sharing are of transformational significance.

At the same time, efforts to answer the knowledge question in cyberspace can make a distinctive contribution to the general anthropology of knowledge and the clarification of anthro-knowledge in particular. A Foucauldian-engendered preoccupation with the knowledge question in anthropology intersects directly in Haraway with the broader implications of new, technoscience forms of knowledge.

The approach outlined in this chapter is anthropological (ethnologically informed ethnographic) study of already existing knowledge networking, especially those practices in which people justify what they hold to be true. As an experiential way of knowing, ethnography has developed ways to hold both people's practices and their cultural constructions of them in the same frame (Hakken 1999a). By refusing to subordinate either the moment of action or that of cultural construction, ethnography constitutes an appropriate approach to the study of knowledge networking. One does ethnography by first identifying an issue of importance (e.g., knowledge networking) and then locating a site (e.g., organizations) where one can observe and participate in the practices of interest and talk about them. Reconceptualizing contemporary knowledge discourses is an area of potentially important anthropological contribution, both to the specific issue of making organizational knowledge networking more effective and also to the broader question of how to use the information we generate with AIT more effectively.

Applying Anthro-knowledge *of* Knowledge Networking *in* Cyberspace

As noted in chapter 3, "thin" knowledge networking practices in 1990s organizations, practices that failed to come to terms explicitly with the social, tended to be merely slightly more complex, even merely renamed, forms of information or even data sharing. Such difficulties were foreshadowed by previous attempts to conceptualize knowledge in purely technical ways, as documented by Dubinskas (1988) and Forsythe (2001). In brief, the systems' failures are often a consequence of their creators' lack of sensitivity to their own actual knowledge networking at work, which tend to be ignored by formal representations of work and therefore of attempts to use these to automate. More generally, despite all the cyberspace hype, careful studies of computer-mediated communication are as likely to document their continuities with pre-AIT communication as some new social dynamic (Hakken 1999a, pace Sproull and Keisler 1995).

Even organizational systems that do attempt to simulate real life tend to conceptualize the knowledge networking too narrowly. Instead of trying, like Heaton's Japanese CSCW, to create virtual face-to-face practices, systems might aim to discover new, AIT-based practices, ones that support collective production and effective dissemination of knowledge in distinct ways that take advantage of new, AITed capabilities. I have made a similar critique of attempts at community computing (e.g., Shuler 1996; Rheingold 1993), which in my view too quickly presume that the goal of community AIT networks is to reinforce and rebuild place-based local or physical community (Hakken 1999a). The alternative (or rather, the complement) is also to use technology to build communities based on relationships different from those of place, or what Mizuko Ito (no date) calls "networked localities." Similar conceptual difficulties are encountered when one tries to deploy notions of "communities of practice" (Lave and Wenger 1994).

As suggested in chapter 2, it is possible to see the modernist and the postmodernist conceptions of the data/information/knowledge interface not as competitors but as complementary. To do so, the modernist notion of "raw" data must be recognized as a culturally constructed silencing. Latour and Woolgar's (1979), Traweek's (1988), and Heath's (1994), as well as Forsythe's (2001), ethnographies all uncover ways in which data are produced socially, the particular contexts in which they are initially situated and from which they are abstracted, and how these abstractions are "disappeared" in subsequent "raw" constructions. However, it is also important to acknowledge that scientific knowledge is still made useful in the world, and that this is done by resituating it in existing sets of social relations and practices, like labor processes. Existing ethnography shows that, far from being either a unidirectional abstraction or concretization, knowledge emerges out of a networking dialectic that moves between various moments of abstraction and concretization.

Can knowledge networking systems incorporate technology to support practices delinked in space and time and still give appropriate attention to all the moments in the complex, neither strictly modern nor postmodern, dialectic of knowledge? Are the parameters and dynamics of such systems distinct enough from existing systems to justify arguments for a fundamental shift in the character of knowledge?

While answering forms of the knowledge question in cyberspace like these is my personal knowledge project of preference, the goal of this chapter has been to focus on a broader set of knowledge issues, those confronting the disciplined social practice called anthropology. My ability to help answer the knowledge question in cyberspace depends upon the work of the colleagues discussed here who are attempting to develop a new anthropology of knowledge in response to the current challenges to anthropological knowledge.

Their work justifies further basic research on knowledge networking itself. Moreover, fundamental change in the character of such networking is correlated historically with profound social transformation. Yet because they existed before automated IT, the current knowledge forms have already influenced AIT's development and continue to affect their use. Is the character of knowledge deeply affected when KNing is mediated by automated IT? How do changing forms of knowledge and its diverse modes of apprehension affect the ways in which IT is implemented?

Before such questions can be answered, the specific conceptualization of knowledge in the professional practices most closely connected to AITs need to be examined. It turns out that they, like the social sciences, have a complex history of knowledge-related notions that significantly influence, often misshaping, their ability to answer the cyberspace knowledge question. It is to this history that I turn in the next chapter.

Chapter 6

A Social Informatics Approach to Knowledge in Cyberspace

Many computing activities depend on distinguishing form from content. For example, once data are transformed into information by representing them in any form compatible with the TCP/IP protocol, the information is, irrespective of content, transmittable from any one TCP/IP-enabled machine to any other.

This form/content dichotomy, central to Cartesian modernism, also figures strategically in accounts of computing's alleged social impacts. For example, under peculiar but highly influential constructions of patent law (Pfaffenberger 1999), standardizations of form like TCP/IP justify claims to own—"proprietizing"—*any* aspect of computing, not just hardware or software, but data, information, and even knowledge. Such constructions extend considerably the terrain on which capital might be reproduced. The "new economy" manifest in the new ways to own knowledge, often described as enabled by them, is also said to have transformed the dynamics of social reproduction (Giddens and Hutton 2000; see chapter 11).

Other analysts reject increasing what can be owned as the chief benefit of computing-enabled form/content contrasts. They instead stress network effects, essentially computers' ability to make connections (Agre 1996). In their accounts, while computing does make more content accessible, what is really important is an increase in the number of people to whom the content is accessible. Rather than an increase in the ambit of capital, pride of place goes to the practical effectiveness and democracy promotion of new knowledge-making structures—like those of the Internet Society, for example—that computing enables.

The first of these two social impact stories could serve as a "legitimating myth" for Microsoft, while the second lends credence to the open source and free software movements discussed below. Here the point is that creation of new forms of knowledge, forms that can be articulated more or less independently of content, is central to the practice of computing and to most accounts of its social consequences claims. The stories equally highlight the importance of the focus of this chapter, the intersection of professional informatics and knowledge.

Pragmatist philosophy and the "new" anthropology of knowledge both offer ways of thinking empirically about knowledge that have more both specific and general validity than scientistic modernism. The approaches of the previous chapters acknowledge that knowledge networking is a moment characteristic of the reproduction of all social formations, but they also stress the necessity of acknowledging the thick connections actual knowledge networkings have with the specific contexts within which they occur. They thus imply caution about just how much form can or should be separated from content, yet, as in stories like those recounted above, radical disjunctions between forms and contents are central to the alleged quantum change in knowledging.

Given their role in design of automated information technologies, it is professionals in informatics whose practices actually separate knowledgings' forms from their contents. Much of what has gone before in *The Knowledge Landscapes . . .* has cast doubt on claims that AITs usher in a truly new cyberspace "knowledge society," while still recognizing its possibility. This chapter argues that a reconstruction of professional theorizing about knowledging is necessary for this potential to become real.

Much is at stake in how adequately professionals theorize the relationship between knowledge and AITs. This chapter outlines a perspective from which it is possible to think clearly in informatics terms about knowledge. It first illustrates the centrality of knowledge activities to theorizing in informatics. Next, it shows how inadequacies in dealing with the social dimensions of knowledge tie together informaticians' multiple failed attempts to engineer it. Finally, the chapter highlights a more viable, alternative approach within the discipline, social informatics, one that already affords the kinds of knowledging outlined in the preceding two chapters. This informatics approach to knowledge is the third intellectual justification for thinking of cyber-knowledge in terms of knowledge networking.

Why We Need an Informatics of Knowledge in Cyberspace

The two preceding chapters addressed knowledging in any and all social formations. This chapter returns more closely to the chief focus of *The Knowledge Landscapes . . .*, knowledge questions in one kind of social formation, cyberspace. At first reading, the notion of an informatics of knowledge *in cyberspace* might seem redundant. If "informatics" is taken to be just the study of automated information technologies, as it is in computer science, and "cyberspace" is a social space produced by such entities, *any* informatics of knowledge would already be an informatics of knowledge *in cyberspace.*

This chapter contains the more complex notion of "an informatics of knowledge in cyberspace" in its title for several reasons. One is an anthropological/cultural relativity point, that all human social formations, not just contemporary ones, have technologies for representing and communicating information. Thus, to treat computing as the only information technology is ethnocentric, inadvertently privileging one social form over all others. In the phrasing of Lewis Mumford, each social formation type has its own technics of information, so it would be equally appropriate to study informatics in all social formations.

Thus I insist, as argued for more fully in *Cyborgs@Cyberspace?*, on using the phrase "automated information technologies" (AITs), as opposed to just "ITs," to refer to computer-based information infrastructures. In asking the cyberspace knowledge question, we are inquiring into how much knowledging changes in the substantial presence of computers. What distinguishes computers from previous information artifacts is not that they are information technologies, but the *kinds* of information technologies they are—that they are both programmable *and* run in relative autonomy from an active human operator. It is these factors that one must feature if one wishes to argue for computers' transformative impact on social formation reproduction.

The more complex phrase is also a useful reminder of the centrality of knowledge to popular accounts of what is distinctive about cyberspace. Recall, for example, chapter 1's presentation of the centrality of change in knowledge to Jon Katz's "Geek Ascension," in which Geeks become the "new ruling class." Katz's is just one of many accounts in which a purported change in professional knowledge practices is discursively linked with the profound social transformation pointed at by knowledge society–type labels. The phrase "the informatics of knowledge in cyberspace" critically links together professional practice, knowledge networking, and theorizations of social change.

A final reason for the longer phrase "an informatics of knowledge in cyberspace" is that it points attention to a specific professional practice. Strategically central to any answer to the cyberspace knowledge question is understanding of the specific professional practices through which knowledge is operationalized in the supposed emerging life world (Habermas 1987) of cyberspace. Those who would "engineer" knowledge are of obvious importance to any informatics of knowledge in cyberspace.

Why "Informatics" rather than "Computer Science"

The issues addressed in this chapter are typically claimed by the academic discipline called "computer science" in the United States. Ownership of these issues is also asserted by "information science," a field previously referred to as "library science," as well as by "information systems" in business. Why follow the lead of Nordic scholars like Pelle Ehn (1988) and use "informatics" as the label for this intellectual terrain, when there are already three claimants?

My reasons have in [illegible] with my "forced" induction into the teaching of computer science, as described in *Cyborgs@Cyberspace?* My objection to the "science" part of the "computer science" label derives from my colleagues' difficulties in specifying a proper scientific object for their emerging field. Typically, they first drew attention to a machine, the computer. However, any science program articulated around an artifact is derivative and likely to be somewhat tendentious, like "automotive science." It is thus better framed as "engineering" or "technology" than "science."

If not an artifact, what else might be computer science's object? Were it the *practices surrounding the use* of a recently invented artifact (the computer), the field would not be a very "natural" science. To be so, it would have to be distinct from mathematics, a field whose "naturalness" had long been a matter of dispute. If, as others maintained, it was about mental processes, such as "computing" or "intelligence," in what way was its object distinct from that of psychology, or philosophy? Why should it be thought of as "natural" anyway? Didn't it make more sense to construct it as one of what Herbert Simon was calling the "artificial" sciences?

"Computer science" is today still a misleading label. It was of tactical advantage within the academic politics of the United States in the late 1970s and early 1980s. Organizing themselves as an independent "department of computer science," as opposed to a bunch of scholars trained in diverse natural and social sciences and humanities who shared only an interest in programming, my colleagues at SUNY Tech left the school of arts and sciences. They joined what became known as the School of Information Sciences and Engineering Technologies, freely acknowledging that "bigger salaries" was a chief motivator.

What is clear is that computer scientists *do* things with and about information: They design automated information artifacts and systems, write programs to get these systems to do desired things, implement the systems, and/or engage in analytic activities necessary to performing these activities successfully. They are skilled at conceptualizing optimal solutions to information problems within complex, externally defined problem spaces and adapting devices to embody these solutions. This engineering quality of their activity is frequently reflected in their terminology: The most frequent term applied to the knowledge preoccupations of informaticians is "knowledge engineering," while "software engineering" is a common course title.

The centrality of such practices means most computer scientists are more similar to engineers and technologists than to scientists. We have in English a grammatical tag that implies engineering-type practices, the "ics" suffix of mechanics, ergonomics, logistics, (at SUNY Tech, photonics), and so on. Thus, because "informatics" more accurately reflects the character of informaticians' preoccupations and typical practices, along with academics in several European countries, especially the Scandinavian ones, I choose to employ it rather than a "computer science" label.[1]

A History of Knowledge Approach in Informatics

There are reasons internal to informatics for pursuing a renewed understanding of knowledge. In what follows, the history of certain discourses in informatics is outlined in a way that gives prominence to moments in which knowledge was of recognized importance. Analysis of explicit readings of "knowledge" in the field demonstrates that the creation, adaptation, and sharing of knowledge have long been goals of informatics. While not always the immediate or acknowledged objective, the changing character of knowledge was initially and has periodically reemerged at the center of the informatics agenda.

While the field has been preoccupied with the knowledge question from its beginning, informaticians have seldom treated knowledge in terms like those argued for in the preceding chapters. There is a direct connection between informatics' knowledge constructs and the relative failures of multiple agendas in the field, including the one that spawned the AITed knowledge management craze chronicled in chapter 3. The discipline's multiple knowledge failures are grounded in a knowledge construct that is insufficiently social. This defect has hampered the field in establishing appropriate engineering agendas. Fortunately, a more effective alternative discourse, social informatics, has existed in the field for long time. In concert with nonessentialist philosophical and new ethnology of knowledge perspectives, social informatics can provide an adequate foundation for a more valid informatics of knowledge in cyberspace.

The Initial Knowledge Program

The computing systems so ubiquitous in contemporary social formations trace their ancestry directly to the ballistic and cryptological machines developed during World War II. Those developing the latter machines, especially Alan Turing, were taken by their inherent flexibility and thus their vast potential, particularly the possible universality of what is now called a "Turing machine." The potential was identified with the capacity to program these machines and then use them to represent what could be abstracted from other, more concrete artifacts and practices as information.

Consequently, it became standard practice in informatics to define computers as programmable electronic devices for the representation and storage of data and the consequent construction of information via data manipulation. This information-focused conceptual practice, like informaticians' activities in the real world, is clearly framed within the modernist knowledge progression (from data to information to knowledge) discussed in chapter 2, and the scientistic modernism critiqued in chapter 4.

In the modernist telling, however, the point of having information is so that it can be used to create knowledge. One person who linked AITed information strongly to knowledge was Vannevar Bush. Bush was perhaps the single most effective advocate for the institutionalization of the "big science" that was the hallmark of the immediate postwar era; he is acknowledged as the progenitor, for example, of the U.S. National Science Foundation. Like the philosophic empiricist Bertrand Russell, who strove to systematize all learning, Bush dreamed of what he called the Memex, a notional machine to store and provide rapid access to all human knowledge. A national science foundation would be as concerned about reproducing knowledge in new forms and sharing it as it would be about fostering its creation. An important part of Bush's agenda for science, Memex is considered by many scholars (e.g., Nyce and Kahn 1991) to be the inspiration for what became hypertext.

Bush felt that a main problem in making knowledge useful was making it accessible, a problem that grows exponentially with the complexity of big scientific projects. The problem with traditional knowledge archives (storage systems), such as paper files or books, is that they must be placed. To deploy multiple copies increases the size of the archive enormously, and multiple copies require changing/updating. Moreover, you are in even more trouble if you forget where the particular place of storage is. Cross-referencing and indexing systems can help, but these require mastery of indexing conventions whose utility also falls off rapidly with increased size (Pfaffenberger 1990).

In contrast, Memex would use computing systems to construct an electronic memory index. An item of knowledge could still be stored in one place, but locating it would be instantaneously possible from any of several directions/access points. Since there would be no point in creating a public bureaucracy to support big science if the knowledge produced was not accessible, something like Memex was in Bush's view a necessary complement of an NSF. The idea (ideal?) of the multiply indexible, easily accessible, knowledge systematizations of Memex inspired numerous subsequent computing agendas, including the World Wide Web (see Tim Berners-Lee's current preoccupation with a universal semantics 1999). The construction of Memex-like devices, second-order tools for referencing and storing the knowledge purported to result from information manipulation and representation of all sorts, was arguably the foundational task (Abbott 2001) in automating information technology and thus a central preoccupation of those creating the new discipline.

The Knowledge Dialectic in Informatics

In a memorable phrase, Mitch Kapor once referred to the World Wide Web as being like a huge library where all the books have been dumped on the floor (quoted in Dery 1994). The comment nicely indexes the extent to which informatics has been and continues to be preoccupied with variations of Bush's knowledge task. Kapor's ironic stance nicely parallels that of the computing center technician's famous "GIGO" phrase ("garbage in, garbage out"). The "GIGO" quip, delivered with shoulder shrug, underlines, albeit offhandedly, how the point of computing is "really useful" information, information acknowledged to be valid, or "knowledge." Linguistically, "GIGO" presumes its inversion, the proposition that storing data and manipulating it for information are only means that require a broader end.

Predictably given Bush's approach, an early metaphor for the computer was the "electronic brain." The continuing centrality of his goal for the new field is indexed by the institutionalization of the "Turing test." Roughly stated, this was the idea that if an outside observer could not differentiate systematically between the results of the actions of a computer and a human being, that computer was intelligent.

The brain metaphor strongly afforded approaches like the "strong program" in artificial intelligence (AI), that the point of informatics was development of computer systems capable of achieving real, human-type knowledge, not just the rough analogue of an index. Albeit having the artificiality of any engineered solution, AIs, were to be functionally homologous with actual human thinking. AIs of the 1960s gave informatics an agenda quite in keeping with Bush's earlier notion.

The term used most frequently to refer to those directly taking up each new version of Bush's challenge is "knowledge engineer." As all these discourses reflect, concern with the broader context within which information is to be interpreted was both an early and a continuing preoccupation of the conceptual space eventually "disciplinized," as Foucault would say, as computer science.

While periodically submerged, the e-brain metaphor regularly resurfaces. At a general level, one can perceive a characteristic oscillation or dialectical movement in the field, from moments of strong confidence in computing's knowledge capabilities, to a knowledge-doubting emphasis on "just data" or "mere information." Strong knowledge programs appear in informatics, draw substantial disciplinary attention, lose their appeal, and then reappear again in new garb (Engelbart 1963; Poitou 1996). Indeed, whether in the form of ongoing programs like AI, more specific technologies like expert systems or, currently, "knowledge mining" (as opposed to "data banks"), to engineer knowledge is arguably the most consistent underlying dream of informatics. It might make more sense to refer to informatics as something like "knowledgistics." Were one to insist on a "science" label and also follow the convention of naming a field after its goal rather than its chief means, it should be called "knowledge" as opposed to "computer" or "information science."

The fact that this is not the common practice suggests that "computer science" is a label that is not only bureaucratically convenient but also helps keep the field's difficulties with knowledge out of focus. Indeed, the history of informatics is littered with knowledge engineering failures, from cybernetics and operations research, through the strong programs in AI and experts systems, to, most recently, knowledge integration. Rather, the great accomplishments of informatics—such as the adaptation of mathematical algorithms to the task of programming, the development of internal memory, the creation of the Internet and later the World Wide Web—are clearly more triumphs of "information" than of "knowledge" engineering.

The repeated failures of knowledge engineering projects have not led to the abandonment of Bush's dream. As long as information remains ultimately only a means, the sometimes hidden, sometimes overt knowledge dialectic will remain central to informatics. The appeal of a notion like "knowledge society" makes deeper understandings of this dialectic an essential preliminary to answering the knowledge question in cyberspace. Such understandings are equally important to improving the informatics contribution to knowledge networking that is a goal of *The Knowledge Landscapes of Cyberspace.*

A Critique of a "Standard" Informatics Account of Knowledge

As a social scientist of knowledging long involved in computing, I have watched this knowledge aspect of computing with particular interest. The phrase "knowledge engineering" suggests the positivist, rationalist assumptions built into the way computer scientists have thought about knowledge. The broad critiques of formalism outlined in chapters 4 and 5 suggest the predictability of as well as plausible reasons for the difficulties encountered on the road to artificial intelligence and expert systems. As with the knowledge management debacle outlined in chapter 3, the main problem with such programs is their weak knowledge construct, one particularly unable to deal adequately with the social and cultural dynamics at the center of all knowledge networking, whether mediated by computers or not.

As an example of these difficulties, consider Tom Stonier's efforts to construct an essentialist informatics of knowledge in *Beyond Information: The Natural History of Intelligence* (1992).[2] Under the rubric of "machine intelligence," Stonier outlines a "strong" knowledge program for informatics. In common with most conceptual approaches within informatics, Stonier from the outset frames his in "computer revolution" terms. The hypertransformative, revolutionary social effects of informatics are explicitly presumed:

> The emergence of machine intelligence during the second half of the twentieth century is the most important development in the evolution of this planet since the origin of life two to three thousand million years ago . . . [being] analogous to the emergence . . . of complex, self-replicating molecules..[M]achine intelligence within a human social context has set in motion irreversible process which will lead to an evolutionary discontinuity. (1992, 1)

Stonier's performance of the standard computer science knowledge engineering trope is particularly effusive in its deployment of "species revolution" rhetoric (Hakken with Andrews 1993), but it is by no means unique. Although in this quotation he references the humanly social, Stonier's theory of intelligence is one in which human entities have little agency. In distinguishing human intelligence from machine intelligence, Stonier labels the latter "pure intelligence," suggesting the former to be "soiled." Particularly fecund, machine intelligence

> represent[s] a qualitatively different form of organization of matter, energy, and life. [It] . . . presages the progression of the human species . . . into a form which, at present, we would not recognize as human. Knowledge engineering . . . contains the seeds of our successors . . . the question is not whether intelligence will supersede life, but how fast? (1)

Artificial intelligence, at the time he was writing the regnant program in informatics, is the engine of this new computer revolution:

> [T]he proper development of AI will allow us to test out various scenarios and simulate various options . . . [P]owerful tools to augment judgment . . . [will i]mprove the quality of life in a myriad of others ways..[with] long term implications: . . . [W]ithin a few generations, humanity will evolve beyond what we currently consider (though we do not understand) to be human. (2)

Pursuit of Stonier's strong AI program leads to a "fledging new scientific discipline—*knowledge engineering.* [This is the] science of information. [We] need scientific information about AI so that we can design large, usable, and reliable expert systems" (2). As a science declaring itself, knowledge engineering needs a theoretical basis. In articulating it, Stonier reveals a strong commitment to the modernist, positivist knowledge progression:

> [A] computer, no matter how sophisticated, is useless if you have . . . no rules to drive the inference machinery designed to process . . . data. Knowledge Engineering attempts to codify and distil information so that it can fuel the increasingly powerful logic machines processing the data stored in their memories. Intelligent knowledge-based systems (IKBSs) and expert systems are computer-based systems which take information and process it in an 'intelligent' manner. (3 and 4; scare quotes in original)

Stonier is quite aware of the thin conceptual assets with which his approach begins. In his view, the field does not even have an adequate theory of information; it merely has one of communication (his negative characterization of Shannon's "nonnoise" approach to information theory[3]). He also draws attention to "the irony . . . that . . . the Knowledge Engineer . . . has virtually no understanding of the phenomenon which we call 'intelligence.'"

Thus, Stonier's Knowledge Engineer is in the embarrassing situation of understanding neither the input (information) nor the process going on inside a computer (intelligent operations) that creates machine knowledge, but she is nonetheless in the process of reinventing life! Despite her enormous intellectual deficits, Stonier asserts that the Knowledge Engineer can safely ignore the enormous philosophical literature on epistemology in constructing an account of intelligence. Atypically for this genre, he explains why. Because they are so suggestive of the intellectual irresponsibility at the root of so many technical discourses on knowledge, Stonier's grounds for this claim are worth examining in detail.

He first makes an apparently substantive claim, that philosophical knowledge of intelligence is irrelevant. He argues this by analogy, saying philosophy can be ignored for the same reason that physicists can safely ignore the literature of natural philosophy prior to Galileo: "Although it is of interest to the history of thought, such physics literature lacked sufficient experience with physical *experiments*" (6; emphasis in original). Apparently because philosophers don't perform experiments, "Any epistemology before the advent of computers can only be of limited value."

However, most informaticians don't do laboratory-type controlled experiments either; they do design based on trial and error.[4] To supplement his case, Stonier switches from the substantively temporal to the topical: "The same can be said for literature not conversant with the constellation of modern cognitive sciences, neurophysiology, and ethology" (6). Of course, ethology is not based on experiment nearly as much as on field study.

It is thus understandable that, having forcefully asserted a (logically weak) "intellectual revolutionist" case for ignoring the work of other disciplines, Stonier would shift intellectual ground again: "It is not that these philosophical analyses have no value, but[,] given the depth and the scope of this literature, it is simply not cost effective for the present author to try to cover it" (p.6). The problem with this weaker justification is, of course, that one cannot know the intellectual relevance, let alone the "cost effectiveness," of coming to terms with philosophical epistemology without having first already done so.[5]

In such rhetorical moves, the modernist voice of positivism becomes regnant by fiat. Stonier's argument for ignoring epistemology comes down to a bald assertion of the hegemonic rights of disciplinarity. He goes on to applaud the increasing number of technically minded philosophers who are becoming professionally involved with AI. Apparently, they recognize that "the train is leaving the station" and you'd better get on, if you want a bigger salary, and so on.

In regard to the acknowledged impossible breadth of the kinds of knowledge necessary to address intelligence satisfactorily, Stonier evokes Schrödinger's intervention as justifying the "new" beginning of his projected three books. The first was an argument for the physicality of information. The third is to be a "real" theory of information. The second text, *Beyond Information: The Natural History of Intelligence*, the one being discussed here, is about the evolution of intelligence from protointelligent systems to the most intelligent systems conceivable. Like the substance of the first book, the title of this one reveals Stonier to be firmly among those for whom informatics is a natural rather than artificial scientific activity. The human-made machines responsible for future intelligence are to be understood as part of the natural world, but one separate from the human. Stonier would dissolve the cyborg into the animal (Hakken 1999a).

To avoid having to base his informatics on communication à la Shannon (1995), Stonier opts for organization. (It is striking how much the problematics of computer science and organizational studies overlap, presumably because both disciplines take the large corporate entity, their usual client, as the prototypical ground for theorizing; see chapter 10.) Unfortunately, Stonier defines organization and information recursively:

> [A]n organised system (OS) is an interdependent assembly of elements and/or organised systems. Information is that which is exchanged between the components of an organised system to effect their interdependence. [This approach is] applicable to biological, social, and other systems engaged in information processing.
>
> [A] system contains information if it causes the system itself or some other system to become organised. (10)

If there is a discernible argument here, it is really a tautology: the way we know that a system is interdependent is by the presence of information, but we can only know if there is information if we already know there is an interdependent system. Stonier later asserts, just as opaquely, "Information is the common coin, patterns of organization are its product; and the nature of these patterns is determined by the properties of the system acted upon by the information" (10). That is, the product of the coin is the coin's determination!

Again, Stonier resorts to analogy—"What mass is to matter, organization is to information . . . [T]he more intrinsic information a system contains, the greater *and more complex* will be its organisation" (emphasis added). But of course this is not a good analogy, because mass is just a measure of scale, not quality. Having more mass means a thing just gets bigger, but Stonier wants to implicate a change in quality (complexity) *as well as* in scale (more information). Most crucially, he wants to distinguish between intrinsic or stored information, plus information added without a basic reorganization of system, on the one hand, versus, on the other, added information which *does* change the system. This somehow leads to the logically nonsensical as well as peculiar comment, "A computer storing information is still a computer irrespective of whether it is engaged in storing a lot of information or none" (11).

Acknowledging the shallowness of his argument does not prevent Stonier from performing a strong measurement trope: "We are still having problems quantifying organization. However, changes in entropy, which can be measured quantitatively in physical systems, can be used to measure changes in the organisation of such systems." There is apparently in his view something like a loss of complexity that follows from a loss of information. This latter loss can conceivably be measured, if it could be specified, in the way that a loss of energy can be measured. However, greater complexity is a qualitative difference; greater information does not automatically mean greater complexity. Moreover, whatever we are measuring in regard to organization, measuring it gets "increasingly difficult when we move to biological and social systems. When we get to human systems, [this is a] knotty problem indeed," because we then "tend to confuse 'information' with 'meaning.'"

Here we get a hint of the central move to come in Stonier's theoretical strategy, constructing a way to talk about intelligence (and knowledge) that excludes meaning (perhaps an ironic synecdoche for intelligence testing in general!). Again, by analogy, "The information contained in a book is there irrespective of whether anyone reads the book, or is unable to decipher the meaning because it is written in a foreign language." Here is Stonier's theory of knowledge, one that by eliminating any dimension of meaning reduces knowledge to information. Knowledge/intelligence is for him a physical entity, one measurable independent of what it means: "The idea that information has physical reality (and is not related merely to human information activities) is axiomatic to creating a general theory of information. It is also a prerequisite for analysing the phenomenon of intelligence."

To prove that information is not dependent upon human mental activity, Stonier resorts to analogy once more, arguing that since DNA has been around for perhaps a thousand million years, we must conclude that "information" has been around for a lot longer than humans.

> If something as simple as an inorganic crystal may possess information, . . . is it unreasonable to postulate that all organised structures contain information? It is therefore axiomatic: any system exhibiting organization contains information . . . [I]f one adds information to a system one has a common coin, like energy, which may be extracted to perform work, to organize another system, or in some other manner, be transformed or transferred. (6–10)

In this way, Stonier creates an abstraction for the essence of organization. This allows him to separate knowledge/information from meaning and still assert that he is holding on to something he calls intelligence. He acknowledges that we tend to view intelligence as starting with the ability to understand, capacity to comprehend, or perception of the significance of an act or a communication. However, we "must go beyond this" (11), presumably for the positivist paradigm to work. To decouple knowledge engineering from the sociality inherent in the notion of constructing meaning, he turns to learning: "In the ability to learn, we see glimmerings of intelligence. If we concede some sort of intelligence to animals . . . then there may be a spectrum of intelligence."

But any such spectrum would be multiplex, not have only a single dimension. Indeed, Stonier comments that "The more *advanced* the intelligence, the more complex it becomes" (12; italics added). Through the magic of language (that is, via a peculiar use of "complexity"), this time quantitative change has become qualitative. On the basis of his previous argument, one might have thought that "more complex" merely meant "more information," but now systems with more information have become "advanced," arguably the prototypic qualitative judgment. The shift to qualitative change is again justified by analogy: "Advanced forms of intelligence must be judged by an increasing number of independent parameters. For example, colour is more complex than light, in that it involves hue [wavelength] value [intensity] and chroma [degree of saturation of a distinctive hue]" (12–13).

Again, avoiding having to invoke humans and their culture, Stonier finds a "natural" basis for the qualitative differences, the multiple dimensions versus a single one, that he wants to call intelligence: "For any given organism, the nature and sophistication of its intelligence is determined by its environment. Intelligence is a complex, infinitely multifaceted spectrum" (doesn't he mean matrix?). The "fuzzy perception of subtle patterns is characteristic of advanced intelligence systems."

This leads Stonier to what he labels his basic definition:

> Intelligence is a property of advanced information systems, a product of the evolution of information systems, a state or condition ascertained only by observing the dynamics or behavior of [not the meaning associated with] a system. Intelligence is required to retain a system's integrity by adjusting it to changing environment. Intelligent activity, mostly, involves the ability of a system to analyze its environment and then make an intelligent response.

This last element, as an autodefinition, adds nothing, nor does his (nonjustified) behaviorist comment regarding methodology. Stonier apparently feels anxious about identifying intelligence with any response to a stimulus, so he adds additional criteria (the addition of none of which is justified theoretically): Intelligent response involves "enhanced survivability, reproducibility, and, if goal oriented, goal achievement" (14–15). He also recognizes the existence of "borderline phenomena, sub- or proto-intelligent," which can organize, process information, or learn. These are only intelligent, however, if they fulfill one of his conditions: enhance survivability, reproducibility, or ability to achieve their goals (but only if they are goal oriented!) Again, tautologically, they are intelligent if they possess "the ability to modify intelligent activity on the basis of previous experience (real or vicarious)."

Knowledge, stripped of meaning in such ways by Stonier, is equated with information, information becomes intelligence, intelligence becomes energy, complexity is both a purely quantitative and a qualitative dimension, and we don't have to pay attention to humans when we think about intelligence. Advanced information systems are intelligent by definition.

To say that Stonier does not take the social dimensions of knowledge seriously is not to say he ignores human events. Rather, he deploys a theory of human phenomena that radically marginalizes anything like human agency. Thus, for him, increased IT storage capacity and increased access to the information stored there are on their own equivalent to humans having increased substantially their effective intelligence. If one accepts this case, it is not necessary that anything actually happens as a consequence of increased information in the world in order to assert profound social impacts. The most profound transformation since the emergence of life need have no discernible manifestations!

In its Herculean effort to marginalize the social in pursuit of a knowledge program in informatics, Stonier's argument is by no means atypical. In its dogged adherence to the tenets of scientistic modernism, it liquidates the problem of knowledge networking. It is not difficult to see how such an approach to knowledge leads to faulty systems of knowledge representation, and thus to knowledge management fatigue syndrome.

The Definition of Information

As acknowledged by Stonier, one of the key contradictions of computer science (if not an index of its "prescientific" stature) is its failure to theorize clearly an obvious initial scientific object, "information." While Shannon (1995) described an interesting way to measure information—the noise to signal ratio of any communication—his approach doesn't articulate what it is about one wave that makes it a signal, for example, information, while another one is noise. We have to know already what information is in order to use his approach.

My critique indicates that Stonier's approach is no more satisfactory. The underlying problem of standard theorizations of information in informatics is their commitment to separating information, whatever it is, from its social context. This move is arguably the most basic one in scientistic modernism.

The necessity of abandoning this approach is paradoxically manifest in Keith Devlin's recent attempt to do it once more, in *Infosense: Turning Information into Knowledge* (2001). Devlin does an excellent job of explaining how we need to separate what we know from the ways we represent it, a lesson he clearly draws from the failed "knowledge banks" described in chapter 3. He also properly acknowledges the importance of maintaining a theoretical separation between information and knowledge, not dissolving the latter into the former.

Devlin unfortunately tries to use a reading of recent "situation theory" in linguistics to accomplish this information/knowledge differentiation. His approach would have the former represented in abstract formal terms and thus make it amenable to the scientific manipulation characteristic of mathematics. To do this, Devlin, like Stonier, must attribute substance to information, making it a thing. While asserting the intellectual value of this move, his book does not show how this attribution helps to solve the practical problems of organizations that are his concern. To posit information as "stuff" in informatics is analogous to the presumption of the "ether" that preoccupied pre-relativity physics. Like ether, the various forms of "info-stuff" will turn out to be other misleading myths.

While he argues for defining "information" in a way that separates it from its social context, his basic definition of information, "representation + constraint," actually violates his own stricture. This is because he uses "constraint" to mean the kinds of links between representations that must exist for information to be used, this latter being his conception of knowledge. To understand links we need to understand context. Consequently, the difference between knowledge and information is not the absence of context in the latter, but rather the absence of clarity regarding what that context is and what it means. It turns out that, in situation theory, information is situated data, while knowledge is information whose situation is known.

Summary of the Critique

The problems of implementing strong knowledge engineering programs based on conceptualizations like Stonier's are predictable. At base, they are the difficulties inherent in any totalizing program to formalize knowledge. Such efforts invariably involve separating what is known from the social conditions of its construction, reproduction, or sharing—dividing the knowledge from the networking.[6]

A large body of social science research documents the negative consequences of such approaches for informatics. Bryan Pfaffenberger (1990), for example, addresses a case where a Stonier-type rationalist program, one that aims deliberately to develop machine intelligence that circumvents human cognition, fails on its own terms. He traces the inapplicability of the positivist approach to the technologizing of information science, as in online searching. Essentially, the programmers made inevitable compromises with other professional groups (especially librarians) in order to make systems that people would use. Consequently, in order to strike a good balance in an online search between getting the best references for your topic and keeping the number of references manageable, you need the assistance of a good reference librarian. As a result, library information systems end up reinforcing the centrality of humans rather than replacing them by machines.

Diana Forsythe (2001) documents numerous examples of a complementary situation, in which computerization fails because of a failure to perceive important social dimensions in attempting to automate human processes. She identifies in particular a number of the problems of the formalist approach to development of expert systems. A common problem with such systems is the tendency of programmers to ignore the informal work typically done by experts, work that is therefore not included in the automated expert system. No wonder such systems fail to work! Forsythe opines that such failures can be traced to the trained tendency of program designers to be oblivious to the informal aspects of their own work. I say "trained" because, as documented by work sociologists building on Kusterer (1978), occupational systems that reward workers primarily for their formal credentials downplay, by not registering its existence, the substantive but uncredentialed activity that makes work possible. Here, a power-indexing cultural predilection, to see only those aspects of work which legitimate the professional, leads not only to inexpert systems but also to making opaque the sources of both nonfunctionality and nonexpertise.

The kinds of silences to which Forsythe and Kusterer draw attention are conspicuous in informatics. Many knowledge engineering initiatives, like strong AI and expert systems, have been either abandoned or greatly modified, most generally by greatly narrowing their scope. However, one seldom finds reference to them in the formal computing literature (e.g., in grant applications)—my experience in three years on the NSF Knowledge/Distributed Intelligence panel.

Professional informatics is replete with examples of knowledge problems revealed in "everyday" practices. In *Plans and Situated Actions* (1987) Lucy Suchman describes the inutility of cognitive science models in explaining the actual activity of systems development. More recently, Gary Downey (1998) and Stephen Helmreich (1999) trace additional contradictions in knowledge programs in informatics that continue to follow from inadequate awareness of their sociocultural context.

An Alternative Conception of Knowledge for Informatics

One need not base doubts about rationalist knowledge engineering programs like Stonier's on social studies of automated information technology, however. In critiquing Stonier, I used the tools of standard philosophical argument, "logic" with a small "l." Before proposing a solution to the problems of the standard, scientistic modernist account of knowledge in informatics, of which Stonier's approach is only one instance, I present some internalist critiques of problems following from such approaches.

Critiques of this approach have long been a part of the work of informaticians, especially those interested in how the technology is actually used. Consider, for example, the introduction to *Readings in Knowledge Representation* (Brachman and Levesque 1985), a collection published some forty years after Bush first broached the idea of using computers to engineer knowledge:

> The notion of the *representation of knowledge* is at heart an easy one to understand. It simply has to do with writing down, in some language or communicative medium, descriptions or pictures that correspond in some salient way to the world or a state of the world. In Artificial Intelligence (AI), we are concerned with writing down descriptions of the world in such a way that an intelligent machine can come to new conclusions about its environment by formally manipulating these descriptions.
>
> Despite this apparent simplicity in general goal, the research area of Knowledge Representation has a long, complex, and as yet non-convergent history. Despite the fact that just about every current AI program has what is called a "knowledge base" containing symbolic descriptions represented in some "representation scheme," there is still a vast amount to be understood before, for example, the knowledge in one system can be shared with another. There are tremendous subtleties in the notions of "representation," "knowledge," and their combination in AI.
>
> The papers in this volume have been collected to allow the reader . . . to begin to explore these subtleties. (xiii; emphasis in original)

Maybe handling knowledge by trying to ignore its social dimensions, or separating form from content (or object from process?) and ignoring the latter, are not good ideas; that at least with regard to knowledge, software is equally implicated in all of these. It may be that different understandings of functionality in software would provide a better key to developing a more satisfactory approach to the problem of mediating knowledge networking via AIT. The goal should be to find AIT-mediated forms of knowledge representation that functionally equal the sociality of existing KNing.

A Current Example: Cantwell Smith and Objects

In *On the Origin of Objects* (1996), Brian Cantwell Smith examines the sources of computer science's fascination with object orientation and the related preoccupation with ontology. His critique lays the basis for a shift in focus back from ontological content to epistemological process.

Smith first identifies the wide range of practical problems that drive programmers to think about ontologies. Basically, these follow from the need for entities to be stable and unambiguous if they are to be representable in the formal information systems designed to be run on computers. To represent it as data, one needs to know that what exists doesn't change.

Echoing Whitehead, Cantwell Smith then connects difficulties in informaticians' ontologies to the shortcomings of their modernist program:

> Perhaps the most interesting thing about [computer science's] ontological effort, moreover, has been the ways in which it has failed. The problem is that, upon encounter with real-world problems, it is hard for practitioners to avoid realizing that such traditional ontological categories as discrete countable objects, clear and precise categories, and other products of received ontological myth, are both too brittle and too restrictive. (1996, 45)

To replace such myths, Smith points to a process-oriented, situated philosophy of knowledge. Something of what his theory strives for emerges in his comments on the role of "objects" in recent computer science efforts at representing knowledge:

> As much as twenty years ago [i.e., the mid-1970s], members of the knowledge representation community in artificial intelligence, and also a number of researchers working on data bases, began to wrestle with many of the same problems as now face object-oriented system designers. In part as a reaction to the insuperable difficulties they encountered, many people in the knowledge representation community abandoned the idea that a system's ontological categories (those in terms of which it deals with its primary subject matter) should be explicitly represented at all. Instead, they viewed them as emerging in constant and dynamic renegotiation with the environments in which these systems play or are deployed. It is interesting to speculate on how the mainstream programming community will rise to this challenge of developing external, social and negotiated categories. (48)

Here Smith lays down a challenge for knowledge informatics (as well as philosophical epistemology): the creation of approaches to ontology that account for objects in ways more sensitive to the processes of their creation and reproduction, especially the social dimensions that are a part of their relationships to their environments. Such ontologies would necessarily be permanently provisional, for if anything, the artificial tends to change more quickly than does the natural.[7]

The Social Informatics Alternative

Indeed—as currently with the notion of "communities of practice"—it is *awareness* of culture rather than abstracting from it that has inspired the most successful approaches to knowledge in computing. Part of the reason for the theoretical "non-convergence" noted by Brachman and Levesque, on its face a failure of "one best way" rationalism, is the existence of a clear, socially oriented alternative to the standard approach to knowledge in informatics. Important early works include those by Joseph Weizenbaum (1976) and G. M. Weinberg (1971), as well as a number of works stressing the use context by Börje Langefors (1993) and the teaching of Kristen Nygaard.

Below I outline the features of this alternative social knowledge program in informatics, showing how it supports nonessentialist, ethnological informatics. The important point here is to understand that this different approach to knowledge emerged early and is a distinguished part of informatics. Terry Winograd and Fernando Flores published *Understanding Computers and Cognition: A New Foundation for Design* (1986), perhaps the most sustained critique of rationalist knowledge programs, six years before Stonier's book. In an important sense, the phenomenon really needing explanation in informatics is the persistence of the formalist program in knowledge engineering, which has continued in the face of both a long string of failures and a clear alternative based on substantive critique. Knowledge engineers have apparently been as oblivious to complexity and contradiction in their own field's knowledge as they were to that of other fields. (I offer a political economic explanation for this situation in chapter 10.)

The alternative practice of informatics is more compatible with the nonessentialist and ethnological alternative outlined in the previous two chapters. It begins with a different sense of what computers are. For formalists in general but especially for the mathematical realist, representation of phenomena via numbers is "natural," so it is "natural" to call machines for processing representations "computers." For scholars like Winograd and Flores (and Scandinavians like Peter Boegh Andersen, Bo Dahlbom, Niels Ole Finnemann, Randi Markussen, and Lars Mathiessen), computers are more general-purpose symboling machines, numbers being only one of the kinds of symbols which they can manipulate. Had these scholars been the namers, "computers" might have instead been called "symbolers," emphasizing how their use as calculators is only one of their many uses.

Another difference between the mainstream and alternative social views in informatics is the centrality to the former of identifying a formal logic to explain computers themselves (creation of a distinct "science" of computers). The alternative sees such a project as subordinate to larger goals, like the general identification of ways to use computers to support the essential human activity of communicating or, more strongly, as means to attain commitment to shared understandings, as means to create knowledge. Similarly, one can differentiate between those for whom simulation via computing is a means to understand a determinant structure hidden beneath the everyday, and those for whom simulation is a means to construct useful plans for collective action.

Social Cognition in Social Informatics

Perhaps the key difference is in the basic attitude toward the social. Positivists like Stonier and Devlin see the central theoretical task of informatics as finding ways to model intelligence, information, or knowledge that marginalize social processes like the perception of meaning. For informaticians like Winograd and Flores, coming to terms with the social is central to informatics. They characterize the rationalistic view, what I call scientistic modernism, as one that

> At its simplest . . . accepts the existence of an objective reality, made up of things bearing properties and entering into relations. A cognitive being "gathers information" about those things and builds up a "mental model" which will be in some respects correct (a faithful representation of reality) and in other respects incorrect. Knowledge is a storehouse of representations, which can be called upon for use in reasoning and which can be translated into language. Thinking is a process of manipulating representations. (1986, 73)

Winograd and Flores reject this "naive ontology and epistemology." It is not the case that "'things' are the bearers of properties independently of interpretation . . . . [W]e are always concerned with interpretation" (73). They are not naive idealists who believe we simply make up our own worlds, but their "central insight is that this world, constituted as a world of objects and properties, arises only in the concernful activities of the person."

What then, is the domain of the minds producing knowledge? It is

> a domain of description . . . in which it is appropriate to talk about the correspondence between effective behavior and the structure of the medium in which it takes place[. However,] we must not confuse this domain of description with the domain of structural (biological) mechanisms that operate to produce behavior. . . . [Actual human cognition, distinct from the analytic act,] is not based on the manipulation of mental models or representations. In saying that a representation is present in the nervous system, we are indulging in misplaced concreteness and can easily be led into fruitless quests for the corresponding mechanisms. (73)

Winograd and Flores question "the distinction between a conscious, reflective, knowing 'subject' and a separable 'object.'" Representation is a derivative phenomenon, which occurs only occasionally, when there is a breaking down of normal concernful action.

This is perhaps the key difference between the dominant and the alternative conceptions of knowledge in informatics. For Winograd and Flores, knowledge is a part of being in the world, not a reflective representation; it is situated in, not abstracted from, actions taken in the context of collective awareness. They contrast this perspective with the "representation hypothesis" central to computer science. While accepting that such a hypothesis may have value in limited circumstances, they critique its generality, pointing out how approaches based on it are severely limited.

Contextualizing Computing

In *Computers in Context* (1993), Bo Dahlbom and Lars Mathiessen outline the theory of knowledge for informatics that inspired the "postmodern regression" outlined in chapter 2. For them, computing inverts the modernist progression, involving instead "[t]urning knowledge into information and information into data." This is "a difficult task as soon as we aim beyond anything but the most formalized and routinized type of knowledge. Sometimes it is quite impossible" (33). They argue that such a difficult task involves development of theory and, in strong contrast to Stonier, draw on large elements of the philosophy of knowledge to develop it. For example, they, like Whitehead, employ a contrast between Platonic concepts of knowledge, which revolve around "perfect instances situated in the world of ideas," and Aristotelian concepts, in which *to know* involves identification of the basic or "primitive" elements that make up an entity. Developing the latter kind of knowledge, an especially key element in object-oriented programming, is very difficult, because in real life "concepts are not primarily packages of information defined by structured complexes of more elementary concepts. . . . [T]hey are primarily defined by our practices and . . . we communicate about them primarily by prototypes or illustrative examples" (35). The act of computing involves transforming a Platonic world of knowledge into artificial, Aristotelian information, but to do this alone would inevitably mean distorting human knowledge:

> One of the challenges of systems developers is to understand and respect the Platonic nature of human knowledge and communication, and to understand

> the computer not only as a machine for processing data based on Aristotelian concepts but at the same time as a tool to support human beings in using and communicating Platonic concepts.
>
> As systems developers, we have to accept the fact that mechanistic thinking, so powerful in producing and characterizing the machine, may hamper us when we are trying to put it to good use. (37)

Dahlbom and Mathiessen reject scientistic modernist computing. Instead, they call for "accepting the challenge of the philosophical position of romanticism, of taking more seriously the actual nature of organizations and their information use" (38):

> Knowledge is not a commodity to be collected under controlled conditions, to be bought and sold on a market; knowledge is subjective enlightenment and edification, difficult but possible to share, *provided there is mutual respect and a sincere attempt at understanding.* (221; emphasis added)

At the heart of this approach to knowing are acts of communal coupling, or "discursive redemption" in the Habermasian quasi-religious language that has particular echo for me. They make the following comments about how to deploy the soft systems perspective, an informatics perspective that they see as central to modeling thinking:

> The method of the soft systems approach is interpretation. We are encouraged, by this method, to consider different perspectives; the claim is that to learn about the world we have to understand, express, and debate a variety of radically different perspectives. (55)

They couple this with the philosophical practice of hermeneutics or "deep interpretation," from which "requirements specification and systems development methods will not seem very important. It is the process of creating the requirements specification and the process of interpreting and changing it that is important" (225).

For them, systems design is a learning activity. Thus, they reject the purely rationalist view that the job of the systems developer is to design and implement a "turn key" system based strictly on her own professional knowledge. They identify an alternative, evolutionary view of systems development, in which multiple iterations of support systems are fashioned and experimented with, modified, and redeployed, until a satisfactory system is coconstructed.

Dahlbom and Mathiessen are at pains to make it clear that there is room for substantial moments of relative rationalism in their approach, that computers are mechanisms for automating activities. The important point is to recognize that these mechanistic moments in the process are not the ideal toward which the process is oriented but periodic events in a dialectic of interpretive coconstruction.

A few additional examples clarify their strongly situated approach to knowledge in informatics. They do advocate quality and quality control in software engineering. However,

> Striving for quality does not simply mean striving for objective criteria and standards of measuring quality. . . . It rather means striking a sensible balance between the most objective and the most subjective, of knowing when to strive for agreed upon criteria and evaluation procedures and when to listen to the idiosyncrasies of colleagues and customers. And it means knowing how to create opportunities for the involved actor to communicate, negotiate, and agree on the quality of the computer system they are engaged in developing. (154–155)
>
> A good project and a good computer systems is such that it makes possible communication, understanding, and rational discourse. (194)

As noted above, they acknowledge Habermasian notions of ideal speech situations as inspirations. They also deploy Habermas's theory of the different reasons for constructing knowledge or "knowledge interests," which includes emancipation along with (rationalist) technology and (romantic) understanding. Indeed, because the systems developer is always involved in changing organizations, she has no choice other than to articulate an ultimate value orientation. This leads them to identify a basic contradiction at the heart of informatics as an engineering practice:

> In contradiction to its essentially bureaucratic [because, e.g., machines are mechanistic] nature, our discipline is facing a fundamental requirement to deal deftly with change. Computer systems are rarely introduced in order to support a traditional way of operation . . . [but i]nstead . . . to change traditional ways of doing things, to solve problems and improve efficiency. (229)

Because the mechanistic approach is inclined toward thc reductionism of abstract rationality, while application requires being situated in the concrete, a communicative, social conception of knowledge, one capable of coping with contradiction, must be at the center of practice.

Finally, they draw on Levi-Strauss's notion of "bricoulage," or tinkering, to undercut further and ultimately displace rationalism or scientistic modernism as the overarching approach to engineering. Under rationalism,

> Computerizing an organization, we tacitly assume, is pretty much like building a highway or fertilizing a field; a matter of introducing the right cause in order to get the desired effect . . . [Levi-Straussian] Structuralism wants us to replace this notion of engineering with tinkering, and it wants us to apply it both to ourselves and to our customers.
>
> [Unless the] . . . material structures already in place in the user organization . . . are changed too, the introduction of the new computer system will scratch only the surface . . . The systems developer has to accept the fact that users will apply tinkering to turn the technology they use into whatever they think it can best be used for . . . (245–246)

This inevitable impact of users on the system is something to be related to, not to be feared or combated. Dahlbom and Mathiessen advocate critical theory as a basis for a general emancipatory project in informatics, while using structuralism to identify strong forces (as is often the case with user impact) that complicate the emancipatory project. In a phrase, computing is a great deal like life, an issue to which I return in chapter 11.

Situation Theory and a Social Informatics

While above I was critical of Devlin's appropriation of situation theory (e.g., Barwise and Seligman 1997), I think a properly social reading of it can make an important contribution to an alternative theory of information. One can begin, as this theory does, with a notion of information as involving both a representation and a set of links, constraints, and environments that together constitute the context relevant to the interpretation of the representation. Devlin would engineer knowledge by stripping these context factors from the representation, because the "problem is that you can't encode the context within the information. The most you can do within the message is try to provide an indication of what the context would be" (2001, 48). His position—that such encoding is not possible, while an indication of it is—is weak and not well argued. For Devlin, separating information from context is a theoretical imperative, derived from his desire to use the transcendent formal representations of math, which by definition demand abstraction from context. To the extent to which "an indication of what the context would be" is already embedded within a message, however, formal abstraction would mean destroying even this indication.

Chapter 7 explains how the "Knowledge Networking Knowledge Networking," or KN^2, project arrived at a very different imperative, that the main task in designing an AIT system to support deplaced knowledge networking is precisely to find ways to embed as much of the context within the formal representation as possible. It is the capacity to support just this which is the technical advantage of the current "bleeding edge" database products, in combination with "tagging" protocols like XML.

In a more social reading, unlike the individual/psychologistic one insisted upon by the standard approach in informatics, knowledge can be said to exist when a group of people acknowledge that something is known. In a non-modernist account, information can include representations that enable us to recover, if we wish, an understanding of the knowledge networking that leads to particular acts of acknowledgment. Devlin is correct to present face-to-face conversation as the means by which we normally engage in such "recursive redemption" in communication. The task of knowledge engineering is to find ways to use AITs that enable different forms of such redemption for people unable, or unwilling, to engage in face-to-face talk.

Examples of Applied Social Informatics of Knowledge: Participatory Design and Computer-Supported Collaboratory Work

Socioculturalist approaches to knowledge in informatics like those sketched out above have provided important inspiration for computing practices as well as theories. Dahlbom and Mathiessen are closely identified with User Participation in Systems Development or Participatory Design (PD), as this approach is more frequently labeled in the United States. As implied by the titles, this approach to the development of computing systems aims to maximize the involvement of users in their creation. It is a way to take seriously the knowledge of the users, finding means—for example, via conversation—for users and informaticians to share their domain and craft knowledge and integrate these into the development of systems. Its goal is to construct understandings shared by all of what the system is for and the particular choices made to implement it. Thus, there is a strong emphasis on multiple iterations in prototyping as well as considerable user participation in implementation, evaluation, and maintenance (Hakken 1999a).

Winograd and Flores are most closely identified with Computer Supported Collaborative (or Cooperative) Work (CSCW). The aim here is to recognize that most work takes place in groups, whereas most computing systems are framed in terms of individual users with single terminals. CSCW aims to come up with computing systems that support groups. Its products run the gamut from relatively simple systems for coordinating the calendars of several workers to complex "ubiquitous" computing environments. These latter provide an array of computing modalities to facilitate collective thinking in both placed and deplaced arenas.

It is of course possible to take a user participatory approach to the development of collaboratory systems. There is an extensive literature on these approaches that it is not necessary to reiterate here. The main point is that the existence of PD and CSCW practices in computing document the viability of approaches to knowledge in informatics that are socially and culturally sensitive rather than dismissive.

How "Irrational" is the Social Informatics Approach to Knowledge?

Does this mean no room for rationalism? One could read Dahlbom and Mathiessen's contention, that at the heart of informatics is a contradiction between bureaucratic automation and the need to cope with change in organization, as a requirement for maintaining some element of a rationalistic approach in informatics. It would be possible, although not easy, to combine knowledge created in this way with user domain knowledge in participatory design. It might be even easier to use knowledge obtained rationalistically to design information systems to support collaborative work. Chapter 2 refers to the possibility of a pragmatist approach to the knowledge dialectic, one that combines modern and postmodern moments. In practice, to what extent is a socio-cultural approach to knowledge displacing the scientistic modernist one in informatics?

In Section III of *The Knowledge Landscapes of Cyberspace*, I will show how I use the theory of knowledge networking developed in this section in my research and practice. I focus separately on each of three moments or aspects of knowledge networking: that of creation, reproduction, and sharing. I wish now to use the distinction between knowledge creation and knowledge reproduction to help focus this discussion of the necessary extent of the displacement by the sociocultural of the modernist rationalistic account of knowledge in informatics.

Networking to create knowledge is, like all knowledge networking, an intentional practice. In addition to creating new knowledge, it can also aim to discover knowledge that in some sense existed but was not known previously to exist. Knowledge reproduction networking, in contrast, is more concerned with applying existing knowledge on new scales or to new circumstances. It is not too misleading to see the former as the preoccupation of organized scholarship (for example, National Science Foundation funding) and the latter as the preoccupation of organizations, like those on which chapters 3 and 8 focus, both for-profit and not-for-profit, public and private.

This distinction allows me to pose the above questions in the following way: Might one acknowledge the greater relevance of the sociocultural approach to knowledge reproduction (i.e., in an organization context) but stress the rationalistic approach in the moment of knowledge creation? Would it be legitimate to differentiate among the moments of knowledge networking in terms of greater or lesser requirements for rationalism and romanticism? Conversely, must one, in order to pursue a knowledge program in informatics, in effect *replace* the ineffective scientistic modernist conception of knowledge with a socioculturally more robust one?

The Open Source/Free Software Movements and the Kind of Knowledge that Code Is

One of the pleasures of being a social scientist, as opposed to being a philosopher, is that it is often possible to find situations where the character of social formation reproduction itself poses and sometimes even provides answers to your questions. Today, the phenomena of "open source" code and "free" software development are exploring how deeply social the character of knowledge should be taken to be.

"Code" in informatics is the instructions which programmers write that ultimately tell the computer what to do. A program is a set of statements in code which, when properly entered into a machine, results in the desired symbol manipulations and storing. "Source code" is the original code included in software, the basis of the processing activities carried out by a machine. If one wishes to know *how* a program works, one needs access to the source code. Easy-to-find source code is "open," while code about which one can talk freely and even alter is "free" (as in both "freedom" and "free beer"; Williams 2002). Conversely, source code hidden by other code, or about which one is not be allowed to talk, let alone alter it, typically for value reasons, is called "proprietary code."

Bill Gates has shown how to make a great deal of money by controlling access to and/or use of valuable code. This is the kind of point with which the chapter began, about the important intersection between economics and accounts of knowledge in informatics. Acknowledging that limiting access to code is very profitable, Microsoft justifies closing and proprietizing code as creating the huge incentives that make possible great leaps forward in the creation of new computing knowledge. Central to this argument is the idea that knowledge production networking is essentially rationalistic, mostly to be produced by large bureaucracies, a managed division of labor, and so on. Current patent law interpretation reflects this view of most informatics knowledge as economically "valuizable."

The existence of substantial open source code and free software are a rebuke to claims like these (Moody 2001). In effect, they assert an alternative model of best knowledge creation networking in informatics, one whose dynamics Eric Raymond characterizes as more similar to the functioning of a bazaar than like building a cathedral (1999). The best code is produced when many independent informaticians collaborate to produce it, not under the domination of a single large hierarchical organization. The way to create better code, and to do it most rapidly, is to publish the existing source code rather than keeping it secret. That way, anybody interested can examine your code, see what you are trying to do and why, and offer evaluations and suggestions. With multiple knowledgeable networkers, good solutions will emerge; or, according to Linus Torvalds—after whom the Linux shell, the open source code which he "keeps," is named—the basic philosophy is: "With a million eyes, all software bugs will vanish" (Torvalds and Diamond 2001, 226).

In organizational development terms, the open source and free software movements rely not on the "hygiene" factor of wages and good working conditions but on the intrinsic motivators of commitment, craft pride, and desire to contribute to a common good. The spread of Linux, the Gnu suite, and free software was arguably the most innovative development in informatics in the 1990s. Linux, for example, is the software of preference for the servers that provide access to the majority of webpages (Moody 2001). When one considers the significant economic power wielded against this model of knowledge creation networking, its spread is quite remarkable.

As examined more thoroughly in chapter 10, the ultimate resolution of the open source/free versus proprietary code contradiction is dependent upon the ability of capital to bend state functioning to its reproduction. As Pfaffenberger (1999) makes clear, the decision of the courts to allow corporations like Microsoft to defend their proprietary rights under patent rather than copyright law is the kind of intervention which puts open and free sourcing at risk.

Whatever the ultimate fate of these movements, three conclusions relevant to informatics knowledge are already apparent. Their challenge to proprietary approaches in software development is a powerful demonstration of the relevance of diverse concepts of knowledge to the practice of informatics. There exist at least two viable, socially very diverse constructions of its knowledge. Whichever prevails in the long run will have done so because of the allies it has been able to recruit to its technology actor network, not because of some ONE BEST WAY inherent in information qua information.

Second, like participatory design and computer supported collaborative work, open source/free software are demonstrations of how, even in informatics, mobilization *of* sociocultural awareness, rather than abstracting *from* it, can be very effective. Again according to Linus Torvalds,

> The theory is that the strongest arguments will develop when the largest number of . . . minds are working on a project, and as a mountain of information is generated through postings and repostings . . .
>
> It's a wrinkle on how academic research has been conducted for years, but one that makes sense on a number of fronts. Think of how this approach could speed up the development of cures for disease, for example. Or how with the best minds on the task, international diplomacy could be strengthened. As the world becomes smaller, as the pace of life and business intensifies, and as the technology and information become available, people realize the tight-fisted approach is becoming increasingly outmoded . . . (Torvalds and Diamond 2001, 226)

Here, Torvalds describes a "wrinkle" which deserves consideration as one argument that there is a very different dynamic to knowledge production networking in cyberspace (see chapter 7), one that really would be "postcapitalist":

> The project belongs to no one and to everyone. When a project is opened up, there is rapid and continual improvement. With teams of contributors working in parallel, the results can happen far more speedily and successfully than if the work was being conducted behind closed doors.
>
> It's been well established that folks do their best work then they are driven by a passion. When they are having fun . . . [t]he open source movement gives people the opportunity to live their passion."
>
> It's like letting the universe take care of itself. (2001, 226, 227, and 229)

Finally, as on Torvalds's reading, it really does make sense to think of open source/free software as social movements as much as technical practices. This is despite (or, more accurately, because of) its many splits and controversies.[8] For

Torvalds, computing is a way for human beings to pursue their basic desires, one of the most powerful of which is to establish social relationships: "There aren't very many things that man [*sic*] is willing to die for, but social relations is definitely one of them." In the open source movement, computing supports the building of social relations, which simultaneously support computing in a virtuous cycle. Like any social movement, open sourcing is a form of knowledge networking which succeeds to the extent that it can create mutual couplings among independent, voluntary participants. The success of open computing movements is the best evidence yet that cyberspace contains substantial affordances to more egalitarian social reproduction. Practices like this are very suggestive of what Dahlbom and Mathiessen called "emancipatory informatics."

Conclusions

The goal of Section II of *The Knowledge Landscapes of Cyberspace* has been to ground firmly an alternative approach for addressing the knowledge question in cyberspace. This knowledge networking alternative has been justified via dialogue with the practice of three academic disciplinary/intellectual traditions. Each dialogue contained certain criticisms of the discipline's past approaches to knowledge.

This chapter's goal has been to demonstrate that the dominant discourse on knowledge in informatics causes certain problems across the entire range of the field as a form of knowledge networking, as well as in other fields that rely on it. At the most basic level, I hope to have convinced the reader that this "reality" should be recognized as a problem, especially when one tries to apply automated information technology to support knowledge networking, and that there is a viable alternative way to think about knowledge in the field.

Of the three disciplines addressed in this section, informatics has had the largest impact on how we think about knowledge in cyberspace, yet its dominant paradigm has changed the least. Indeed, this skewed paradigm is in an important sense responsible for the peculiar form in which the cyberspace knowledge question has to be raised, on both intellectual and popular agendas.

Or course, informatics' slowness to change has to do with its being an engineering field, as opposed to philosophy and anthropology, which are more purely academic. As fields committed to "basic" research and conceptual practice, anthropology and philosophy count conceptual schemata as their primary "product," so criticisms of such schemata can be deeply unsettling—even producing "paradigm revolutions" (Kuhn 1970). While one can find influential examples of the application of both anthropology and philosophy to practical problems—including how to construct cyberspace—these fields' application influence pales in comparison with that of informatics.

As an engineering discipline, informatics aims to produce systems that work, and many AIT systems have been of great utility. Further, I accept that, like all "practical" practices, what it means for a system to "work" is not an "either/or" matter, but rather a matter of degree, spread out, moreover, across multiple dimensions of effectiveness and efficiency. For all these reasons, engineering fields are full of examples of systems that worked well despite shaky conceptual foundations. One may not really need a good answer to the cyberspace knowledge question in order to build a workable system.

While I have had the benefit of doing much applied work in my career, I have seldom had the responsibility for making a system with a large technical component "work" in the sense of a new information system. Besides, informatics is not my field; when the opportunity came to switch, I stayed in anthropology. Perhaps this has something to do with the fact that formalist moments are central to informatics in ways that they need not be in philosophy and anthropology. In the latter, formal argument is a tool, whereas the majority of informaticians still try to build systems on formal bases.

For all these reasons, my aim in this chapter has been somewhat more limited than in the previous two. I hope those developing AIT systems will recognize formalism as inevitably problematic. Having done so, one can draw on the approaches described in the latter parts of the chapter to develop systems both more effective and more efficient. The history of computing in this chapter overstates the degree of uniformity in both dominant and alternative approaches to knowledge in informatics. Nonetheless, it does acknowledge that alternatives, ones which rejected the formalism of AI and Expert Systems and which, like Participatory Design and Computer Supported Collaborative (or Cooperative) Work, accepted the sociality of work, did indeed develop *within* the field. These alternatives may have failed to displace the mainstream, but not just because of academic politics. Lorna Heaton's work contrasting CSCW systems in the Nordic countries versus those in Japan is in its own way as suggestive of why as is knowledge management fatigue syndrome.

Awareness of the cultural differences at the heart of these examples increases our sensitivity to how difficult it is to bring about desired change in cyberspace knowledge networking. Recognizing the centrality of the sociocultural to any project to engineer knowledge is just a first step; the hard work of changing practices comes next. Trying to do so has been a central aim of my work over the last two decades. Section III recounts several attempts to apply the approach to knowledge outlined in these first two sections.

Section III

Ethnographic Perspectives

The Dynamics of Knowledge in Cyberspace

Chapter 7

Creating Knowledge with Automated Information Technologies, or Knowledge Networking Knowledge Networking

Summaries and Introduction

Summary of Section II

Before we can say how much knowledge changes in cyberspace, we need a clear conception of what knowledge is—its attributes, characteristics, and dynamics. Section I of *The Knowledge Landscapes of Cyberspace* introduced a conception of knowledge as networking that makes it more possible to answer the knowledge question in cyberspace.

Systematic grounding of this conception began, in Section II, on the knowledge landscape of philosophy. Attention was drawn to the etymological history of the word to suggest that, at least in English, conception of "knowledge" as abstract content was actually *preceded* by awareness of processes of acknowledgment, like an overt, public recognition of the "truth value" of purported knowledge. Even if for purposes of argument the Western "content" notion of "justified true belief" is accepted, chapter 4 showed how the knowledge networking conception could be separated from the Cartesian, modernist presumptions about the centrality to justification of an introspective search for essences. Instead, the chapter advocated a "turn to the social" in conceiving of knowledge, centering thoughts about knowing on comparative study of multiple forms of actual knowledge networking. Understanding the social practices central to acknowledgment across different social formations becomes the starting point of a new philosophy of knowledge. The chapter drew on pragmatism to provide a framework for merging modernist and postmodernist discourses on the relationships among data, information, and knowledge, to integrate positivist and deconstructionist moments into a single networking dialectic.

Chapter 5 on anthropology described parallel developments encouraging new, practice studies of knowledge, both ethnographic and ethnological. Concentrating on the social relations within which knowledge is created, reproduced, and shared, new knowledge anthropologists make thick connections that ground content in specific processes. Their studies grow out of the recent pluralization of anthropology's sense of its own knowledge, a consequence of its "crisis of representation." Their practice approach to knowledge also centers logically on knowledge networking as a moment characteristic of the reproduction of social formations. It follows that any general characterization of changing knowledge dynamics, including the idea that a "knowledge society" emerges with automated information technologies, can only be justified in relation to comparative empirical study in multiple social formations.

Chapter 6 focused on the discipline of informatics and its long but often backgrounded preoccupation with knowledge. It demonstrated how constructions of knowledge in this field were generally framed in formalist terms, tending toward the simplistic. Such constructions generally direct analysis and design away from the social practice concerns raised in anthropologies and philosophies of knowledge and in so doing impede creation of a real "computer science." However, an alternative social informatics tradition offers a more useful basis for a knowledge informatics. Social informatics, a tradition based on recognition of computers as machines for general-purpose symboling, not just tools for counting, treats knowledge as complex. The emerging open source and broader open computing movements in informatics are two of the many ways of doing computing which, by taking the social seriously, generate technical infrastructures with more potential to support truly new forms of knowledge networking.

Introduction to Section III

To illustrate further the potential of the networking approach to knowledge, Section III of *The Knowledge Landscapes of Cyberspace* shows how it is being used already in ethnographic research and practice. The chapters describe my personal efforts to answer the cyber-knowledge question on academic, organizational, and educational terrains, on each of which I have both studied knowledge networking mediated by AITs and worked to incorporate specific AITs strategically. The chapters are similar, in that they all present the cyber-transformationalist cases typically made on the relevant terrain, describe research and practice relevant to it, and use these to evaluate the cases.

Unlike most popular trade books on computing (e.g., Kidder 1981), these chapters contain few heroic tales and do not fit the "success story encouraging emulation" conventions of airport business books (e.g., Hammer and Champy 1994). In conventional terms, my interventions have had at best only modest results. Instead of aiming to inspire, the chapters provide cautionary guides regarding what to keep in mind when preparing for the "next go." I choose this approach for several reasons:

1. Any general transformation of social reproduction must be complex, relevant to multiple social terrains.
2. My activities mean I have something concrete to say about several kinds of AITed knowledge networking.
3. My stories illustrate the "ethnographic imagination" as it emerges when following the arc of praxis, moving back and forth between the construction of thick understandings of social worlds and efforts to act in those worlds based on these understandings.
4. It is these experiences that led to the knowledge approaches presented in the preceding sections, so they offer additional reasons for thinking about knowledge in these ways.
5. While a storytelling "moral" is honored more often in the breach than in the observance, I really do feel (fortunately for me) that we learn as much from failure as success.

A Framework for Comparative Research on the Forms of Knowledge Networking

A basic purpose of comparative study is to establish the forms of a practice and the extent of their diversity. It is common for comparative researchers to proceed by first identifying some potentially different types of the practice in question, locating examples of each, and then examining the similarities and differences among the examples.

In the West, at least three moments in knowledge networking—creation (or production), adaptation (or reproduction), and sharing (teaching and learning)—are separately institutionalized. It makes sense to use institutionalized differentiations like these to structure initially inquiry into the diverse forms of knowledge networking, so each is the focus of a chapter in Section III.

Differentiating the ethnography of cyber-knowledge networking into moments of creating, adapting, and sharing is only a first step toward a robust theory of knowledge networking. Any typology is subject to revision; it is fundamentally only a means to initiate analysis of underlying processes. There may be additional forms that need to be explored, or some other differentiation might make more sense. Still, a focus on the institutional differentiations among research, business, and educational organizations highlights important differences in the forms of knowledge networking. Once in a position to say something substantial about the degree of similarity or difference in the broader correlates of computing in these forms, we are better able to address the general validity of the "knowledge society" idea.[1]

Moreover, this typology of the main kinds of knowledge networking is robust enough to have important practical implications for AIT design. It allows investigation of, for example, whether the forms of knowledge networking are similar enough to warrant construction of a single, generic type of AIT infrastructure, or are so different as to necessitate design of multiple, distinct systems. If distinct systems are needed, how many and of what kind should they be? Systems to support many different kinds of knowledge networking would have to be very generic, while systems to support only one form, or a few closely related ones, would be "purpose built." How generic can systems be and still be useful?

As described in chapter 3, design of technical knowledge infrastructures was not postponed until an understanding of the particular range of knowledge networking forms supportable in particular cases had been achieved. "Knowledge management fatigue" suggests inadequate "user-friendliness" as a likely feature of tools too general, their overly generic quality being an important reason for the failure of many 1990s knowledge management systems. More specialized systems require less modification to make them useful, but they are generally limited to the specific situations for which they were designed. The generic bias can be explained as the approach most likely to be taken in social formations dominated by the reproduction of capital (see chapter 11), in which decisions about degree of generality normally have to be made early in a development process. Under conditions of substantial market competition, there was an initial bias toward generic systems, the greater range of potential applications increasing the range of potential markets.

Thus, there is reason to think that insufficient knowledge about the differences among, the relative prevalence of, and the importance of the diverse kinds of knowledge networking are main design problems, particularly central to knowing where to strike the balance regarding degree of generality. Properly designed research could facilitate such decisions, especially if it were able to identify the distinctive aspects of primary knowledge networking forms. Through assessing the degree of similarity and difference in the dynamics of these three forms of knowledge networking, we should have something useful to say about the generic design question. (The fact that much of the research to be described has been pursued from an "action" perspective increases the initial value of this analytic framework while also increasing the chances of its being ultimately displaced.)

Knowledge Creation Networking

The current chapter on knowledge creation asks whether this moment of knowledge networking is fundamentally transformed by AITs. It includes description of my primary research project at the time of writing, "Knowledge Networking Knowledge Networking," or KN^2. The experience of this project, along with other research described in the chapter, gives insufficient support to the idea that knowledge creation networking has already been profoundly transformed by AITs. (The following two chapters will discuss cyber-perspectives on networking in which knowledge is, respectively, adapted and shared.)

Research, intellectual, and academic organizations are the institutions most associated with knowledge creation networking. As will be noted in chapter 8, other organized activities, like product development and customer servicing, do create some new knowledge. However, from a knowledge networking perspective, the primary focus of the organizations within which these latter activities take place most frequently is the adaptation and application of existing knowledge to new but recognized classes of circumstances—in a phrase, the *re*production of already existing knowledge.

Similarly, educational institutions also produce new knowledge. Constructivist pedagogy is the approach to knowledge sharing networking or teaching/learning that strongly informs chapter 9. One of constructivism's central notions is that for students, learning is not simply a matter of passively accepting content from a book or teacher. Rather, in an important sense, a group of learners must actively re-create knowledge for themselves if it is to be truly learned. Still, there is an important difference between the knowledge they *re*create in the process of incorporating it into their own social reproduction and the new knowledge of a chemist analyzing data, or even that of a designer developing new approaches by reflecting on her practice.[2] What is different about the latter is pointed at by the notion of knowledge creativity.

Knowledge Creation and the Knowledge Society Debate

Certainly central to Western knowledge discourses, and perhaps important to others as well, is a notion of real newness, of truly creative knowledge networking. Even when an aspect or quality of some phenomenon may already exist, the recognition or "discovery" of this quality is treated as involving something almost as distinctively new as that which is created "out of whole cloth."

Indeed, many of the knowledge society discourses recounted in previous sections presume (albeit in their own way and often only implicitly) such creation/discovery claims. A large contingent of those who would argue that AITs are fundamentally changing society focus attention on "revolutionary" creation/discovery entities like collaboratories, document management infrastructures, online publication, and the development of new, artificial sciences. The National Science Foundation, universities, state legislatures, and the American Anthropological Association all cite changes in knowledge creation as necessitating fundamental policy shifts.

Thus, at least to a westerner, any truly serious knowledge society argument would have to specify substantial changes in the character and/or social functions of new knowledge production. However, few knowledge society–type claims are serious in this sense. Consider President Clinton's 1998 State of the Union simple assertion that the "information superhighway" was associated with a doubling of the amount of knowledge in the past year. To demonstrate that half of what was known at the time of the speech had been created/discovered in the year prior to its being given *and* that this constituted a case for knowledge creation transformation, one would have to be able to

1. Specify and then apply a rule by which new knowledge could be separated from old
2. Demonstrate substantial differences in the two categories of knowledges' characteristics and/or social functions
3. Explain why these new forms and/or functions of knowledge were significantly, not trivially different
4. Connect these developments closely to cyberspace—that is, specify close links between the cyberspaced new knowledge and AITs
5. Demonstrate why the cyberspace-enabling AITs were primarily responsible for, rather than themselves being a product of, the changes, or both the AITs and the changes a consequence of some third entity, and
6. Repeat such analyses in relation to multiple aspects of knowledge creation networking.

One might then be justified in speaking of really new knowledge creation. Yet even this would not yet establish a "knowledge society," because things could still be different in knowledge adaptation and/or knowledge sharing networking, or in other, as yet to be identified forms of knowledge networking. Instead, knowledge society claims like Clinton's are usually only illustrated, as by allusion to an increase in the number of books published. Foucault's razor must be applied to such claims; for example, they should be attributed to processes like an increased pressure on academics to "publish or perish" before acceding to "change in knowledge's quantity" claims.

Specialization and My Experience Studying Knowledge Creation Networking

Transformative claims about knowledge creation networking are difficult to justify legitimately for reasons beyond their necessarily complex logic. Consider, for example, the analytic difficulties that follow from the specialization characteristic of Western knowledge creation. It is her sense of what is adding to and/or deliberately changing in her specialty's base that is the likely starting point for any Western intellectual or academic who thinks about the implications of AITs for knowledge creation, but this particular privileging has no theoretical standing. Indeed, specialization is so extensive as to make very difficult any meaningful discourse on general knowledge creation.

From another perspective, an obvious initial "candidate" dimension for knowledge differences in cyberspace is change in the degree to which the networking that creates it itself manifests specialization. As argued in chapter 5, the new ethnology of knowledge has heightened attention to variability in knowledge discourse, within but also across disciplines as well as cultures. Attending to specialized forms of knowledge networking aids discovery of the division of work in a social formation—men's from women's, hunters' from farmers', anthropologists' from systems developers', and so on. In an important sense, of course, the very idea of Western civilization is linked to the notion that there is something distinctive about Western approaches to knowing, yet at the same time, each intellectual tradition, academic discipline, school, and the like, depends on social recognition of the distinctiveness of its own "knowledge base."

As an anthropologist, I am inclined to think first about my discipline's knowledge when I think about AITs and change in knowledge creation. Indeed, AITs have substantially aided my participation in constructing my field's knowledge. Tools like e-mail and listservs have been of great value in overcoming considerable isolation (being the only anthropologist on my campus, etc.). Moreover, as recounted in *Cyborgs@Cyberspace?*, I have made extensive efforts over the years to legitimate various aspects of AIT-related phenomena as research topics of importance in my discipline, efforts which have been recognized in, for example, the Textor Prize of the AAA. Interestingly, despite the critical tone of my published work, this activity has led some colleagues to mistake me for a "computer revolutionary." The distinctively Western disciplinary solipsism about knowledge creation apparently leads them to presume that someone who has both benefited from AITs and "entrepreneured" study of their cultural correlates would perforce perform such rhetorics.

Actually, studying knowledge from anthropology has helped me avoid capture by the romance of computer hype. Unlike those in many other disciplines, but like many in STS, every anthropologist grapples extensively with at least two domains of knowledge, that of her own discipline and that of some other practice—for example, that of traditional healers among the Bongo Bongo. I use the term "AIT developers" in a very broad sense, to describe the "other tribe" whose knowledge I have attempted to come to terms with over the last twenty years. Relatively deep study of another knowledge base involving frequent shifts of knowledge frameworks tends to heighten one's awareness of the relativity of knowledge claims, while also opening one to talk of knowledge in general.

At any rate, as a westerner with my particular history, it is through the windows of anthropology and AIT systems development that I tend to view claims that knowledge creation is transformed in cyberspace. Given the anthropological bias toward celebration of the preexisting, and the systems development opposite bias toward performances of projected futures, this pairing turns out to promote a certain balance. As described in detail in *Cyborgs@Cyberspace?*, my basic approach to the study of computing and social change is to do participant observation fieldwork in protocyberspaces, those situations which are arguably prefigurative of what a fully developed cyberculture might be like. I have worked with developers of various kinds of AIT systems. In conjunction with the organizational and educational work described in the following two chapters, I have had opportunities to observe and participate firsthand in the application of computing to knowledge creation in several other disciplines as well.[3]

Focusing on Technoscience in Cyberspaced Knowledge Creation Networking

Serious study of the cyberspace question in knowledge creation networking must go beyond single disciplines. Still, as Marietta Baba pointed out to me forcefully (2002), the scope of disciplined human knowledge is vast, and some focusing is necessary. "Technoscience" is a generic term referring to self-described scientific, engineering, and technological disciplines. For several reasons, technoscience is the chief arena I will consider regarding AIT-induced fundamental change in knowledge creation networking. Technoscience is of perceptual criticality, because many of the transformationalist claims about computing and knowledge creation relate directly to fields within it. Moreover, given their importance to the reproduction of the U.S. social formation (and thus to world reproduction), demonstration of transformation in fields like physics and biology, while not automatically justifying general knowledge transformationalism, would certainly be very suggestive. Conversely, if evaluations of transformationalist claims in such disciplines showed them to be greatly overblown, this would be generally corrosive of the transformationist case.

Therefore, I will focus in this chapter on modes of talking about knowing in "disciplined" technoscience. However, my privileging of AIT-mediated technoscience knowledge creation networking is only methodological. Indeed, many other practices in complex social formations make knowledge claims. It would be a serious violation of the ethnological approach to knowledge at the heart of *The Knowledge Landscapes of Cyberspace* were the inferiority of these other knowledge discourses to be accepted even tacitly, thereby replicating unquestioningly the a priori claim of technoscience to the top spot in the hierarchy of knowledge. The notion "discipline" itself at base only means a knowledge discourse that has found a relatively high prestige institutional home, in opposition to those that have not.

I focus on technoscience knowledge creation networking, then, not because its knowledge is more important or "better" than that of other or less disciplined modes of knowledge creation. Rather, I do this because of the strategic place of the former in the talk about cyberspace and in current social reproduction dynamics. An additional clarification: It follows from all that has gone before in this book that consideration of knowledge shorn of awareness of its social dimensions is intellectually perilous. Thus, I include in technoscience those aspects of disciplines like anthropology and sociology, as well as crossover trends like social informatics, that acknowledge how social dimensions are an integral part of technoscience.

Action Research on Knowledge Creation Networking in Cyberspace

As an anthropologist, my primary way of learning about a culture, including the protoculture I provisionally call "cyberspace," is through doing fieldwork in it. I therefore begin my evaluation of transformationalist knowledge creation claims by describing my own knowledge-related research and involvement with AITed knowledge creation.[4]

The relationship of knowledge and technology has been an abiding interest since the 1970s, when I studied the use of computerized teaching machines (Andrews and Hakken 1977). My first knowledge projects had a more knowledge sharing, than knowledge reproduction or creation, focus. My first extended ethnographic project, during 1976–1977 in Sheffield, England, focused on the relationship between worker education and the reproduction of working-class culture. This project exposed me to the importance of technology change at work as a mediator of what I now call knowledge reproduction networking (Hakken 1978).

Changes in the economic character of the Sheffield region led to my second extended study during 1986–1987, this time of knowledge adaptation. This project was an examination of the relationships among new AITs, regional labor processes, changes in organization, and broader social changes. In this second Sheffield study, I was also interested in an applied issue, the success of attempts to use knowledge *of* the social correlates of technology to intervene *in* their construction, especially to encourage desired correlates and discourage unwanted ones. I concluded that these sharing and reproduction efforts had conceptually interesting but, ultimately, marginal effects, and also that adapting knowledge networkings locally could affect more general social formation reproduction only with difficulty (Hakken 1990; Hakken with Andrews 1993). This conclusion led me to center my 1993–1994 project on national-level production of knowledge about computing, through a comparative study of the cultural construction of computing in Norway and Sweden. An important part of this work involved analyzing the actions of systems developers not only on specific projects but also on national policy discourses and on how computing came to be practiced in organizations (1992).

I integrated the results from these extended field studies into my teaching in the State University of New York Institute of Technology. I also participated in a series of ever more ambitious efforts to foster effective AITed knowledge networking among not-for-profit and public organizations in the upstate New York region around it, and more nationally with regard to policy issues in disability technology. (These activities are described in detail in the following chapter.)

Taken together, this research and practice convinced me of the potential benefits of integrating knowledge of AITs' social correlates into both policy and implementation. Such integration was crucial if computing was to be integrated effectively into organizations and/or social reproduction more generally, but my research and practice gave me a strong, sometimes acute sense of how difficult it is to do this.

"Knowledge Networking Knowledge Networking" (KN^2): Original Project Design

While increasingly less tied to purely "local" knowledge, the projects described above remained framed geographically. However, while all networking is socially spaced, information technologies mean their spaces need not be tied to particular places. Indeed, their capacities to "decouple space from place" is often cited by scholars like Ina Wagner (1989) as one of the most important transformative potentials of AITs.

Thus, my most recent project was not framed geographically. In 1998 I initiated "Knowledge Networking Knowledge Networking," or KN^2. The project involved collaborative design and implementation of technologies to support knowledge creation among ethnographers doing research on knowledge reproduction networking in organizations. Designed to include ethnographers from around the world, KN^2 was deliberately, pace Marcus (1998), "transsited," based on creating a notional social space rather than studying particular places.

KN^2 was an effort to put the broad social consensus that we are in, or at least moving toward, a "knowledge society" to an empirical test via action research. The Project was a reflexive, "meta" attempt to implement the ethnological approach to knowledge outlined in *The Knowledge Landscapes of Cyberspace*. Its design presumed that the creation, sharing, and reproducing of knowledge are profoundly social activities, carried out by human, cyborgic, and nonhuman entities involved in distinctive, often complex social relations. Organizations' late-1990s efforts to use AITs to deal with knowledge provided interesting sites for basic research on these social activities, sites where one could usefully study whether the character of knowledge was changed in cyberspace and, if so, how. In its focus on how ethnographic study produced new knowledge about knowledge networking, KN^2 was a study of AITed knowledge creation networking.

KN^2 was conceptualized in conscious response to the urgings of the AAA leadership that anthros develop larger-scale, longer-term, more technologized projects, so that we could compete effectively in the world of mega-technoscience. In 1999, support for planning stages of the project was obtained from the Societal Dimensions of Engineering, Science, and Technology Program at the U.S. National Science Foundation.

Ethnographic Knowledge of KNing in Specific Field Sites

As action research, KN^2 was also an effort to intervene technologically in knowledge creation networking, to take opportunistic advantage of two events. One was the turn toward anthropologically inspired ethnography in organizational studies, a development examined more fully in the following chapter.

The decision to privilege ethnographic approaches in KN^2 was of course related to the1990s business romance with them, but it was also discipline based.[5] Ethnography's original use and longest sojourn has been in the academic subfield of cultural anthropology. However, as noted in chapter 3, a range of scholars from various fields appropriated various aspects of the ethnographic gaze a key moment in the emergence of the social informatics (Bowker et al., 1997) described in chapter 6. Early social informaticians wanted ways to develop systems that enhance the role of users. A number of early Nordic experiments (Florence, Utopia, DUE—see Bjerknes and Dahlbom 1990) involved anthropologists or anthropological methods. Several distinct systems development approaches (e.g., participatory design, CSCW) owe a great deal to this initial cross-disciplinary encounter.[6]

Ethnography like that described in chapter 3 provides rich understandings of what goes on at work, because they feature a social knowledge perspective. By documenting the substantial working knowledge of employees, workspace ethnography paved the way for organizational interventions into knowledge networking. By the mid-1990s, when organizations turned to knowledge technologies to address these problems, ethnographic study was already well established in organizations, which quickly turned to field study of knowledge management.

Unfortunately, results from the ethnographic study of actual knowledge networking in organizations were seldom integrated into mainstream organization studies or development of AIT systems to support it. Ethnographers had not yet organized themselves into an effective "invisible college." Their separations followed from different national policy or cultural sensitivities, from disciplinary boundaries, and from diverse commitments to basic or applied research or practice interests. Perhaps because committed to postmodern perspectives, many had ideographic, localist conceptions of ethnography, such as those that inform ethnomethodology or similar sociological constructions of "qualitative" methods. For the anthropologists among them, separation followed from their training. Ph.D. anthropologists normally conceptualize their fieldwork independently, identify their own appropriate field sites, enter into a field often far removed from potential collaborators, and execute, analyze, and write up their findings all on their own. Such highly independent researchers are not easily yoked with others, especially those trained in, for example, a natural science worldview.

Organizational knowledge ethnographers did gather periodically at academic and industrial meetings, often when presenting the results of their research to audiences whose members could compare them and attempt their own generalizations. These academic performances, intended for broad audiences, were important in legitimating the still shaky status of ethnography in organization studies. They nonetheless seldom led to the generalizing discourse essential to a shared episteme or paradigm.

Nor did encountering each other's work in academic journals, both off- and online, lead to a sustained discourse. Writing would appear in the same books but as independent essays rather than coauthored texts. Ethnographers occasionally succeeded in creating small organizations of their own and could be found periodically working on shared contracts. The former, however, tended because cross-disciplinary to be very tenuous. The latter actually impeded scholarly collaboration by competition, both professional (e.g., for jobs) and market-based (for clients).

In short, ethnographers of knowledge networking in organizations networked, but seldom both knowledgeably and collectively. Thus, while ethnographers created several interesting descriptive accounts of organizational knowledge networking, new general knowledge, reasonably a prerequisite to substantial practical intervention, was not accumulating. An empirical research base with substantial potential for reconstructing knowledge talk remained largely unaccessed and inaccessible.

This situation seemed to be a real opportunity to use AITs to help knowledge creation. These scholars already shared a sense of the importance of knowledge networking. They also had a relatively high degree of familiarity with the successes and failures of its existing AIT-mediated forms and could be presumed to have at least some interest in making these work in organizations. Third, many of them already participated in initial national and international research discourses on knowledge networking, for a variety of personal and occupational reasons. Individual ethnographers would involve themselves in the project out of intellectual interest and out of a desire to be identified with a key prestige (via NSF funding) project of potential significance to clarifying the roles of the social element in technological processes. A properly conceived intervention should have been able to promote more knowledgeable networking, so KN^2 could serve a real knowledge creation need.

Knowledge Management

The other event of which KN^2 was designed to take advantage was of course the emergence in organizations during the midnineties of several practices explicitly related to knowledge. As described in chapter 3, these included efforts to design, build, and sell knowledge technologies and use these AITs to manage knowledge, protect intellectual property, and construct "knowledge capital." These organizational knowledge developments were clearly tied to technology sales and the more or less continuous computer science knowledge engineering thrusts described in chapter 6, but they also had a clear material base. They grew out of a contradiction, increasingly manifest in contemporary organizations, between increased organizational scale (e.g., "globalization"), on the one hand, and growing commitment to team and other, more worker-dependent organizational forms and structures, on the other (see chapter 8). In part because of the obvious difficulties of dealing with an entity (knowledge) so obviously socially implicated, many of these organizational knowledge activities incorporated qualitative components. It made sense that, by the late 1990s, several organizational ethnographers had been hired to study and/or assist in implementing knowledge-related practices in businesses.

The KN^2 strategy was to deploy AIT infrastructures to record and manipulate accounts of what went on in organizational sites collected in the field. These infrastructures would be used to integrate studies of specific work sites where AITed organizational knowledge adaptation networking was actually testing the promise of ethnographic knowledge creation networking. Working in the mode of participatory action research (Whyte 1991), the intent was to use current, off-the-shelf knowledge networking technologies to construct a "collaboratory." Best practices discovered among ourselves would be fed back reflexively into individual site work. At the same time, best practices from field sites would inform our own designing, prototyping, and implementing work.

Collectively, we would use the collaboratory to explore our own questions (e.g., is it a good idea to think of dispersed teams as "communities of practice"?) as well as those of other fields.[7] Through developing "best practices," the project would also make a contribution to the practice of AITed KNing. As we fed our collective understandings back into the projects in our individual field sites, we would be acting in concert with the cyberspace ethnography ethics outlined in chapter 12. At the same time, KN^2 would test these infrastructures' "bleeding edge" capabilities.

Initial Planning Activities

The first recruits to KN^2 were ethnographers whom I already knew through fieldwork or as a guest lecturer, author, and conference/listserv participant. With them I initiated consideration of the various AIT-supported infrastructures, both virtual and real, that could help it happen. Existing studies of computer-mediated communication (e.g., Sproull and Kiesler 1995) provided additional suggestions about what to build into KN^2. For example, good KNing needs substantial coordinating. For KN^2, this was envisioned as an expansion of the good listserv moderator's role, fostering discussion on both knowledge content issues and on process—for example, development of an explicit project "netiquette." Moderation had to be based on appreciation for the particularities of each participant's field situation, professional objectives, and personal interests. (Field visits, for intensive face-to-face networking with likely core project participants in their regular locale, were an important planning activity.)

For these reasons, considerable responsibility devolved on me as project coordinator. I was in a good position to carry out these responsibilities. For a long time, I had been visibly promoting networking among cyberspace ethnographers. I successfully initiated structures to support such activity (e.g., the Community of Anthropologists of Science, Technology, and Computing of the American Anthropological Association and a similar form in the Society for Applied Anthropology). I had academic standing in relevant areas within my profession (as president of the Society for the Anthropology of Work and an invited co-organizer of the Oslo Conference on Technology and Democracy, for example) and substantial related grant activity. During the planning period, I built on this base by:

- Organizing KNing presentations at the meetings of the AAA, the Society for Social Studies of Science, and the European Sociological Association
- Evaluating relevant grants (e.g., I participated in two relevant NSK panels, that on the Societal Dimensions of Engineering, Science, and Technology, and the Knowledge Networking subgroup of the Knowledge/Distributed Intelligence panel, as well as doing additional ad hoc NSF and Wenner-Gren Foundation reviewing)
- Reviewing manuscripts (e.g., I worked on *Anthropos Tecnika*, conceived of a online journal in the anthropology of science, technology, and computing, turned into a project in collaborative open design)

- Continuing related professional work (e.g., I completed a term on the AAA Executive Board and as chair of the AAA Committee on Scientific Communication, with oversight responsibility for organizational meetings and publications; was an active member of AAA Advisory Group on Electronic Communication; and was a candidate for new AAA Public Policy Committee)
- Planning for the face-to-face KN^2 participatory design workshop

During the planning period field visits and via e-mail, we considered several relevant issues, including:

Appropriate Intellectual Foci for KN^2

The final form of the project was of course to be determined by the group, as per the methodology of participatory action research (Whyte 1991). Action researchers deliberately intervene in practices of interest, learning about them by trying to foster and/or change them. Because the anthropologist learns by attempting to embody the practices of interest, ethnography is normally to some extent action research. Hawthorne or researcher-induced effects are obvious potential drawbacks of action research. These are of less salience when there is need to explore the limits of a new thing and can be of great help when, as with knowledge networking, there is considerable conceptual confusion in existing practices. Action research contrasts with the typical development project, which has the disadvantage of commitment to one predetermined approach.

Research is participatory when those whose practices are of interest have substantial impact on the project, including problem identification, method selection, execution, and/or analysis. While development of a seamless research design is highly complicated by such participation, the latter approach is appropriate when establishing collaboration among researchers has high priority. The KN^2 planning period was intended to foster development of both a stable network of participants and agreement regarding primary project objectives and technological infrastructures.

Thus, the issues addressed in the initial proposal were those which appeared most pressing in my personal work on computing and social change. Others' experience generated different issues or different approaches to them. To what extent did the issues initially outlined constitute an adequate project problematic? How could we network in a manner responsive to what we learned/shared from our individual work while maintaining focus on shared intellectual concerns?

Network "Membership" and Roles

Under KN^2, network members were to explore the important social dimensions of knowledge networking by collectively doing it, using the knowledge gained from their primary sites to inform the project networking, and project networking to inform their site work, whether basic or applied. For the sites, this promised that the project would be a source of additional insight as well as providing an opportunity to reflect on the knowledge generated there.

My own primary site was the networking among not-for-profit and public social service organizations in the Upper Mohawk Valley of central New York State discussed in chapter 8. Other network participants have similar projects in several different kinds of sites. A range of primary sites enabled a discourse over the generalizability of particular cases to emerge naturally out of project discourses.

Knowledge networking activity normally fluctuates in participants' intensity and involvement, but good networking achieves a certain stability among core participants. The individuals initially involved were all organizational knowledge scholars whom I had met and whose work I respected. While such an individually linked network is characteristic of an initial networking state, whom should be included in any expansion? How could new participants be integrated? How large should the network be? At what points and in what venues was it appropriate for us to discuss our shared work? How were we to manage cross-disciplinary paradigm and cross-national/cultural negotiations most effectively? Were we striking the proper balance between participatory democratic decision making and fostering networking, between intellectual and material incentives to participate and respect for individual interest trajectories, and so on? Did it make sense to identify some "core" participants with more specific network responsibilities, say, for fostering networking on a specific topic or coordinating a specific task/writing group? Equally important was identification of satisfactory roles for those more peripherally involved. How should we conceptualize the relationship of those centrally involved in the project to those more peripherally involved? Was the role for me outlined above an appropriate conception of the network convener?

Appropriate Networking Infrastructure

KN^2 conceived of at least four kinds of knowledge networking content: individually created, both think pieces for network participants only and pieces to be made more broadly available; and group created, again internal drafts and those for wider distribution. Codesign and coconstruction of an appropriate information infrastructure to support networking of and around such content was to be a central activity of KN^2. In the planning period, key decisions had to be made. Did the "off the shelf" approach give sufficient scope to "growing our own networking support system"? How should differential accessibility be achieved? How should these electronic arrangements map onto the various roles conceptualized? What were the implications in terms of software and maintenance cost? Which system was most compatible with participants' hardware? What "netiquette" would be most conducive to generating reflexive understanding of knowledge networking? What kind of training would participants need to take maximum advantage of the infrastructure? What kind of technical support would be necessary?

The initial eighteen-month (7/1/2000–12/30/2001) planning grant provided support for travel, network building, initial design at a distance, and a final preimplementation decision-making/training workshop for core project participants. During this period, further funds to support actual project implementation were sought; because of my prior experience, the Knowledge Networking segment of the National Science Foundation's Knowledge/Distributed Intelligence initiative seemed a likely source for a three-to-five-year effort, which would:

- develop and maintain a substantial electronic Project presence (e.g., a broad knowledge networking portal)
- generate means to study its own networking activity (both electronic and human/ethnographic),
- promote extensive communication and travel,
- represent the project in appropriate venues,
- support joint writing, and
- include additional IRL meetings.

Summary of Original Project Goals and Activities

In sum, KN^2 aimed to create new knowledge on the basis of a well-constructed and broadly shared problematic, sense of appropriate network roles, and infrastructure. KN^2 was to add significantly to our understanding of what separates knowledge from mere information networking, when it is appropriate for AIT-mediated knowledge networking to aim to simulate in-real-life knowledge networking, and other possible and/or unique contributions of AITs to networking. In the long run, the project was intended to clarify the extent to which mediated knowledge networking changes both the character of knowledge and what it can contribute to solving organizational development issues in the current era. Ultimately, it aimed to contribute to the creation of a general framework for understanding of what separates knowledge from other forms of networking. This means addressing not only the wide variety of knowledge networks but also questions like the extent to which it makes sense to act as if knowledge is separable from its networking infrastructure. It was to do this by placing AITed networking systematically in broader contexts, comparing it with various other forms of knowledge networking, both in other cultures and in different modes of social formation reproduction.

First Project Rethink

One does planning grants because big projects generally have to be modified. Compelling change for KN^2 were:

1. The emergence of the "knowledge management fatigue syndrome" described in chapter 3, manifest in the winding-up or substantial modification of several of the organizational knowledge projects of ethnographers initially enthusiastic about the project. Those employed as consultants were forced to find new projects (and often, new employers), and those who were academics to find new field sites. Thus, several individuals who had indicated an initial interest in participating had less opportunity or willingness to contribute to it, normal difficulties in fostering communication among consultants whose competition with each other was being exacerbated.
2. A program change at NSF, the intended funder of the full-scale project, from an interdirectorate initiative (Knowledge/Distributed Intelligence) to a technoscience directorate-based initiative (Information Technology Research Based in Computer and Information Science and Engineering). A preliminary approach to this new program was rebuffed, probably because the project's focus was perceived as insufficiently technical.
3. Technical/administrative support glitches at my home institution, despite the fact that this branch of the State University of New York proudly quotes its rating as one of *Wired Magazine*'s 100 best. These were particularly ironic, given the project's general theme. The first campus person contracted with to design an initial website for the project disappeared, while State University of New York's

> draconian limitations on consulting agreements and intellectual property compelled the second to quit. In order to use the open source software of preference to my third scripter, I had first to negotiate an unprecedented agreement with the university's lawyers. In the midst of project design, the sociologists in my so-called joint department decided to eliminate anthropology and marginalize me, and they successfully recruited local administration to their purpose. Sent to the woodshed despite the fact that I was also nominated for a distinguished professorship, I no longer had sufficient institutional support for a large project

To cope with these developments, the project planning stage was extended, ultimately for two years to allow for redesign. While the overall scholarly intent, research approach, and use of AITs remained the same, for practical reasons (e.g., access to sufficient sites and participants) the target knowledge contexts had to be changed. The rapid rise and fall of knowledge management is typical of a business organization environment, but such fashionable trendiness (Hanseth 2000) is not easy to fit to the longer cycle necessary to building collaborative scholarly research. To stick to a narrow business organizational site would have required the ability to predict the next "organizational hot ticket," but knowledge management fatigue syndrome suggested that this was unlikely to have a knowledge focus. KN^2 needed greater distance from the dynamics of for-profit computing and competitive careers, of both chief knowledge officers and organizational ethnographers.

Second, preliminary planning, like many other knowledge management exercises, had been impeded by multiple property/proprietary imperatives. Indeed, the frequency of this occurrence suggests that is may be an important source of KM's failure to succeed. It is obviously difficult both to maximize information flow and maximally protect "knowledge capital." If dispersed knowledge networking is to compensate for its loss of sociality compared with face-to-face knowledge networking, it may need to distance itself from the imperatives of commodification. There may in practice be a contradiction between the use of proprietary software and the aim of maximizing knowledge creation.

The free software and open source movements provide tools to cope with these problems. My growing interest in "open" perspectives has two sources. One is the specific question of whether overtly nonproprietary approaches to information system development are fundamentally more supportive of knowledge networking than proprietary ones. Also, do they create knowledge in substantially distinct ways? The broader source is our interest in the obvious importance of "open," shared, public, democratically arrived at standards, such as TCP/IP or those enabling the World Wide Web. What are the appropriate standards for knowledge networking, and how can they be developed in a manner that maximizes the knowledge networking potential of cyberspace? Concern with such broader implications for knowledge of the question of openness means recognizing that threats to privacy and efforts to use the patent process to frame code as property have obvious implications for the future of knowledge networking.

Third, in analyzing the sources of the knowledge management fatigue that blocked the original conception, I began to wonder if the initial knowledge management tools were too generic. This led to the attempt described above to differentiate among basic types of knowledge networking, and to the idea to build exploration of this framework into project design.[8]

Second Project Plan

Planning funds were therefore redirected to prototyping two noncommercial knowledge networks, each with reasonable prospects for developing further on their own. One initiative, in the anthropology of knowledge/anthropological knowledge, was to be a knowledge creation initiative. The second, to support the Harvesters Project of the Episcopal Church, was to be about knowledge sharing.

After initial discussion of the technical prototype architecture,[9] system specification decisions about the prototypes were made at a workshop in the fall of 2001 at Manchester Community College in Connecticut, where the technical structures were to be supported. That the two prototypes shared an initial infrastructure was central to the new research design. Their subsequent developmental trajectory was to have illuminated the issue of the proper typology of knowledge networking, as well as the generic/specialized design issue. If the prototypes developed in different directions, this would tend to support the value of separate support infrastructures for knowledge creation and knowledge sharing networking. In contrast, similar developmental trajectories would support more generic design protocols. (An alternative, more minimal interpretation would be that this result would not be supportive of making the creation/sharing divide outlined above a focus point for either theorizing or differential design.)

There were many reasons to believe that prospects for successful pursuit of such an approach were good. Many organizational ethnographers have interesting stories to tell about KM fatigue, but several organizations (including ethnographers' employers) remained financially committed to selling KM products. Hence, a need remained for an independent venue in which to process recent experience, place it in broader contexts, and figure out what it means. This could be done more easily under an explicitly anthropological rather than an informatics rubric.

The prototypes themselves were not the ultimate goal. Rather, their study would provide initial steps toward creation of a much more generic, nonproprietary system to support dispersed knowledge networking via AITs. Sciences, disciplines, and organizations all face the challenge of finding ways to support dispersed knowledge networking. In the future, these and many other activities/forms will continue to adapt to the potentialities of AIT while adapting AITs to their needs. Experience with the prototypes' development would suggest the outlines of a larger project on the relevant forms of AITed knowledge networking. Development of the most effective open source, free software–based system on the basis of the experience with the prototypes would also be an objective of the broader implementation project.

The most significant feature of the prototypes might have been the scalability of their infrastructure. In development, high priority was to have been placed on maximizing extensibility, so that networking activities could increase in scale and take new forms without the demands of periodic platform switching. These features, along with the quasi-hypothesis testing quality of the revised research design, make the project a better candidate for funding to support continued development.

Second Project Rethink and Third Project Plan

Despite the efforts of several individuals, neither of these prototypes got off the ground. AAA staff were unable to take part in the Manchester workshop because of September 11, 2001, travel problems, and in any case the staff concern had become more archival than knowledge creation. Nor did proffered interest by the AAA leadership have issue. The Episcopalians, for their part, had to deal with several internal organizational issues that displaced technoenthusiasms.

Remaining project funds were thus again redeployed to develop and implement a generic open source tool suite. These tools were used to build a demonstration promotional website, www.knowledgenet.org. As of writing, these tools are to be used by student members of the Society for Social Studies of Science and the Community of Anthropologists of Science, Technology, and Computing to support their own knowledge networking activities.

Action research on their use and elaboration forms part of my new project, a comparison of open computing in the "Christian" North Atlantic and the "Muslim" Island Southeast Asia—that is, if the recent tragic events in Bali and SARS do not prevent cyber-fieldwork with the latter comparand.

Preliminaries to Analyzing AITed Knowledge Creation Networking

The research trajectory described above gave me an intimate and detailed understanding of certain aspects of the knowledge/AIT nexus. Little happened during the KN^2 experience, for example, to incline me to abandon my skepticism toward the idea that AITs have already transformed scientific knowledge creation networking. The experience revealed as many technology-related impediments to knowledge creation (e.g., commoditization and formalization) as supports for it. Perhaps the strongest conviction arising from my experience is of the large gap between the rhetoric used to promote knowledge AITs and their actual dynamics in use. It is not that the interventions have no correlates, but that the discernible changes associated with their introduction do not converge in the transformative dynamic that the rhetoric evokes.

Of course, deciding whether knowledge creation has generally been transformed is a complex task, clearly not to be resolved by even the most detailed single research effort. To bridge to the broader question, I want to place the KN^2 experience in the context of knowledge change in technoscience. This will explain further why I think the transformation case regarding knowledge creation has not yet been demonstrated. This is not because I am ignorant of AITs or refuse to see their potential for transformative change,[10] but because careful efforts to engage transformative arguments on empirical grounds, like KN^2 just do not support transformationalist claims.

A Technoscience Knowledge Creation Baseline

To decide if the knowledgescape of technoscience is transformed in cyberspace, one must be able to compare cyber-networking with how technoscience knowledge was created previously. Prompted in part by the energized informatics visions of artificial intelligence outlined in chapter 6, a new awareness of the process of technoscience knowledge creation, as well as new ways to study it, emerged in the 1960s. However, actual computing had only begun to make substantial inroads in most technoscience arenas. Thus, despite some significant problems in this scholarship, the image of technoscience networking it created can serve, when appropriately critiqued and modified, as a baseline model of precyberspace technoscience knowledge creation networking.

The most influential study in the tradition I have in mind is Diana Crane's *Invisible Colleges: Diffusion of Knowledge in Scientific Communities* (1972). While the term "diffusion" in the title seems to suggest reproduction or sharing more than knowledge creation, this work is actually about growing scientific knowledge.

For Crane, the entities that produce scientific knowledge are dispersed circles of individuals who interact indirectly, without formal leadership; indeed, authority structures are contrary to their norms of interaction. Initially, their knowledge creation networking is organized around a shared set of concerns and ways to address them ("a paradigm"), and an addressing social circle, centered on one or at most a few high prestige central figures and their closely linked students.

In their second or "takeoff" stage, these circles develop sufficient density in their networking to constitute "invisible colleges." These are dispersed communications solidarity groups, similar to western extended families or social movements. They build morale and maintain motivation via frequent communication and information transmission. They allow criticism, and regular contacts lead to common standards and norms for handling conflict. The structure of an invisible college is not that of an interest group as normally understood. In social network terms, the structure is instead that of a "centrally noded" network, one focused on a few highly visible, perceived-to-be-productive scientists.

In sum, Crane's model of 1960s technoscience gives central place to social relationships, stressing particular aspects of their social character (e.g., their voluntary character) and their discourse features. It recognizes the need for critical "discursive redemption" in knowledge creation networking, and, at least overtly, it resists privileging some forms over the others. In both the breadth of her gaze and her stress on the social, Crane's description of invisible colleges is compatible with the project of developing an ethnology of knowledge networking.

Critiques of the Baseline

There are some serious limitations to Crane's work, many related to its foundationalist presumptions. Like Derek de Solla Price, the founder of this trend (1979), she takes the growth of technoscience knowledge and its cause (the exploitation of intellectual innovation) to be self-evident. Crane similarly presumes this growth to be a natural phenomenon, one that moreover proceeds inexorably according to a predictable mathematical (logistic) developmental curve. While acknowledging that one could argue theoretically that there were multiple growth mechanisms, her analysis presumes them to be fundamentally the same.

In short, instead of asking whether or why growth occurs, Crane focuses only on the how. If the growth mechanism is the same in all fields, any differences between them must be due to some contextual factor or factors. Methodologically, she operationalizes the above assumptions in the presumption that there are distinct research areas within the broad community of science, and that their distinctivenesses will be reflected in publication differences, especially citation rates and related practices. By comparing how science communities' growth is differentially reflected in indicies, she hopes for insight on how such contextual social relations affect the growth of knowledge. To increase the likelihood of difference, she chooses to compare a "soft" area and a "hard" one, rural sociology and the algebraic theory of finite groups.

Crane finds significant differences between the two fields in her indicators of knowledge growth. Because she assumes growth to follow a regular pattern, the only thing institutions can influence is the pace at which the logistic curve is followed. Following Merton and Sztompka (1996), she offers a social institutional explanation for the differences, differences in an informal and specialized social system, a particular type of community.

While the pace varies, she sees sufficient similarities in the modes of social and ideational development in her two fields to conclude that the same theoretical concepts can explain them. In her conclusion, Crane argues for the more general validity of her model of knowledge creation, that it is not just limited to these two fields or even technoscience in general. She sees it as providing the base for a general sociology of cultural products, of knowledge in general, to include art, ideology, literature, philosophy, political thought, and religion as well as science. Thus, her ultimate analytic gaze ranges broadly, better described as like the European attention to *Wissenschaft* (German) or *Vetenschap* (Scandinavian) than like the narrower American focus on natural science.

Nonetheless, for Crane, technoscientists are the prototype knowledge creators (albeit while including sociologists among them!). Her focus on them reinforced the tendency to privilege science in cultural sociology. This privilege is captured in the name generally applied to the field that developed after

Crane: the sociology of scientific knowledge (SSK). Like the normative view of science practice advocated by Crane's mentor Merton, her views have been much criticized in STS.

Baseline Modifications

While inadequate as an *analysis* of knowledge networking, Crane's *description* offers a reasonable starting point for a model of precyberspace technoscience knowledge creation networking, The invisible-college dynamics she describes, when appropriately modified, can serve as a useful comparative baseline against which to identify any new correlates of AIT-mediated technoscience knowledge creation.

To be plausible, a transformative case must show substantially further departure from this modified Cranian/Latourian world. Some have argued, for example, that, through AITed instrumentalities like digital collaboratories, knowledge creation has become deplaced. However, dispersion was already a characteristic of Crane's networking, as indeed it was of early modern technosciences like Copernican astronomy.

If the growth of knowledge is not only natural but follows a predetermined pattern, all that can vary is pace. AITs could only change knowledge creation in terms of slowing it down or speeding it up.[11] Because of the centrality of communications to invisible colleges, the mechanism to carry this effect would be discourse. Thus, on Cranian conditions, the consequence of computing could only be transformative by something like what Stephen Jay Gould calls "punctuating" an equilibrium (2002). This is an evolution in which numerous quantitative changes accumulate and produce a qualitative change, such as the emergence of a distinctly new species. The analogous point in knowledge creation networking would be new communications and communication patterns accumulating so quickly as to force a radical paradigm shift, likely the emergence of an entirely new intellectual problematic as well as unprecedented new communication patterns.

Crane's inclusion of a paradigm as a regular aspect of an invisible college reveals her Kuhnian orientation. As noted in chapter 4, Steve Fuller has properly emphasized the stability/equilibrium orientation, rather than transformationalism, of Kuhn. Arguably for Kuhn and despite his title, there has only been one real "scientific revolution." Thus, the different citation patterns in different technoscientific communities found by Crane—indeed, most changes within a discipline—are indicative of "normal" technoscience, not transformation. Moreover, one can just as easily imagine new communications leading to information glut, and new communication patterns leading to paradigm confusion.

Thus, on the Cranian model of knowledge creation networking, because Kuhnian, the threshold for a transformationalist case is so high as to be virtually impossible to achieve. The ensuing SSK literature, whether Mertonian or Latourian, operates with this Cranian tradition and offers little support for any transformationalist case. One can, like Merton, see techno-scientific knowledge as generated through an evolving kind of ethical practice, but this practice, too, is one Mertonians project as continuous.

I take from Crane an image of deplaced, dense scholarly social networking as the basis of precyberspace knowledge creation in technoscience. I also share Crane's presumption that creating knowledge is a conscious, deliberate process (e.g., about finding better justifications for better beliefs). I first add, from a Latourian perspective, organizations and artifacts to individual scholars as potential actors in the networking. Apparently conversely, Latourians view technoscience knowledge as an artifact of the hegemony of particular alliances or actor networks of widely dispersed agency. In a Latourian world, conflict among potential knowledge regimes, as well as various rises and falls, as well as attainments of hegemony, are also normal. Any serious Latourian transformationalist account would have to break analytically from this tradition. It would have to develop an account of knowledge creation that did not rely on anything like current technology actor networking. (The famous debate over the agency of nonhuman entities (to which we turn in chapter 12) is not about a movement *away* from technology actor networks but about including these entities *within* networking models.)

Thus, in neither the Mertonian nor the Latourian tradition are phenomena like the emergence of a new research field or an additional invisible college sufficient evidence of transformation, as opposed to normal development or conflict. Indeed, in both modernist (Mertonian) or nonmodernist (Latour) accounts, the presence or absence of AITs has been tangential to the main SSK stories, where emphasis is on institutional continuity.

A presumption that the only possible change in knowledge production is in its rate also means it is very difficult to make a cyber-transformationist case, but this is not a necessary presumption in an antifoundationalist theory of knowledge networking. Any sustained argument for transformation in knowledge creation would probably demonstrate change in qualitative as well as quantitative aspects of social networking. Pace Crane, such an argument would likely argue for change in networks' character and/or social functioning beyond mere pace of accumulation and/or quantity of knowledge content.

To make such a case, one would have to operate outside of the main theoretical orientations in the sociology of scientific knowledge.

Deconstructing Knowledge Creation Networking

In what ways *might* the knowledge creation networking of a technoscience not bound to the presumptions identified above be affected by AITs? How would one identify changes, especially qualitative ones, which *would* count as transformative in relation to this modified Crane/Latour model of knowledge creation networking?

Technoscience has its own, admittedly idealized, version of how it creates new knowledge, the version often taught as research design. This account can be decomposed into distinguishable moments (really, sub-submoments of knowledge networking in general). Among the moments typically included in this process are:

1. Formulation of a new knowledge question
2. Execution of a literature review; for example, stating what is already known about how to answer the question
3. Design of new research concepts, activities, and organizations to pursue the new question; for example, outlining a potential new "paradigm"
4. Execution of the new research
5. Analysis of the new results
6. Publication/dissemination the new results and conclusions
7. Working out their implications by (re)building (new) networks/alliances/relevant discourse arenas of relevance to # 2
8. Performance of quality control on "normal" standards—that is, reconstructing all the relevant knowledgescapes

Breaking down knowledge creation networking in this admittedly formalistic and idealized manner makes it possible to ground a better transformationalist argument. That is, any particular transformationalist case can be situated in relation to a particular aspect or aspects of the technoscience knowledge creation process, and its impacts, real or imaginary, can then be examined more concretely.

For example, the general changes in the character of thinking posited by, for example, Stonier (1992) and Bailey (1996) seem particularly relevant to the first aspect. The arguments of those like Lincoln (1992) who hypothesize a radical democratization of scholarship might be seen as most relevant to the seventh. Charles Ess's focus on electronic media (1996) hit the road especially hard at aspects 2 and 6. Bill Bainbridge argues (2001) that "Computer-related developments across the social sciences are converging on an entirely new kind of infrastructure that integrates across methodologies, disciplines, and nations." His examples are relevant to a range of the sub-subaspects, including 2 (in relation to libraries) as well as 3, 4, 5, 7, and 8.

This analytic move provides a framework at least potentially more conducive to transformative arguments than those in the SSK tradition. It also makes it possible to argue not only that computing transforms particular moments in the knowledge creation process, but also that the comparative strategic status of each moment may be transformed as well. However, ethnographies of technoscience (e.g., Suchman 1987; Forsythe 2001; Dubinskas 1988) show that "real" knowledge creation networking seldom approaches anything like this degree of linearity and cleanness. Questions are revised, analytic frames changed through the inclusion of new organizational partners, and so on. There is seldom a single "ah hah!" event, let alone a stage in the process outlined at which these events normally occur, making it the "crucial" one. None of these steps is necessarily more central to new knowledge than any other. Thus, as the more thoughtful authors cited above acknowledge, any persuasive case for transformation would likely address multiple points in the progression.

Transformed Knowledge Creation Networking in Cyberspace?

While there are specific AIT-related changes in particular sub-submoments of this process, these changes are not substantial enough to sustain a general case for transformation. To explain why I hold this position, I need to devote detailed attention to arguments like those of Lincoln, Ess, and Bainbridge. In what follows, I use the pedestrian deconstruction of knowledge creation networking outlined to place the KN^2 experience in the context of other experiences in AITed knowledge creation networking. Because my skepticism is provisional, it is also important to me to specify those arguments that would substantiate a serious transformationist argument. To illustrate, I often present potential anthropological knowledge creation networkings in the relevant transformationist terms.

Formulation of a Question/Research Agenda/Research Enterprise

The emergence of a profoundly different set of research questions, demanding new kinds of knowledge, would likely entail major innovation in each of the technoscience knowledge production stages. Thus, a logical case could be made for this first one as the key stage for the transformationalist case.

One set of the obviously new cyberspace scholarly questions is related to AITs themselves. There was no field of computer science before computers, hence little research on questions with this specific label. Now there is. Is this sufficient evidence of a transformation in knowledge creation? I should think not, else the existence of any new field would be prima facie evidence of transformation.

Nor does the pronounced proliferation of new interdisciplinary fields on its own justify transformationalism. Some (e.g., anato-informatics and physio-informatics, and other new fields within bio-informatics) may really only be pseudo fields, a proliferation of disciplines being an artifact of the specialization dynamic long a part of Western intellectual life. Computing merely provides new terrain on which this process can be worked out.

Many of the problems on the contemporary technoscience agenda require multiple perspectives for their solution. The absence of breadth sufficient to this task in existing disciplines, however, may be evidence of needed changes, but not necessarily transformative ones. Rather, it may be an artifact of prior *hyper*specialization, for which the new fields are a corrective.

The reflexive critique of anthropological knowledge associated with what some call the linguistic turn (Scholte 1986; Clifford and Marcus 1986) might be an example of such a new set of questions. However, there is little reason, despite the efforts of some like Lyotard (1984) to connect the postmodern sensibility to AITs, to view the critique of my discipline's knowledge discussed in chaper 5 as caused by the arrival of new technology. There was calculation in anthropology before computers, but now there is much more. Is the existence of these new activities involving the use of AITs in itself sufficient evidence of a transformation in its knowledge creation? I should think not, since the potential of the machinery has barely influenced the questions we ask.

Computing does allow new agendas to be pursued, like those of the high-energy physicists described by Sharon Traweek (1988). To decide that this justifies transformationalism, the new agendas would have to be indexes of something more than the availability of new tools and techniques; they would have to be manifestations of profoundly new research frontiers. There has long been a dialectical relationship between technological and intellectual innovation in technoscience. As the career of Albert Einstein demonstrates (Fölsing 1997), there are certainly moments when intellectual innovation runs ahead of technical. The opposite has arguably been the case in recent years, as the availability of increasing computing power makes it possible to pursue approaches—for example, simulation—and therefore questions in ways that were impossible in the past. This is normal science. To be convinced of the transformationalist case, I would want to see a number of fields in which not just agendas but analytic approaches and standards were new.

Bailey's notion of computers taking us "beyond thinking," like Stonier's similar case for intelligent machines (addressed in chapter 6), were they persuasive, would begin to constitute a transformationalist case. By failing to maintain a consistent distinction between information and knowledge, and being overly impressed with the size of data sets or speed of calculation, such arguments decline into mere rhetoric, their promises undeliverable in the real world. Such arguments document neither new characteristics of knowledge nor new social functions. Because this kind of argumentation is so fraught with opportunities to confuse causes with effects, I am inclined to apply a Foucauldian razor to transformationalist claims at this stage.

Execution of a Literature Review: What Is Known about How to Answer This Kind of Question/Pursue This Agenda/Create This Organization

Various forms of representation, both graphic and visual—"literature," construed broadly—have long been a part of technoscience knowledge networking. Via interventions like digital libraries, the researcher in an AITed world has potential access to a substantially greater portion of these representations, even a continually increasing proportion of them. She has tools like search engines of growing sophistication to assist in accessing them. Via collaboratories, there is the opportunity for a deeper engagement with a broader range of ongoing research at earlier stages.

Still, some caution is in order before concluding that AITs' consequence is an unequivocal transformation in the quality of the researcher's awareness of existing knowledge or her ability to use it. First, there are the obvious problems of information glut. William Durham, long-term editor of *Annual Reviews in Anthropology*, argues that anthropologists' substantial commitment to e-mail means they read less, especially the formal representations of knowledge (books, articles, films) in the field (2001).

Problems of glut are compounded by problems of access; remember Kapor's comparison of the World Wide Web to a library where all the books have been dumped on the floor. In his study of online searching, Bryan Pfaffenberger argues convincingly that, as in the past and despite all the new tools, the searcher remains profoundly dependent upon search intermediaries, like librarians. This has much to do with the ways in which these professionals' indexing conventions continue to be built, often with good reason, into the new tools. Researchers have limited access to these conventions, to a great extent still embodied in the collective practice of human librarians, but also now additionally hidden in code and sometimes opaque FAQ files. To this extent, the ratio of useful hits to her total hits is likely to remain low (see also Kotter 2001). These were the kinds of considerations that led Geoff Bowker and his colleagues, in pessimistically evaluating the digital library initiative, to question whether knowledge is in any important sense portable (personal communication).

Research Design Activities

Via listserves and even chat groups, the AIT-mediated researcher can extend the peer group within which she designs her research.If research were normally carried out within a well-defined paradigm, along the lines argued by Crane and Kuhn, it should be portable and easily executable. Invisible colleges should grow in size, their development accelerated, even be transformed.

However, there is reason to question the extent to which technoscience research is paradigmatic. The anti-Kuhnian picture painted by STS scholars (e.g., Fuller 2000) is one of much weaker and more conflicting paradigms. Constructing research technology actor networks, in the pattern of Latour and Callon, involves building competitive alliances rather than shared paradigms. On Latour, expanded opportunities for collaboration equally expand the terrain for competitiveness. The Net provides additional, even faster means for *intra*paradigm communication, but this complicates construction of shared research direction/paradigm generalization. At least for less postitivistic technoscience, current institutional arrangements like institutional review boards remain at best an annoyance to collectivizing research design; the revisions proposed by the National Research Council in 2001 would actually complicate online research design. The failure of high-profile efforts like MIT's MediaMoo to create a self-sustaining community of researchers suggests that interest in using AITs for such processes is a novelty (Bruckman and Jensen 2002). After waiting several years for my organization to provide technology for sharing calendars, I for one have gone back to using a paper rather than electronic one.

Research Execution

While the term "collaboratory" has been used in a number of ways, perhaps its most frequent referent is that of a "virtual" laboratory space, spread over multiple physical locations, in which a common research agenda is being pursued. While as in most technoscience there is a division of labor, computer-mediated communication is supposed to keep everyone "in the loop." By enabling further deplacing of research execution, collaboratories theoretically impact the extent to which research can be upscaled in size and made a more complex enterprise. More transformationally, they promise more collective, and more democratic, research.

Before one can claim that these possibilities have been made transformatively real, one must acknowledge that technoscience research has often been, to borrow a phrase from contemporary ethnography, multisited; the shared astronomical observations of Copernicus, Brahe, Kepler, and Galileo are a famous example. Further, the upscaling of research into an enterprise has been a longtime trend. While to a substantial extent enabled by AITs, upscaling is a more direct result of the globalization of the corporate research enterprise. As argued in chapter 11, the search of capital for mechanisms to free its reproduction from, for example, the national form has more to do with the expanded scale of corporate-sponsored research than with some internally driven realization of the technology's potential.

KN^2 illustrates just how hard it is to collaborate "normal" ethnographic research, at least for nonelite anthropologists; this is not to be achieved by simply "building the highway." In 2000, I participated in a workshop on collaboratories at the annual Computer Supported Collaborative (or Cooperative) Work conference. A number of projects were discussed, many of which appeared to make good use of AITs. However, their predominant function of the AITs was to disseminate the results of research conducted by a core group (often in one place!) to a broad audience, not to facilitate collaborative research in a fuller sense. Even those projects that did involve data collection in multiple locations had a "public relations for science" character. Students and marginal researchers were encouraged to contribute data (and, yes, occasionally questions) to the "database" of core researchers, who were primarily responsible for doing analysis. Again, I feel we should use Foucault's razor, to differentiate between efforts to make people "feel like" they are involved in a collective research activity and deplaced but real collaboration.

Bill Bainbridge has stressed (2001) the advantages that the migration to cyberspace offers certain research traditions, as in certain forms of survey research. While potentially drastically cutting some costs and therefore extending its ambit, this migration also brings problems—with access, technical incompatibilities, and an already noticeable decline in response rates. Indeed, many of the very advantages of the migration—its ease of access—

produce substantial negative side effects. That is, the ease of constructing and conducting on-line surveys led to their rapid overuse and was arguably a non-trivial cause of declining response rates. The online environment appears to lessen respondents' tendency to reply in ways desired by the researcher, and online performativity imperatives may also mean responses are less reflective of "real" opinions. While only marginally responsible for invading peoples' spaces, online research may indeed be caught up in the looming privacy backlash, which should add additional representativeness problems. In sum, such switches to new, potentially fruitful ways to execute traditional research also entail relatively unique, non-trivial problems; their mere existent is not a good basis for the transformative case.

Results Analysis

Computers, because of their ability to analyze very large sets of data and mine them for patterns, hold out the promise of revealing deep structures not normally visible to human brains. A group of computing anthropologists (e.g., Read 1991) has made this argument for a trasformation in the analysis of cultural data.

I think most of my colleagues remain skeptical of such claims. They have had little impact on general disciplinary practice, as manifest in, for example, the program at the annual meetings. Admittedly, in the recent period, anthropologists have had their eyes more on postmodern than formalist questions. Having gone some way down Read's road, I find it very difficult to separate this program from the romance of formalism or the empiricist hope that the world will reveal its form to us without our having to formulate questions. One unfortunate side effect of this intervention has been that computing in anthropology has been perceived as allied with the formalist side in our internal science wars.

Distance collaboratory data analysis is clearly facilitated by AITs. However, given the rapid growth of computer memory (e.g., "Moore's Law"), the advantages of sharing analytic tasks across sites have been exaggerated. One of the downsides of virtual analytic models, particularly apparent in climate research, has been "model peonage." That is, use of certain costly models becomes de rigueur if one wishes to engage the discourse in particular invisible colleges. As time goes on, the limits of a model become more obvious, but the professional price of abandoning it rises as well (Lahsen 1995). Other problems of the virtualization of analysis, such as effectively restricting research to costly modeling exercises, also reinforce dominant paradigms at some cost to research creativity. In all these ways, computers can impede rather than promote progress, let alone transformation.

Publish/Disseminate Results and Conclusions

What we think of as the Internet had its roots in the informal communication that emerged around use of an early Defence Advanced Projects Research Agency (DARPA) funded utility for rapid sharing of research results. The communicative potential of e-publishing has long been recognized, and innovative forms, like hypertext, have been the focus of considerable development work. A promise to publish research results on a website is now a regular feature of the many NSF proposals that I read. Medical information is, after pornography, the most frequent reason for going online. Academic libraries are increasingly active proponents of systems to make research free and highly accessible online. There are doubtless now more technoscience research results available to a wider audience than before the Internet.

Perhaps with the exception of medical information, however, the opportunity to publish results online has not necessarily revolutionized dissemination practices in a positive way. There is first the problem of reduced control over the quality of the results reported, which may, like the writer, be a dog. For reasons doubtless associated with the problem of quality control, despite all the interest in e-publishing and numerous experimental e-journals launched, the tendency of technoscience e-publishing thus far is to mimic the characteristics of paper-based publishing, rather than the other way around (Meyer and Kling 2000). These reinforce prior dissemination patterns rather than transform them.

Bryan Pfaffenberger argues convincingly (1999) that the decisions of the U.S. courts to treat software as patentable rather than copyrightable has potentially drastic limiting implications for dissemination. In essence, these decisions make anything expressible in code subject to patent. It is difficult to specify how limiting on technoscience dissemination the imposition of more rigorous systems for protecting so-called intellectual property will be, but surely they are not likely to be more democratic. Neither my own experience on KN^2 nor that of the human genome project inclines one toward transformative confidence on this issue.

It is reasonable to presume continued expansion of for-profit organizations influence on knowledge dissemination, either through their direct ownership of publications or indirectly through their influence via "triple helix" research funding and control (Leydesdorff and Etzkowitz 1998) or "Mode 2" research (Nowotny *et al*, 2001). Consider, for example, the 2001 U.S. National Institutes of Health cautionary view of the high and increasing proportion of medical researchers with a property interest in the results of their research.

It is also reasonable to be concerned about the impact of certain rhetorical conventions of online representation, such as the preference for brevity and simplification, on the quality of that technoscience which is disseminated. While widespread adoption of such conventions would constitute a transformation of a sort, this would be more of a devolution than a transformation in the conventional, positive sense.

As an anthropologist, I am struck by the extent to which the American Anthropological Association annual meeting program remains unaffected by AITed dissemination media. In 2002, for at least the fourth year in a row, the AAA managed to schedule me in two places at the same time. I still had to try to fit the complex argument of this chapter into fifteen minutes, and I couldn't even use Powerpoint™. Except for my fellow panelists, no one in the audience was able to read my paper online before the conference.

In sum, the situation with regard to the impact of AITed mediation on technoscience dissemination patterns is not clear. There are some reasons to expect change, as well as reason to be skeptical of it. Even more difficult to identify is the character of change, which is arguably as likely to be antidemocratic and limiting as empowering and broadening.

Construction/Reproduction of Knowledge Networks/Alliances

My understanding is that the first significant unauthorized talk on the Ur-Internet was about science fiction. Rather than snuffing out this chat as illegitimate, wise grant administrators recognized that system operators were creating a many-to-many communications modality not initially envisioned. Experiences like this have been parleyed by Internet gurus like John Perry Barlow and Howard Reingold into the argument that online communication is inherently communo-generative. My early experience with professional listservs was consonant with this reading. Like Yvonna Lincoln, I experienced a strong sense of relative liberation from the elitist biases of more conventional venues for scientific argumentation, and I was an avid participant. These experiences made me optimistic that online discussion of research results could change the trajectory of disciplinary development in substantial ways.

Further experience with these modalities has sobered me, however. Perhaps it was being personally flamed in a patronizing fashion by Paul Gross and Norman Levitt, or being trivialized by Jay Labinger (Hakken 1995a; Labinger 1995). More generally corrosive was the eventual trajectory of the "netiquette" or distinct networking etiquette that emerged around the disciplinary alliance building, to call it by its more pleasant aspect, that migrated into protocyberspace. As manifest on SCI-TECH STUDIES, the list of the Society for Social Studies of Science, for example, this initially encouraged voluntarily limiting your contribution to only two screenfulls. While conducive to widening the discourse to include others, and allowing one to continue to read what was posted and still have a life, this convention hardly helped raise the level of discussion of complex topics. Moreover, when a group of self-appointed defenders of "true science" decided to ignore the convention in pursuit of their agenda, most of those previously active on the list were reduced to lurking. Perhaps because of events like this, as well as the proliferation of spam and other junk e-mail, this list, like many others, has become more a venue for announcements and queries for help than an arena for sorting out basic issues.

In their studies of who contributes to online publication of research findings, Eric Meyer and Rob Kling (2000) found that the vast majority of postings are by those already at the center, either geographically or in terms of prestige, of the disciplines they studied. Rather than undermining the preexisting social organization, online publishing tends in this regard to reinforce it. More broadly, from the general perspective of what we might call "paradigmization," online communication is as much about reprofessionalization as deprofessionalization. If one uses a search engine to gather sites on a medical topic, for example, the sheer number of hits is likely to be quite impressive. This experience has been used to argue that the Net opens up professional discourse to anybody who wants to participate. Formally, every writer does have a shot at readers' attention. Unfortunately, the way search engines work undermines any potential for democratization here. Various mechanisms, many of them pecuniary, are more available to the well-provisioned to raise the ranking of their sites in a search result. In addition, the sheer volume of sites drives the surfer to rely on well-known names and institutions. The long-run consequence of current activities is to reinforce, rather than undermine, the influence of the well-connected.

In 1997 when he was a student in my "people and systems" class, John Backman carried out an exercise which revealed to me a peculiar feature of what may be described as the web's "perceived topology." He invited into a computer lab on campus a group from local inner-city, minority organizations who had never "surfed the web." After some brief instruction, he helped them get started exploring sites of interest to them. At the end of the day, he asked for their impressions of the experience. One idea widely shared was how, contrary to expectation, African Americans and their organizations were "well represented" on the web.

At that time, of course, the proportion of websites by and about African Americans was tiny. Yet the architecture of the web is such that the surfer's sense of the importance of her particular area of interest is easily magnified. Is this "exploded" importance best understood as a distortion, or should one trust the intuition of Backman's informants?

I can see the possible validity of a much weaker version of the "democratization" argument. Younger, more marginal scholars, because of their positive experience in discipline-oriented listservs, chat groups, e-publication discussion groups, and the like develop a level of self-confidence more quickly than might have otherwise been the case. They venture more quickly into making scholarly claims, and at least some of these have an impact sooner than might otherwise have been the case. The result is a moderate broadening of those whose contributions to disciplinary discourse are taken to be substantial.

Would this mean a less hierarchical dynamic for invisible colleges—for example, one less centered on one or two central teachers? Possibly, but it should also be noted that this real effect would follow from a probable *mis*-perception of the actual significance of the students' Internet interventions. It is difficult for me to accept a transformationist case based on the systematic misrepresentation of the actual situation.

Perform Quality Control/Reconstruct the Knowledgescape

As the participants in particular disciplined discourse communities reach individual and collective assessments of the veracity/significance of any new initiative, a period of more general reassessment ensues. How does the dominant (do the existing) paradigm(s) need to be adjusted? How should these adjustments be implemented in terms of new or altered standards of discourse and/or models of exemplary practice? What new topics/research areas now appear to be "bleeding edge"?

One of my early frustrations with academic work was with the character of the print versions of our typical "product," the book or article. The stability of the printed page contrasted strongly with the fluidity of my still evolving thinking on most topics, while the individuation of authorship (admittedly pronounced in anthropology) violated my sense of the much greater importance of representing what "we" rather than "I" think. The flexible symboling of the computer held out the promise of forms of representation which would allow better tracking of the developing trajectory of a line of thinking as well as be a tool to encourage collective rather than individuated cogitation. This could be accomplished via continuous circulation of disks or webpages of text and graphics both inherently more alterable and supported by software capable of tracing the alterations. Would the new communication technology mean that the "combat" style of academic discourse would be displaced by a more collaborative approach?

There is little basis for arguing that disciplinary reconstructive dynamics have altered fundamentally. In my experience, those online academic journals that have survived have done so by modeling themselves closely on their offline predecessors. Instead of becoming more transparent, standard practices have arguably become more obscure. Partly this is the result of the need to master broad arrays of new computer-related technical skills to be an effective standard promoter. Computerization of data means in many cases the analytic processes that led to a particular conclusion are harder to recover, perhaps because they are buried in closed source code programs that few have access to, even if they could understand it. Web publication has turned out to be a nightmare for archive librarians. The particular representations on a webpage today were not necessarily those there yesterday. Does this mean each webpage needs to be archived every day? It's been difficult to identify conventions for citing webpages. At a minimum, a much more complex sense of time, far beyond year of publication, has to be developed. It was developments like these that led Langdon Winner to describe the age of the computer, not as a "knowledge society," but as the age of "mythinformation" (1984).

Again, I am not convinced that mediation by computer has transformed this aspect of knowledge creation. How then to account for the widely shared sense that things really are different? Several critiques (e.g., Fujimura 1999) have suggested one fundamental change in the disciplined knowledgescape, what I think of as its colonization by computer science. As medicine becomes bio-informatics, this may not represent merely the adoption of some new tools by an old discipline. Ethnographers like Forsythe found themselves in situations where they were deskilled, as each systems developer became her own ethnographer. As outlined in chapter 6, an aggressive knowledge program is alive and well in computer science. While at the core of each discipline is one form or another of the belief that its worldview is the "most fundamental one," putting knowledge creation online may provide unique opportunities for one disciplinary project's massive self-aggrandizement. Academics are familiar with the battles that take place over disciplinary turf, but seldom is there wholesale colonization like that now taking place. Again, this is transformative in a certain sense. However, it is not of obvious general positive benefit to humanity, nor, given the previous failure of knowledge programs in informatics, is it likely to experience long-term success.

Metanalysis, Alternative Perspectives, and Implications

While the image of AITs' impacts presented above is complex, the overall argument is that computing has not, at least yet, shifted fundamentally the general landscape of technoscience knowledge creation practices. To use them to address the broader question of general knowledge creation, such summative analyses need to be more deeply contextualized, subject to the reflexive process sometimes called metanalysis.

For example, there are problems with privileging technoscience, even relatively as this chapter does, as a vehicle for generic discussions of change in knowledge creation. Technoscience tends to privilege a rather Taylorist notion of research. On this view, knowledge creation is presumed to be normally carried out in large groups, over a number of sites, and in a sequential fashion—step one is completed, followed by a discrete step two, which is then completed, followed by step three, and so on. The analyses offered presumed that one could decompose technoscience knowledge creation into discrete steps and examine them independently. This model of knowledge creation is similar to the discredited "over the wall" conceptualization of software development.

There are additional problems with taken technoscience as the paradigmatic tranformationalist case:

1. It privileges big over small science, experimental over theoretical physics, laboratory over field science, positivist over ethnographic epistemologies, and so on.
2. It privileges "science" over engineering and technology, both in that the latter are much less rationalistic and more opportunistic.
3. It tends to background the tighter, dialectic relationship between science and technology implied by the term technoscience.
4. It overemphasizes the extent of discreteness between various moments in the techno-scientific enterprise, which often take place more simultaneously that linearly in time.
5. It may imply that proof of transformation at any one stage would be sufficient, whereas for transformationalism to hold, it would have to be present at more than one if not several stages.

For all these reasons, an approach to evaluating the transformationist case based on technoscience may not be the best one.

Is there an alternative strategy for assessing the correlates of computing in knowledge creation networking? Still following Crane in accepting that knowledge creation is an institutionalized social process, one might make a transformationist case via a more limited number of more holistic structural arguments. These might include:

1. A change to organizational context argument—for example, that with AITs the central locus of knowledge creation moves, from universities, research centers, think tanks, and R&D centers to interorganization networks, triple helixes, collaboratories, and so on
2. A networking change argument—that knowledge creation networking moves from discipline-based, elitist discourse communities to more democratic, market-based flexible networks and/or mass discourse spaces
3. A content argument—that knowledge creation is no longer about "knowledge for-knowledge's-sake" but moves, for example, to commodification and/or dumbed-down "information candy"

It is to such more sweeping sorts of claims that the Stoniers, Druckers, Barlows, and Tofflers are drawn, perhaps because they reduce the need for complex empirical argument. One of the salutary consequences of taking on transformationalist claims on the turf of technoscience's view of itself is to make plain how any sustained argument for transformation would have to be complex. Besides being mutually inconsistent, the structural hypotheses outlined above are simplistic. Their contradictory character reinforces my view that transformative arguments must reach a relatively high threshold; it also explains why I do not take each of them on at this point.

The Limitations of Casting Knowledge Creation Networking as "Community"

Having used Crane's notion of "invisible college" to frame description of the social relations of the typical disciplined knowledge creation network, I conclude by pointing out an additional limitation of existing theoretical approaches. Most readers of this book will have grown up in cultures that privilege science by treating it as the most central form of knowledge networking. For a long time, much of science has involved deplaced networking—at universities, research centers, corporations, and home studies at some considerable distance from each other.

Its normally dispersed character means that some celebrations of unity—Festschriften, banquets, award presentations with large elements of collective self-congratulation—are periodically an aspect of the disciplined creation of technoscience knowledge. Unfortunately, the term "community," as in "community of practice," while appropriate to description of celebratory performances, has come increasingly to be used to describe all organized practices of knowledge production. This metaphoric extension unnecessarily complicates empirical study of knowledge creation networking.

On a postmodern reading (e.g., Cohen 1985), performances of community involve collections of people evoking symbols of what they share. These are often performances of boundary construction, of separating an "us" from a "them," of celebrating what unites "our" identities. Thus, in performances of community, an ethic of radical egalitarianism is often presumed to exist among those sharing the presumed identity, a concomitant ethic of difference being imposed on those outside the identity boundary.

Knowledge networking does invoke shared symbols, like those that stand for the shared set of standards by which we collectively distinguish "mere information" from what we together "know to be the case." Its shared symbols, unlike those of community, are tied to a metadiscourse (often explicit in knowledge creation networking) about what constitutes knowledge. One might call this reflexive networking talk an ethnophilosophy, with its own notions of ontology (what is to be known) and epistemology (how it is to be apprehended).

Knowledge networking also definitely involves boundary work, as when participants assert what separates their knowledge from that of others. This leads to a dynamic very unlike the notion of community, however, for it involves divergences recognized as being within the network. Knowledge creation networking especially evokes differences of view. As network participants disagree about each other's assertions or dispute their right to make them, there emerge competing ideas, in relation to both what standards should be shared and how they are to be applied in concrete cases. Thus, knowledge networking involves a rougher sort of equality than that of community, one more like the multiple oppositional cultures characteristically developed by "inmates" in "total" institutions (Goffman 1959). Performance of differences internal to networking is not congenial to "community"; it is more characteristic of social movements.

Donna Haraway has recently provided what in many ways is a useful summary of this chapter's notions about how to study knowledge networking

in technoscientific organizations. The identity of the "'knowers' of scientific knowledge claims" is very social; they are neither individuals, nor "no one at all." This recognition leads Haraway to call for a "different kind of strong program in science studies, one that really does not flinch from an ambitious project of symmetry[,] that is committed as much to knowing about the people and positions from which knowledge can come and to which it is targeted as to dissecting the status of knowledge made" (1997, 36–37).

However, Haraway characterizes the "knowers" of such claims opaquely as "social communities," a characterization from which I dissent. In my view, a knowledge network is not a special kind of community. Knowledge networking is better understood as a distinct kind of social activity.

It is in this sense that Bryan Pfaffenberger's study of online searching (1990) provides a better picture of the networking that actually creates knowledge than either "community" or "invisible college." Pfaffenberger addresses why early efforts to create a "radical information democracy" via remote, electronic searches of databases of library-type references did not succeed. In his view, this is because of the character of the indexing practices integrated into these systems. These practices are a hybrid of mindless "text crunching" programs that indexed every word and the indexing conventions of skilled reference librarians. For the typical user, effective use of the online databases still requires a knowledgeable guide; the tepid quality of my students' online searching on their own indicates that this is still largely the case.

This chapter has examined similar processes that share this quality, where knowledge is "really" shared only to the extent that those in a knowledge network share rules for establishing both epistemological and indexical status. Networkers embody such knowing insofar as they participate in the establishment of the rules. The "trust" involved in a knowledge network is conditional on such participation, not "trusting." It is based on social experiences—often of compromise, of negotiated optimality, of experience-based confidence that rules will be applied with similar senses of scope and coverage.

What science requires, in addition to occasional performances of Cohenian community, are institutions that simultaneously promote and manage conflict. It especially requires mechanisms that support often conflictual metadiscussion of the criteria by which to judge which kinds of data and arguments are to be found telling.

Consequently, a major task of AIT systems to support knowledge creation networking must be to enable substantial degrees of such participation. This is likely also to involve promoting awareness of both standards of just how "known" something is and what it represents. Much more important to this

form of networking than the occasional Festschrift is the considerable non-, even *anti* community performance: conflict over ideas, competition, back-stabbing, sabotage, and so on. Thus, a metaphor like "the scientific community" is largely misleading and design dependence on it distorting.

Deplaced AITed knowledge creation networking in disciplines has to be as supportive of conflict as of community. To describe a collaboratory as a "virtual community" is to emphasize that aspect of knowledge creation networking which humans find relatively easy to perform, celebration of those things that we share. The other, more difficult, conflictual aspects have to be "engineered for" (not engineered) as well—how does one organize contexts that allow performance of community but also support the pursuit of multiple, often contradictory knowledge projects? Other than in KN^2 little attention to this problem has been given by the AIT interventions discussed in this chapter, but it was very much in mind in the constructing of the project featured in the next one.

In short, to be considered transformative, change must be cumulative, substantial, and explicable. My hope is that those wishing to argue a transformationist case are motivated to make it in comparably complex terms. The absence of such argument remains perhaps the strongest reason for skepticism.

Moreover, even if sustained and substantial in the creation aspect, change would also have to be equal or even greater with regard to knowledge reproduction and sharing. Further general argument is best postponed until the transformationist case has been addressed in relation to the other aspects of knowledge networking (reproduction and sharing) addressed in the following chapters of this section, as well as the political economic issues, like commoditization, addressed in chapter 10.

Chapter 8

Knowledge in Organizations and Knowledge Networking Oneida County

An Ethnological Alternative to Knowledge Management

In the future, will the transformative potential of AITs be realized in organizations, or will this potential remain blocked? Despite its criticisms of knowledge management, chapter 3 acknowledged that major knowledge challenges still face contemporary organization. Fatigue with knowledge management developed not because the challenges disappeared but because KM systems failed to meet them. Would taking an alternative approach make a difference?

This chapter shows how the ethnological concept of knowledge networking can be applied to organization. Combining ethnographic study of knowledge in organizations with AITs based on an alternative theory of knowledge could increase the chance for transformation. This means a focus on knowledge adaptation/reproduction, the part of knowledge networking most characteristic of organizations.

The chapter begins with a discussion of why organized knowledge networking focuses so much on adaptive practices. It focuses on one indicator of whether AITs change organizational dynamics, the extent to which they "resocial" work—that is, make work substantially more likely to produce new and/or stronger social relationships. Since fully resocialed work really would be work that is fundamentally changed, prospects for resocialing are a central concern of cyberspace labor process studies (Hakken 1999a).

The arguments for and against the resocialing impacts of AITs are illustrated through recent case studies, including my efforts to promote fulsome

knowledge networking in Oneida County, located in the Upper Mohawk Valley of upstate New York. These efforts were based in part on the idea that, contrary to the implicit presumption of organization studies, research on public and not-for-profit organizations provides better information on the knowledge potential of AITs than on private ones. Studies in private organizations tend to misrepresent AITs' impacts on knowledge because of these organizations' greater dependence on the reproduction of capital. Moderate success networking Oneida County knowledge enabled clearer specification of the social forces with which a truly coherent effort to change the dynamics of organizational knowledge reproduction has to contend. What was learned through this and other interventions is critically integrated with recent work in organization theory, itself largely stimulated by efforts to come to terms with knowledge (Pitt 1998). The result is an approach to knowledge in organizations that is truly supportive of resocialing and could actually "engineer " transformative organizational change.

Organizations, Organization, and Reproduction

Their basic dynamics make organizations' primary relation to knowledge one of reproduction, as opposed to either its creation or its sharing. Lucy Suchman nicely describes the replications and moderate refashionings that are typical of organizational knowledge networking:

> Within organizational settings there is an intimate relation between forms of discursive and material practice, and action's rational accountability. Learning how to be a competent organization member involves learning how to translate one's experience, through acknowledged forms of speaking, writing and other productions, as observably intelligible and rational organizational action . . . Demonstrations of competence are inseparable, in this sense, from artful compliance with various professional and technological disciplines, reflexively constituted through those same demonstrations. At the same time, artful compliance necessarily involves endless small forms of practical "subversion," taken up in the name of getting the work of the organization done. (2000, 313)

What, then, makes organization? Informing the anthropological theory of social networks described in chapter 5 is a conceptual continuum of all social interaction spaces that includes organizations. On one end of this continuum are nonorganized spaces, like the informal structures of band societies, as well as the random and anonymous public places in complex societies. Networks

are in the middle of this continuum, exactly where depending on the complexity of their patterns and their degrees of relational connectedness and overlap. On the other end would go bounded entities of interaction, ones with notions of "membership" and explicit or at least articulable rules.

Organizations fit at this more organized end of the continuum of social interaction spaces. Among early anthropologists, the lineage was the organization or "corporate group" most often studied. In sociology, business, and the overlap field I will call organization studies (OS), a particular type of organization, the limited liability corporation (LLC), is similarly privileged. OS emerged out of Alfred Chandler's interest in how the newly created, huge LLCs of the early twentieth century managed their considerable problems of scale (1990).

Initially, this privileging was so complete as to make the LLC the sole object of study. With the apparent emergence of a permanent public stake in the economy associated with post-Depression Keynsianism, however, OS's rhetorical focus broadened. The field began to call itself "organizational science" to imply a more generic inclusion of public as well as corporate entities. However, that the "natural" forms of both types of organization were fundamentally similar has been assumed rather than demonstrated. While this proposition is open to question, the tendency of the field has been merely to extend theorizing based on private cases to the public sphere.

In early OS studies, voluntary organizations were also marginal to the field, considered elsewhere beneath rubrics like, in sociology, "social movements." Under the pressure of those working in the study of development in what used to be called the "third" world, however, a third type of organization, the nongovernmental organization (NGO) has also been added to the OS agenda. A parallel practice has emerged in studies of the "first" world, including as organizations the not-for-profit (nfp in the United States) or voluntary (Europe, Canada) organization.[1]

Sometimes associated with this additional broadening was a second renaming process, from "organizational science" to "organization studies." Users of this latter term often wish to convey an analytic loosening and broadening, from a focus mostly on formal structure (the organization as a form) to include practices (the organizing) that produces the dynamics of interest. This "poststructural" shift (Clegg et al. 1996) is associated with the influence of Anthony Giddens's "structuration" theory (e.g., 1987).

The move from focus on a thing (the organization) to the process of organization parallels a move toward institutional, resource-based theories of the firm as opposed to neoclassical, market theories (Nahapiet and Ghosal

1998). Resource-based theories of firms emphasize that their internal characteristics are as determinant of their actions as is their market position (Pitt 1998). This formalist to substantivist shift is manifest in related fields like sociology. Writers like Castells (2000a), for example, argue that the "network societies" in which we now live derive their new dynamics from a change in the characteristic processes of organization. That is, the chief organized entity is no longer the single firm; this has been replaced by the interorganizational network (Powell 1990). Network society advocates similarly claim that in cyberspace membership in formal groups (e.g., organizations) is less important than is participation in multiple, much less tightly bounded networks. Held to be a related phenomenon is the self-help described by authors who urge workers to accept corporate and public organization downsizing and give up their addiction to jobs, these being no longer the primary, let alone necessary, means of access to the means of existence.[2]

In short, the original problematic of organization studies was, legitimately, critiqued for presuming that there is only one basic kind of organization, of which the large, for-profit, jobbed entity is the prototype. The consensus that the field's task is to create a foundational "scientific management" (Taylor 1998) also no longer holds (Clegg et al. 1996).

Yet while occasionally drawing attention to large public bureaucracies and NGOs/nfps, the OS approach still privileges *Fortune* 500s—the big, job-based bounded entities. Little attention is given to small organizations, nor has enchantment with inter-organizational networking yet led to study of entities with principles of membership or recruitment other than paid labor.

These biases toward the big and the jobbed limit organization studies' ability to theorize organization in general. Because the job remains individuals' primary way to access the means of day-to-day reproduction, and organizations remain the primary providers of jobs, this privileging is understandable. The dynamics of job-based organizations may be shifting, but these organizations, private, public, and not-for-profit, retain their central place in the reproduction of capitalist social formations. They must have, for now, a central place in our study of organizational knowledge networking. In this chapter, acknowledging an empirical centrality without taking it as the theoretical one, we concentrate on organization in jobbed entities.

Organization and the Reproduction Aspect of Knowledge Networking

The rethinking of organization from a resource rather than a market perspective follows from thinking about organizations from a knowledge perspective.

A resource perspective follows from viewing jobbed organizations as being more about "reproductive" than "creation" or "sharing" submoments of knowledge networking. Arguably some of the activities of organizations—for example, a research department—involve the production of new knowledge. However, as reflected even in the dual name of the functional units typically charged with creation tasks—"research *and development*"—this is a minor part of what most organizations do. Similarly, while internships and apprentice programs—knowledge sharing submoment activities like those addressed in the following chapter—are common in organizations, these also are seldom their primary concern.

Most commonly, organizations adapt knowledge. They reproduce it, sometimes exactly. More often, they apply already existing knowledge to new, varied situations, or apply newly acquired knowledge produced elsewhere to already encountered conditions. Typically in adaptation, the content of knowledge is of less concern than is the form or style of its expression. Knowledge's representation is altered to fit different, specific, concrete circumstances.

In many systems of social reproduction, the outputs from one period become the inputs to the next. Organizations are a sociality type that can adapt themselves, change themselves deliberately in order to compensate for different inputs or changed contexts; indeed, this is the chief benefit of organization. Moreover, as argued long ago by Karl Marx (1871), under capitalism reproducing social systems over time tend to change themselves. This is because it is highly *un*likely that inputs will remain constant, given the frequent output altering changes in the system's context.

Organizational change is normally more adaptive than transformative. Because temperature has declined, the thermostat in a central heating system causes a furnace to ignite. So in organizations, the point of the system is more likely to be something similar to homeostasis—to keep things the same—than to change them substantially. Similarly, in neoclassical economics, when supply rises to meet demand, this change is seen as supporting equilibrium, admittedly a new one in some formal sense, but substantively a lot like the old one.

Occasionally, however, the small, normal accommodations that are part of adaptation or "normal" reproduction of a social system accumulate in such a manner as to create a fundamentally different dynamic. Such an occasion, as when organizational knowledge networking is truly transformed, would be like what Gould (2002) calls a "punctuated" equilibrium in regard to physical evolution. Marx differentiated such transformative reproduction, which he called "extended," from "simple" reproduction, or the replication of a system's basic characteristics (Hakken 1987).

In sum, if knowledge AITs *were* radically to extend organizational reproduction, this would likely be primarily because they change the way organizations *adapt* knowledge. Further, since organizations are not generally about the business of creating new knowledge, the transformative changes associated with AITs would be manifest in changes in the *social functions* of knowledge as well as in *new types*. Since employment-based organizations are the distinctive feature of employment-based social formations, fundamental changes in organization knowledge adaptation will doubtless be a necessary component of any general Computer Revolution. If one is interested in whether AIT-induced changes in knowledge networking transform social reproduction to create a knowledge society, as we are in *The Knowledge Landscapes . . .*, it makes sense to pay attention to organization.

A Research Agenda for the Ethnography of AITed Organized Knowledge Networking

The following set of questions was first articulated at an early stage of the KN^2 project. This agenda emerged from extended field studies of sites where people were actively networking knowledge in organizations (e.g., Sachs 1994). Chapter 3's account of the failure of knowledge management was based on the ethnographic experiences described here in Section III. That chapter's account contrasts strongly with those that have dominated the literature on AITs in organizations. Once the striking-because-rhetorically-hyped-but-ultimately-thin veneer of cybertalk was peeled away by careful empirical study, the dynamics of early computer-mediated organizations were seen to be only moderately different from their predecessors. Like Attewell with his productivity paradox (1994), I reached this conclusion in my initial (1980–1995) ethnographies of AIT in organizations (Hakken with Andrews 1993; Hakken 1999a; pace Sproull and Kiesler 1995). In part, my alternative analysis derives from an important aspect of organizational ethnography, its effort to study organization from multiple perspectives, to study "up" as well as "down."

Knowledge and Extended Organization Change

Indeed, concern about the absence of expected changes was one impetus to the development of knowledge- and resource-oriented constructions of organization. For example, earlier AIT systems for deplaced organizations aimed to simulate existence "in real life" (IRL). It became clear that the knowledge networking aimed simply at "replication at a distance" was too narrow. Instead of trying, like Heaton's Japanese CSCW, to create virtual face-to-face practices, knowledge systems needed to foster new, AIT-based practices, ones that support collective production and effective dissemination of knowledge in ways that take advantage of the new technology's distinctively different capabilities.[3]

What specific developments would be most indicative of fundamental changes in the ways AITed organizations reproduce knowledge, not indicative of "normal" reproduction but of some fundamental, "extended" transformation (new evolutionary stage in social function)? Do current changes indicate something beyond merely new but still quite comparable points of equilibrium? Is there reason to think that newer versions of knowledge-oriented IT systems promote really new dynamics?

The Forms of Organized Knowledge

In those organizations in which her position was initially created, the chief knowledge officer was insufficiently distinct from the previous chief information officer, the individual with responsibility for the organization's electronic information system (*Information Strategy* 1998). In an important sense, such organizational interest in knowledge grows directly from awareness of a need to do something more than just share information. Like more general discourses, the CKOs responsibility was generated and communicated in two very different epistemological traditions, yet it was presented as unitary. Organized knowledge is informed, but only implicitly, by both modernist and postmodernist conceptions (see chapter 4). On the one hand, numerical, often statistical information developed in the manner of science is essential to much modern production. On the other, much of the knowledge necessary for success in the market is understood to be literally embodied in staff, a collective resource deeply situated in workspace culture, to be best communicated in existentially compelling stories (Nonaka and Takeuchi 1995). The coexistence of such different kinds of knowledge "content" produced an important underlying tension in early knowledge management systems. In such ways, the undertheorizing of knowledge networking's problems undermined the systems' performance.

As argued in chapter 4, it is possible to see the modernist and the postmodernist conceptions of the data/information/knowledge interface not as competitors but as complementary when sufficiently incorporated into a common frame. This begins with recognizing how the modernist notion of "raw" data is a culturally constructed silencing. Latour and Woolgar (1979), Traweek (1988), Heath (1994) Forsythe (2001) and other science, technology, and society ethnographers and historians illuminated the many ways in which technoscience data are produced socially, the particular contexts in which they are initially situated and from which they are abstracted, and how these abstractions are disappeared in subsequent "raw" constructions. To make scientific knowledge maximally useful in the world, one needs to create framings in which modernist and postmodernist knowledge can coexist, simultaneously resituating relevant relations and practices, like labor processes. Neither a unidirectional abstraction nor concretization, knowledge emerges out of a networking dialectic that moves among various moments of abstraction and concretization.

Can knowledge networking technologies be designed that embrace this dialectic? Can systems be implemented that support practices delinked in space and time while still giving appropriate attention to all the moments in the complex, neither strictly modern nor postmodern dialectic of knowledge? Would the dynamics of such systems be distinct enough from existing forms of knowledge networking to support transformationist ideas about the social functions of organized knowledge?

Knowledge and Organization Ethics

The creation and spread of AIT pose a series of ethical problems, for example, in relation to privacy, security of data, and the several dimensions of integrity (Hakken 1991, 2000, and 2003; Sproull and Kiesler 1995). Chapter 12 argues that the development of denser knowledge networking in organizations would ratchet up many of these issues.

Do the new approaches to knowledge ownership (Coombe 1998) developed in law and organizations suppress rather than support creative adaptation? What kinds of approaches to informed consent have sophisticated knowledge networking organizations developed? Do these approaches adequately enhance human dignity? Reflexively, from which stances, in what kinds of positions, do ethnographers find themselves studying knowledge networking in organizations? How do they manage the potentially conflicting problems of contributing to enhanced networking while remaining responsible to a broader scholarly knowledge network? What research ethics safeguards should be built into knowledge networking in organizations?

Knowledge and Organized Sociality

The globalization/team contradiction in contemporary organizations described in chapter 7 helps explain why knowledge networking remains a key point of AIT intervention in organization. The problem of knowledge in organizations is very basic, and therefore attempts to deal with it are inevitable, while the very failures leading to knowledge management fatigue provide a basis for creative rethinking. As described in chapter 3, the core problem with the first generation of self-labeled knowledge management systems was their failure to come to terms explicitly with the social. These systems were "thin" because they were designed technically and managerially, not socially. Indeed, some were merely renamed media for information or even data networking. A long tradition of social thought has posited a close connection between experience at work and such broader social dynamics. Some optimists, following the Blaunerite analysis (Blauner 1964) discussed below, have argued that AIT at work will tend to reverse the degrading, deskilling trajectory of work evolution (Braverman 1974), resulting in both more workplace democracy and more democratic society (e.g., Toffler 1983; Rheingold 1993). I have summarized these arguments in the notion of a dynamically new, cyberfacture stage in the evolution of the labor process (Hakken 1999a). Pessimists, however, see even greater control and deskilling at work, and/or greater unemployment, perhaps a devolution to a previous stage in the labor process (e.g., Rifkin 1995).

With which view are the dynamics of "cutting edge" organizational knowledge networking most compatible? Do second-generation knowledge-engineered organizations create new social dynamics? Is there more sociality, as manifest in, for example, more trust, community, or sense of common purpose? How do they create sociality, and how well? To what extent do organization knowledge networking initiatives create alternatives to—as well as simulacra of—trust, shared cultural and social constructions, and so on? To what extent do they develop alternative, technology-based means to achieve social relationships with these characteristics? What separates successful efforts of this sort from unsuccessful ones? What new strategies are suggested by the study of such efforts?

The Framing Analytic Issue: Is AITed Work Resocialed?

This research agenda can be summarized in terms of the presence or absence of a new social dynamic. Do AITed organizations manifest substantially more sociality than their predecessors? A historical perspective suggests a rephrasing: Do AITs *re*social work?

Managerialism and Desocialing Work

This rephrasing is compelled by the manipulation of work sociality—indeed, the attempt to eliminate certain forms of sociality among workers that created distinct worker knowledge—central to the notion of organized management. The necessity of management was the fundamental premise of much of late-nineteenth and early-twentieth-century reform of work organizations. This presumed necessity was manifest most clearly in Taylorism, but it is still implicit in, for example, most industrial engineering. Under Taylorism, making the management function distinct is presumed to be necessary to bringing production under the control of science, to achieve "scientific management." For production to be managed, it was necessary to seize control of the activities through which commodities were produced. The basic strategy for achieving management control was to individuate work. By reorganizing work into tasks carried out by discrete individuals so that how each worker performs the new tasks could be controlled, both workers' collective and their individual control of work was to be obviated. Any social relations, but especially those developed "spontaneously" by workers, were to be eliminated wherever possible. Managerialism meant *de*socialing work.

In other words, there is an inverse, dialectical relation between knowledge at work and managerialism, and its cousin, technicism. Incomplete understandings of knowledge at work derive in large part from the more or less purely managerial perspective of organizational science or human relations (Clark 1995). These inadequacies also follow from conceptualizing knowledge in overly technical ways, a point well documented ethnographically by Frank Dubinskas (1988), among others. His work in organizations neatly complements, for example, that of Diana Forsythe's (2001) on the informal work ignored by formal systems. As described in chapter 6, Forsythe explained the general failure of "expert" systems, and, by extension, any "strong" informatics program, in terms of their creators' insensitivity to their own informal but still important work processes, rather than to technical design problems. The tendency of technical and managerial approaches to misunderstand knowledge at work is a legacy of a deliberate silence about this organization knowledge politics.

Managerialism and Organization Studies

Managerialism and its desocialing strategy remained an important implicit assumption of post–World War II organizational science. In the 1960s, documenting the sociality that persisted at work (despite managerialism) once again became a prime activity in social science. Workplace ethnographies like those completed by recipients of the Society for the Anthropology of Work's Arensberg Award, for example, Louise Lamphere (1979), Karen Brodkin (Sacks 1982), and Fred Gamst (1995), made a substantial contribution to the renewed interest in the actuality of the social at work. Their work studies, a presence, for example, in the 1979 Andrew Zimbalist collection, *Case Studies on the Labor Process*, complemented the work most influential in reviving the sociology of work, Harry Braverman's *Labor and Monopoly Capital* (1974).

Perhaps the key contribution of this new work ethnography was to justify a key Bravermanite methodological presumption, that of the relative autonomy of the labor process itself. New work ethnography documented in particular the sociality surviving despite Taylorist attempts to expunge it. Actual activities on the job are related, significantly and dialectically, to socioeconomic and technological as well as managerial contexts. Since the dynamic of a labor process is not generally reducible to any one context, it operates with relative autonomy from all of them. An important implication of the relative autonomy of the labor process is its centrality to the broader dynamics of social formation reproduction and hence to the social analysis of current social formations. No longer viewed through a technicist, Taylorized lens, actual work reemerged as a theoretically privileged moment in social reproduction.

Braverman was responding to Robert Blauner's 1964 classic, *Alienation and Freedom.* Blauner acknowledged the debilitating alienation of work under mechanized mass production or machinofacture, the dominant form of industrial capitalism in the Taylor period. In contrast, Blauner felt that newer workers, their labor process transformed by the new, continuous-process, "high" technologies emerging in chemical production and oil refining, were freed from alienation. Continuous-process technologies transferred the onus of worker activity from mindless physical movement to responsible mental attention, from habitual enactment of embodied manual skills to varied application of mental knowledge of science-based processes. Blauner's arguments, transmitted via the sociologist Daniel Bell (1973) and the management guru Peter Drucker (1970), laid the basis for much contemporary popular wisdom about knowledge; for example, the notion that the information society is a knowledge society, that knowledge is now the key factor of production, that brain has replaced brawn, and so on.

Deliberately subtitling his book *The Degradation of Work in the Twentieth Century*, Braverman rejected Blauner's optimistic contentions. In contrast to Blauner's causative stress on technical relations of production, Braverman's basic argument was social. In brief, he argued that owners of capital in the long run do not select those technologies that are most efficient and effective.

Rather, they choose technologies conducive to work arrangements that reproduce their social power, even when these mean inefficiency and smaller profits. In similarly stressing sociocultural rather than economic or technical relations, workplace ethnographies of the 1970s and '80s are properly perceived as more supportive of Braverman than Blauner.

To be optimistic about the transformative implications of AITed work means to affirm a Blaunerite conclusion in spite of Braverman. Like Blauner, I see liberatory *potential* in AIT-enabled changes in workspace knowledge networking. Standard pro-AIT analysis, in the Blauner tradition, attributes changes in contemporary work institutions directly to new production technology, or to similar new technologies of organization that are more participatory, "Japanese," matrix, virtual, and so on. The failure of first-generation knowledge technologies in organizations means we must reject simplistic Blaunerism, however. Since the ideology of managerialism was responsible for much of the initial desocialing, it is doubtful that real AIT induced workplace transformation is compatible with continued managerialism.

Is managerialism being marginalized in organization studies by perspectives more conducive to revival of old forms or the emergence of new forms of sociality? While presumed to reverse Taylorist individuation, do AITs so substantially rekindle sociality as to set off a fundamental transformation *in actuality*? Moreover, before it can be argued that AITs transform knowledge networking, work's persisting sociality must first be recognized. Since the work ethnographers documented the continuing presence of substantial sociality at work in spite of Taylorist managerialism, the mere presence of sociality in AITed work does not in itself demonstrate fundamental change; we would need to see new forms of sociality.

Transformation probably also means supporting worker knowledge, not ignoring or suppressing it (Orlikowski 2000). As illustrated in chapter 3, however, advocates of AITs (e.g., Sproull and Kiesler 1995) claim they automatically increase social connection in the labor process. (These arguments parallel the view that AITs are inherently democratizing as critiqued by, for example, Winner (1994).) To illustrate, a prime argument for installing new computer-based technologies is their capacity to expand interactive communication. Increased bidirectional or "many-to-many" communication serves and is served by denser social relations. To automatically resocial work, AITs would have to reverse or at least undermine any individuating/desocialing remnants of Taylorism. There is little to be gained by computerizing communication tools if communication itself is still discouraged, or by expanding communication capability while simultaneously discouraging sociality.

The section immediately following presents the case for resocialing. The next section outlines the opposite, more consistent with Braverman argument, the "continuing degradation" one. Data for the following considerations are derived primarily from my study of AIT-mediated labor processes in the United States, England, and the Nordic countries. They include basic research on technological change and economic decline, national difference and the institutionalization of technology, and change in knowledge and technological change. My applied research in this area has focused on various attempts to intervene in the technology/social change nexus, at work group, organizational, community, local, state and national levels (Hakken 1999b).

The Case for AIT-Induced Resocialing

"Resocialed" work would be at once both substantially more social (less individuated) and would have its sociality legitimated, neither being ignored (as in human relations) nor suppressed (Taylorism). The analytic case for resocialing is based on the idea that, as in other time periods, a series of contradictions has once again increased the relative autonomy of the labor process. In such moments, innovation at work is more frequent and the labor process evolves along a relatively more independent trajectory.

Because it increases labor process autonomy, computerization has a resocialing impact. The increased autonomy can be theorized as resulting from a contradiction between, on the one hand, the "globalization" of production (what Rothstein and Blim (1992) refer to as a "the global factory") and, on the other, the increased dependence upon specific local forms of sociality at work for production of surplus value. Several increasing scales of production are typically referred to collectively as globalization. Production increasingly takes place within larger organizations, each of which is more likely to include multiple locations, many of which in turn have different regional, national, and cultural locations. Moreover, more permeable organizational boundaries mean production occurs within technical and social networks which cross company cultures. A weakening of local, regional, national, cultural, and corporate regimes of influence and control is complemented by the increased power of larger units of capital reproduction, like Microsoft and the Wallenberg family holdings now only loosely based in Sweden. The accompanying increased ambit of the anarchic tendencies inherent in capitalism is manifest in, for example, the rapid spread of recent recessions. (I discussed these developments at length in *Cyborgs@Cyberspace?* [1999e], and chapter 10 explores the problems for those who like Castells [2000] locate their source too deeply in technology.)

A correlate of rising scale and greater anarchy is an increased need for coordination, often across different languages, communities, and cultures, in the labor process. The recent shift in organizational talk about computing, from stress on computer-mediated activity to computer-mediated communication, is partly reflective of the urgency of this globalization steering problem (Habermas 1990). That is, first-wave computerization tended to focus on, say, transferring control of a single machine tooling operation from a human to a database. In contrast, second-wave computerization focuses typically on running multiple machines, often manufactured and/or operating in several different countries. We are fast approaching a situation where the majority of machine activities can be computerized and the results of this computerization communicated anywhere. (From the perspective of a labor theory of value, this could lead to the reemergence of a crisis in value realization.) In a phrase, there is a decoupling of the work*space* from any particular work*place*.

Such developments mean even greater difficulties coordinating production. It is predictable that such coordination difficulties would foster a renewed discussion of how to organize actual work processes. Some of this talk, stimulated also by knowledge management failure, stresses development of even more complex, formal systems of control. The bulk of contemporary organizational talk, however, has taken a very different turn. The rhetoric of team has replaced that of military command. My colleagues in our Business School at SUNY Tech now stress facilitation, no longer decision making, as the chief role of management. Downsizing and the flattening, even the elimination, of large sections of bureaucracy and hierarchy also increase organizational dependence upon workers' initiative in the actual labor process. Phrases like "knowledge capital" reflect how the knowledge of workers, based on their capacity to create and reproduce social relations, is increasingly evoked as a source of an organization's ability to compete. Such knowledge is applied best when applied spontaneously, as in cooperative, self-managed groups, rather than via hierarchical command structures.

Body-to-body, face-to-face interaction is the condition under which such group formation and action normally take place. Hence, the contemporary workspace manifests a delicious contradiction: work becomes more dependent upon workers' abilities to create close social relations at the same time as the globalization process that creates this dependence inhibits their construction. The shift in attention from using computers for data and information to knowledge networking is an outgrowth of this contradiction. Many recent initiatives, like those funded under the knowledge networking component of the U.S. National Science Foundation's Knowledge/Distributed Intelligence program, address an obvious problem: how to replicate or at least simulate the embodied social interaction on which team relationships and voluntary social cooperation depend.

Such means are necessary if capital in its globalized form is to be able to reproduce itself. The shift in relative importance at work from computer-mediated activity to computer-mediated communication is itself a harbinger of a broader reappropriation of computing technology. One unfortunate consequence of the initial public uses of computers in scientific contexts is their relatively vapid social construction as fancy machines for calculating, a social construction that misrepresents computers' much broader potential as general symboling machines. While previous language innovation processes have held referent constant while altering code, or visa versa, computers make dual innovation possible. As such, they are unprecedented machines to support cultural innovation, including ever more virtual cultural performances.

Thus, the conjunction between more computing and expanded attention to the cultural dimensions of contemporary production is not happenstance (Greenbaum 1995). The contemporary labor process is unique in the extent to which it is marked by a second, cultural parallel process (Hakken 1999a). The first parallel process emerged in the late nineteenth century. In the then new era of machinofacture, an effort was made to replace the skilled coordinating workers of complex hand production (manufacture via detailed division of labor) by machines. A separate management hierarchy of engineers, clerks, secretaries, and technicians controlled these machines. These new workers carried out a paper-based representational process that paralleled and was perceived as controlling the physical process of production.

The second, new parallel process of today adds another layer to the first one rather than displacing it. In the new parallel process, staff members are in workshops and training—a seemingly endless round of meetings and rallies whose actual relation to "real" work seems even more distant than that of paper shifting. Performances of the second, cultural labor process may achieve at least one, perhaps strategic function: both customers and employees, dazzled by the performance, become convinced of the validity of the commodities that result. (As pointed out by Richard Sennett 1999, it becomes increasingly difficult to distinguish the commodity from the performance!)

As multipurpose communicating symbolers, computers are vital to this second, cultural, parallel labor process. They are thus the appropriate symbol of the potential new stage in the evolution of the labor process, cyberfacture, of which the cultural parallel process may be the distinctive feature. The symbolic power of computers as communicators was obvious in the recent dot-com moment of commodity fetishism, in which Internet companies were worth billions (but only for a while!) even though they never turned a profit.

In short, the case for resocialing gives priority to forces that increase the likelihood of a socially revived workplace, one where innovative practices and groups foster new forms of socio- and ethnogenesis. The new forms grow in the space created by the contradiction between globalization of organizations and the relative abandonment of degrading forms of work organization. In the resocialing reading, the individuation of work demanded by Taylorism—that is, the constructed isolation of the worker from her fellows and the effort to achieve detailed control of work through control of each movement of her body—is necessarily a casualty. At places like Sun Microsystems, for example, the principle underlying new architecture is never to allow work to be carried out in private. Lounges, cafeterias, "smart" library tables, and conference rooms replace individual offices. At least half of every telephone conversation is overheard. And at the same time, workers are encouraged to communicate with each other and with customers via e-mail rather than telephone, resulting in an easily searchable electronic record of all communication.

One can derive support for the resocialing hypothesis from various additional sources. One is the apparently increasing importance to work of what are now called "communities of practice," a notion articulated through interdisciplinary collaboration at the Institute for Research on Learning in Palo Alto (Barley and Orr 1997; Orr 1996; Lave and Wenger 1994). Another is the idea that networking relations among organizations are supplementing, if not indeed replacing, competition between them (Powell 1990; see also Castells 2000a). These perspectives place substantial emphasis on an increase in the importance of the more social qualities of the relations within which work is embedded.

This increased scope for the social at work is a real possibility. In Richard Lee's terms (1998), it is a potentially significant locus of (relatively) autonomous work. It thus provides at least a vision of the separation of work from conditions of labor, at most a vision of a transformed society. If it were to come to pass, it would be a more substantive manifestation of a "postcapitalist" social formation than those conditions described by Drucker (Marquart and Kearsley 1998).

I find persuasive the ideas that globalization has proceeded too far to be reversed, and that those organizations more able to get the social right are, mutatis mutandis, more likely to survive increased competition. There is a real affordance for resocialing work. In protocyberspace organizations, are new forms of sociality displacing the colonization of work by the formal? Is the result a more genuine organizational culture? The possibilities should be taken seriously (Hakken 1999a).

The Case against Resocialing:

The resocialing case may be vulnerable on several grounds. One is that the desocialing of work attributed to Taylorism was more celebrated than accomplished, that it was an essentially failed project. To assert that AITs at work are strongly transformative merely because they increase sociality, Taylorism needs to have actually desocialed work in the first place, to have been relatively successful. Conversely, were Taylorism a more ideological than real desocialing practice, AIT-related sociality would be less generally transformative.

Opinions differ among scholars as to how successful Taylorism was, the extent of desocialing being a key point of argument. (Compare Braverman 1974 with Wood 1982, for example.) A long line of work ethnography, including the contemporary interest in apprenticeship and skill (see Barley and Orr 1997) but traceable at least from Ken Kusterer's *Know-how on the Job: The Working Knowledge of "Unskilled" Workers* (1978), documents empirically the continuing importance of worker knowledge despite Taylor. Notwithstanding efforts to deskill work, much production remained centrally dependent upon worker knowledge, largely constructed in groups. Management's inattention to its dependence on worker knowledge for value is a function of ideologically (self-)imposed ignorance rather than any real independence. Wood and his contributors (1982) made similar critiques on more theoretical grounds, using them also to attack Braverman's deskilling notion. From such perspectives, the apparent resocialing of work is nothing more than acknowledgment of a sociality always present. While such recognition may have symbolic significance, it should not be confused with actual transformation of the labor process (postmodern continuity).

A second critique focuses on the politics of the resocialing hypothesis, that the apparent liberation of increased sociality actually masks a shift in the locus of control, from a supervisory force embodied in foreman and machines (machinofacture) to a more abstract workspace culture. Gideon Kunda's account of *Engineering Culture* (1992) at "Tech," a high-tech computer design firm which many take to be modeled on Digital Equipment Corporation, makes this case. In his view, while a matrix structure has replaced the overt hierarchy of organizational reporting structures, management control is still retained, but via "normative" means. Control is accomplished by a "laying on," via the second, parallel labor process, of an overt corporate culture, literally called "the culture" at Tech. This culture sufficiently colonizes engineers' consciousness so that they normally choose "spontaneously" to do what is in the interest of capital reproduction. Their social psychology is so dominated by the organization that they can no longer distinguish between their personal and the corporate interest. Along similar lines, Jan English-Lueck's and her colleagues' study of Silicon Valley (2002) can be read as documenting the colonization of home by work. There is a similarity between this vew and sociologists' notion of post-Fordist hegemonic control or concertive control.

Kunda is pessimistic about the long-term prospects for democracy in a society where the most prestigious workers, like those at Tech, no longer have a life away from the workspace. Any "autonomous social zone" constructed by such workers is more delusion of false consciousness than resource for liberation.

More generally, factors like the strong potential for and actuality of surveillance in high-tech worksites can be used to argue that the apparent new worker culture is spurious, not genuine (Sapir 1990). Computers' very flexibility as machines of representation can be bent to the dynamics of capital reproduction, their enticing capacity to constitute an apparent "second self" an illusion of independence which obscures the reality of increased dependence. Far from demonstrating a renewed relative autonomy of the labor process, the second-wave computered worksite, with is complex communicative loops and number of peripheralized telecommuting workers, is less the site of real networking among autonomous coproducers of knowledge than Brave New World. While from the resocialing perspective, cyberfacture constitutes a distinct potential for a more humane working life, from this critical perspective it is a further, perhaps even more insidious because more easily hidden, form of degradation. It is more like nineteenth-century manufacture, with a small group of information-rich skilled workers that is surrounded by a larger group of the unskilled, information-poor. Both groups may be being used by the very machines they like to think they are using.

In short, the resocialing issue focuses study of the transformative implications of AITs for organization. Through fostering a resurgence of sociality at work, AITs could displace one hundred years of managerialism and thereby enable broader social transformation. If merely creating an illusion of sociality while more effectively masking external control of work, however, AITs at work are primarily an ideological means to perpetuate the organization status quo.

Relevant Practice: Knowledge Networking Oneida County

Phenomena like knowledge management fatigue syndrome suggest that, so far, AITs at work fall far short of the first, transformative alternative. Yet later technological interventions can be informed by history and be based on a more nuanced understanding of the real place of knowing at work. Will next-generation knowledge technologies support real resocialing, not merely its illusion?

The Relative Autonomy of Knowledge Networking in Public and Not-for-Profit versus For-Profit Organizations

The Knowledge Networking Oneida County (KNOC) project was an intervention to explore the limits of knowledge adaptation via computing. The location of the KNOC project in the public/not-for-profit sector of the economy was an important part of this limit-testing capacity. One of the problems of studying knowledge reproduction in large private firms is the ever-looming presence of the "bottom line" as the standard of success. Knowledge management projects are seldom evaluated from any other perspective, such as the quality of the adapted knowledge, just as information technology purchases are seldom the object of cost-benefit analysis (Hakken with Andrews 1993). Neither the knowledge "managed" nor the social relationship created is examined independently. Rather, their quality is presumed to be reflected directly in profit or loss.The indexical quality of profit or loss is, like the necessity of a distinct group of managers, an assumption seldom questioned in the organization studies literature by any of the organization studies writers, even those praised below like Davenport and Prusak.

The arguments for resocialing suggest instead that the structures presumed by such assumptions stand in the way of truly extending knowledge adaptation networking in organizations. While the "bottom line" orientation does interfere with evaluation of knowledge engineering in the for-profit firm, this is less the case in the public or not-for-profit organization. Since they have no "natural" bottom line in the sense of profits, publics and not-for-profits cannot rely on this to measure knowledge networking success. To mimic the standard OS testing of knowledge networking, these organizations must self-consciously invent quasi–bottom line indicators. There is no compelling functional reason to try to run these organizations along capitalist lines. While some do try to construct a shadow notion of "profit" as a standard against which to measure success, these efforts are generally too hackneyed to be very credible. They are appropriately related to as mechanical, ideological "lay ons," forced applications of an outmoded organizational science paradigm.

Publics and not-for-profits are less in thrall to capital. They are in general less preoccupied with its reproduction and more with other things, like keeping politicians popular. Indeed, in a democracy one expects government to serve the citizenry and voluntary organizations to serve member or client, rather than corporate interest.

For these reasons, publics and not-for-profits are of at least as much empirical interest to the knowledge ethnographer as large private ones. Theoretically, such organizations present a good empirical terrain on which to decide what works and what doesn't in using automated information technology to facilitate knowledge networking. By picking her sites carefully, the knowledge ethnographer can gain entrée to processes of greater interest. This has been one reason for concentrating my personal efforts to apply my theory of knowledge networking in publics and not-for-profits. In the Upper Mohawk Valley, these were also the sites where I have had the best opportunity to do so.

The Upper Mohawk Valley and Its Social Service Challenges

Investment in sophisticated information technologies makes little sense if only used to store bad or the wrong information, or if organizations don't know how to use the information gathered. Organizational redesigns that stress teams, virtual organizations, or interorganizational networks will contribute little if staffs don't have access to the information they need to make good decisions. Similarly, the development by a county mental health office of integrated (e.g., "special needs") plans in order to decentralize control will have little lasting impact if communities cannot apply locally meaningful standards to evaluate intervention outcomes.

For more than twelve years, I worked on such AIT issues with publicly and volunteer-funded disability, mental health, social, and employment services in Oneida County, New York. Like those in many other U.S. regions, central New York publics and not-for-profits must overcome substantial challenges if they are to provide effective services efficiently. However, several factors specific to this county provide both unique advantages and disadvantages in using AIT in this effort.

The economic history of the Upper Mohawk Valley is shaped by its settlement by Europeans just after the Revolutionary War. Oneida County, and its largest cities, Utica and Rome, played central roles in nineteenth-century U.S. development. Utica, for example, began to grow because it was the key point of transfer from the Erie Canal to stagecoach transportation for the main western migration; hence it became the origin point for the Pony Express. During the Civil War, both U.S. senators representing New York were from Utica.

One legacy of this period is a larger-than-average infrastructure of public and voluntary social service organizations. The Utica Psychiatric Center (opened in 1844 as the first New York—and second in the nation—lunatic asylum) and the not-quite-as-old Rome Developmental Center were outstanding examples of the former, just as the Neighborhood Center, one of the first settlement houses in the country, is an example of the latter.

In contrast to this nineteenth-century experience, the twentieth century was largely one of economic stasis and cultural marginalization. When facing typical urban problems, the region drew on its substantial organizational resources. In the 1960s and '70s, patients at the Utica Psychiatric and Rome Developmental Centers were deinstitutionalized. The organizational legacy of the nineteenth century made it possible to implement, under a Community Services Board, a broad system of community support services for the majority of thc 12,000 residents who resettled in local communities.

Another positive resource for meeting human needs was the ability of publics and not-for-profits to work together. An example was execution in the late 1980s of a transorganizational, participatory social needs assessment, staffed by the SUNY Institute of Technology Policy Center (O/HHNA Report; Hakken 1991). This needs assessment led to adoption of an action plan, broadly supported by private and public organizations, which set infrastructure change as the highest priority. In brief, rather than the largeish number of historical legacy organizations, fewer and more cooperative networks of organizations were needed to meet more complicated contemporary social needs. More flexible structures could address multiple problems in a coherent fashion. Building on the community services network, area social services and health care providers entered into more formal networks aimed at greater efficiency, just as public entities accelerated their efforts to share services.

These efforts could draw upon substantial local AIT expertise as well. In the 1950s, civic leaders accomplished a transition from textile production to light manufacturing. As a result, several of the early UNIVAC computers were built at a Sperry-Rand facility in Utica, and a local firm, Mohawk Data Sciences, was responsible for several important commercial IT successes in the 1960s and early '70s. Some local firms have staff that can be traced to these events, although only one of these, PAR Technologies, has sustained significant size. Several other entrepreneurial AIT initiatives developed from activity at the U.S. Air Force's Rome Laboratories. As with the local IT industry, however, none have flourished here, perhaps because they did not develop a relationship to the public and not-for-profit social institutions at the core of the regional economy.

Until the 1990s, the Upper Mohawk Valley dealt fairly well with the social problems characteristic of other urban regions in the United States. There were some relatively unique challenges, such as the presence of a large number of immigrants, the result of a refugee resettlement effort involving both Southeast Asians and people from Eastern Europe and the Balkans. However, these challenges, not generally being of a massive scale, were not overwhelming. Declining urban population meant that problems related to an aging housing stock remained largely hidden. Expanding public programs and increasing funds for relatively large organizations that got along well meant that ad hoc solutions could be found for most problems.

In the 1990s, however, this approach was less effective. Substantial changes were introduced into social service delivery

1. Decentralization of service administration, for example, through development of teams and branch and satellite offices
2. Privatization of service providers—mostly a shift from public to not-for-profit sectors, although a few for-profit providers (e.g., of alcohol rehabilitation services)—atrophied public infrastructures
3. Penetration of managed care modalities, both in third-party payments and in (largely failed) efforts to create quasi–managed care public modalities, as in "special needs plans" which shift funds from "contract for services" to "fee for services" basis, all of which was supposed to promote competition among providers
4. Declining public funds, what remained being channeled through more diverse local, state, and national streams, and an accompanying shift, both in public and foundation sectors, to grant-based rather than continuous "budget line" funding
5. A preference for legislative initiatives that press populist "buttons" but are not carefully thought out in terms of broader system impacts—for example, regulatory change in key structures, as in the Temporary Aid to Needy Families (TANF) approach to "ending of welfare as we know it"—that subverts local planning structures

While a case for each change could be made on its own terms, their interaction heightened local problems. For example, the shift of services from public to not-for-profit sectors meant that there is no accessible central repository of information on who does what for whom with what result. These developments led to substantial growth in some not-for-profits, but their growth masked a general decline in resources and services. Another result was a ballooning in Medicaid expenditures, for which county government is 25 percent responsible but over which it has no direct control. This problem may have been a direct outgrowth of efforts at the state level to respond to contraction in some federal programs by "Medicaiding" as much as possible.

Unfortunately, the local social formation had to adapt to these challenges with a contracting economic base. Manufacturing shrank through several factory closings. Problems of a declining tax base and challenges to United Way contributions were exacerbated by the shutting down of Griffiss Air Force Base and the elimination of local General Electric defense work in the early 1990s. The region had barely begun to participate in the economic revival associated with the new economy when the dot-coms collapsed. Even the rapid expansion of the prison population, which brought jobs, also meant new populations and service needs.

Service providers and public monitors responded to the new realities. Public and private organizations came together to identify and apply for new sources of funds, such as redirected TANF, savings to help with transitioning off welfare. The new funding patterns created an additional problem, however—a proliferation of ad hoc networks, each of which comes together for discrete funding opportunities. The cumulative effect is to disperse access to information, decision making, and control even further.

Still, ad hoc collaborations around specific cases can conceivably be converted into more durable connections, which through leadership can become durable organization. Organizations recognized that their individual data had to be shared to succeed in the new world of grant-based funding. Might better coordination be built on a good systemwide database? Could a demonstration "knowledge strategy" project, where specific cases of using shared information are shown to reproduce useful knowledge, foster further sharing and prompt necessary leadership?

Initial Attempts: Information Networking Oneida County

To create systemwide data, one must bring relevant organizations to the networking table and support development of network links. As an early step in addressing these needs, the Policy Center at SUNY Institute of Technology tried to implement the needs assessment follow-on in a 1994 project with the Oneida County Department of Mental Health (OCDMH). At the time, I was just joining the Community Services Board (CSB). Public authorities at the state level had decided to channel public funds for provision of mental health services through a managed care–private insurance system. Since it's quite difficult to make money from public provision to poor people, many suspected (correctly) that implementation of this system would fail, yet state Poohbahs articulated no fallback alternatives. I suggested we use computer simulation to stimulate a "home grown" second plan. That is, on the basis of participatory discussion in the CSB about the kinds of local systems we wanted in the long run, we would construct an alternative and be ready once the managed care smoke cleared.

About this time, SUNYIT announced that, as part of a technology transfer exercise, the local Air Force research lab had installed for free use on campus a sophisticated suite of process simulation, decision support software. We liked the idea of using a product developed locally at the Air Force's Rome Labs. Via adapting military command and control software to more positive activity, we could model better ways to bring managed care to terms with publicly funded mental health services. Assured that it could be used to simulate alternative organizational futures, I argued for public use of this tool that the taxpayer had helped to develop, to bring business process reengineering to the public sector.

However, several months into training, the project foundered on the inadequacy of the software. I was finally told by the trainer, "This stuff works best when you sketch your options all out with paper and pencil before trying to put it on the machine." Rather than a tool to discover things we didn't know, this simulation could at best only illustrate what had already been discovered the old-fashioned way. Time as well as $20 million in taxes had been wasted on a graphics package! An alternative effort to adapt Mental Health's existing d-base III system was also unsuccessful.

Later Efforts: Knowledge Networking Oneida County

In 1998, a broader effort, Knowledge Networking Oneida County (KNOC), was initiated at the Policy Center. The first step was to recruit a group of people with relevant technological and organizational expertise as well as a shared interest in information and change in not-for-profit and public organizations. This group strove to develop a shared view of how to prepare organizations for the future, summed up in the phrase, "flexible, participatory codesign." This meant letting participating organizations and members chart the development path by identifying the appropriate goals and next steps. While organizations led, the Policy Center's role was to support their initiative by ensuring access to the relevant AIT technical, organizational, and policy expertise and promoting a high level of interaction between the group and county knowledge networkers.

We began again, this time with an Oneida County Department of Mental Health mandated data exchange project. The first step was to develop "c-info," a customized database using Microsoft Access™, to handle required quarterly reports from not-for-profits contracting with the Oneida MH Department.

Once up and running, based on the design participation of personnel from several local organizations, c-info data were to be merged with other data to support application for grant funds. For example, the New York State Department of Labor issued a request for proposals to increase local capacity to identify individuals who were already clients of the public mental health system and were at risk of losing TANF benefits. Once they were identified, a coordinated set of services could be designed to support them. Identifying these individuals meant altering c-info to amalgamate data from two (state-level) public agencies and from additional local not-for-profits.

In the midst of these changes, the Assisted Outpatient Treatment (Kendra's) Law was passed by the New York State legislature. This law assigns responsibility to the mental health system for administering multiple court-ordered services, up to forcible medication and institutionalization. Additional functionalities and design specifications, to accommodate the substantial exigencies of fast-track implementation of this piece of legislation, had to be retrofitted into c-info.

The Policy Center KNOC group treated all of these events as occasions to support OCDMH as it discovered and implemented its own uses for new and additional information. At the same time, we began trying to align intra- and interorganizational information procedures. As the potential of this locally based infrastructure became more widely known, others quickly visualized additional uses. Combined with information from other sources, c-info information was sought for grant applications. The potential of additional reports stimulated the department to launch a new program. This brought together the Community Services Board, both public and private agencies, and consumers through the Mental Health Subcommittee of the CSB around ambitious plans for new outcomes measurement initiatives. The existence of c-info put Oneida County at the forefront of complying with Kendra's Law–related mandates. The KNOC project was presented twice to the annual New York State Institute on Mental Health Information Management and at numerous other gatherings of state and county bureaucrats.

This experience demonstrated the advantages of a more comprehensive strategic approach in moving from information to knowledge networking. On the basis of both the initial successes and the problems revealed, the Policy Center group developed a plan to expand KNOC. Building on initiatives already supported by the county and local private service providers, we presented a broader community knowledge strategy with the following elements:

1. Local development of a customized, flexible, multifunctional, privacy-protecting, web-accessible database (an expanded "c-info") about recipients, services, and needs

2. Broader and deeper "bottom-up" participation of public and not-for-profit organizations in the expansion and continuous revision of c-info
3. Expanded use of c-info, as a basis for, for example, internal organizational self-monitoring as well as joint grant applications; it could be adapted as a platform for organizational accounts
4. Integration of information derived from this database with that from other regional, state, and national sources, to provide better data for county administrators, policymakers, and community representatives
5. Coconstruction by "knowledge teams," composed of stakeholders in the local social service system, of new knowledge based on this information to address specific informational and/or policy needs
6. Preparation by these groups of more specific policy initiatives, using all these capabilities: for example, to assist the Community Services Board in designing meaningful outcomes measures
7. Promotion of greater cooperation among county organizations in both public and not-for-profit spheres, as well as more effective competition for external funds, as the network of local proficiency in public/nfp organizational development and use of automated information technology expanded
8. Formulation of innovative legislation on the state level; as more was learned about how the public/nfp hybrid system actually worked we would be more able to address problems like the extent to which current legislation and funding practices actually encourage organizations to "capture" clients, rather than reward them for maximizing client independence
9. Nurturance of opportunities for product development and additional private sector job development in software and consulting in the one area, social service, in which its history gave Oneida County an advantage

We pitched a two-year project. In the manner of existing, joint-venture biotech/pharmaceutical firms, we specified a set of "milestones," events whose accomplishment would be taken as indicating that the project was "on target" for the first year; second-year funding would be dependent upon their attainment.

Unfortunately, the county agency heads to whom we delivered this pitch decided it was too ambitious and declined to push its funding. Applications for alternative outside funds were prepared and submitted to the U.S. Commerce Department's Technology Opportunity Program and the NSF's Information Technology Research and Partnership for Innovation Programs. With the assistance of the county executive's office, two requests to finance small pieces of this broader project—an expansion of the data collaboration between Oneida County's DMH and its Department of Social Services, and a model of system incentives for State Senate Social Services Committee policy purposes—were prepared for member item sponsorship by our local state senator, who chaired the State Senate Social Services Committee. We also contacted the Center on Technology and Government at SUNY Albany to explore options for partnering. Consistent development of such a complex strategy, even a participatory one, requires an adequate staff as well as financing base, so I also applied for a one-year researcher position at the Oneida County Social Service Resource Center, housed at the Leavitt Center for Public Affairs at a local private, Hamilton College. The position would have allowed me to pursue coordination of KNOC full-time. Even though the Hamilton position was partially funded by county money and county personnel sat on the selection committee, another person was picked.

All these efforts failed to produce the additional funding for strategic knowledge networking. Other problems with this initiative followed from funding difficulties. Development of the strategy had to rely substantially on volunteer effort, as well as on serendipitously available funds through the Department of Mental Health. While the Policy Center group as a whole ceased to function, two of its members set up small consulting firms. While over $250,000 of public money has been used to purchase equipment, design and implement the c-info database, prepare reports, and process data from other sources, no funding commitment has extended beyond six months. At the time of writing, the c-info was finally moving toward a beta or "real-life" test, but staff members were also quitting to find more secure employment.

Results

The primary activity stimulated by KNOC was AIT systems development. C-info has been expanded into c-info::TxPIM, which includes additional functionalities (treatment tracking and personal information management) and a dual structure or architecture. A knowledge strategy like KNOC depends upon a high level of cooperation from private nfp as well as decentralized public organizations, only some of which can be mandated. An initial reluctance of the chief executive officers of organizations to share fully data on all their clients led to a decision to create separate workspaces in c-info for each organization, as well as one for the entire database, administered by OCDMH. The basic principle was that you got access to aggregate data in relation to the proportion of your data that you put in.

These expansions are only one facet of a most difficult systems development process, called "normalization" locally. Essentially, it has been necessary to construct a system that maintains much organizational autonomy and ambiguity in data fields in order to entice "buy in" from autonomous organizations. For such initiatives to work, organizations have to get more useful information without large extra cost. Technically, this means a web interface, a step which itself involves development of complex security arrangements, integration of these arrangements with existing county firewalls, and substantial staff training.

As discussed in chapter 6, this approach is an advance over standard system specification. Still, progress toward a viable information infrastructure has been effectively stymied by the need to adapt constantly to new demands without sufficient staff. The infrastructure has also been changed to separate mental health from other county data, and to this extent KNOC had become part of the information segregation problem it set out to solve.

Better knowledge networking would help these organizations be more effective, as well as those like me, in my role as chair of the Community Services Board, trying to make sure services meet community's needs. With development of an appropriate array of outcomes measures, we would be able to help the recipients of services use the knowledge reproduced to make more informed choices about the organizations from which they will receive services, to actually exercise the consumers' power which was to be the chief benefit of privatization.

Unfortunately, creation of structures for active knowledge reproduction has been even slower than information infrastructure development. Knowledge adaptation depends on prior consensus regarding desirable outcomes measures, as well as support for transorganizational knowledge building teams to guide the development process from the bottom up, but the bulk of funding has gone to the technology. C-info has become an elaborate management information system, albeit one with potential to be quite successful in the long run. Still, it is not clear that first priorities—maintenance of work to date by documenting c-info, training permanent staff, and working out a coordinated plan for regular expansion—have been achieved. This can only be accomplished as part of the development of a broader knowledge plan for the County.

Long-term success for KNOC-type initiatives depends on real policy innovation. At a local level, this depends on some initial, high-profile successes. We were hopeful that one of these would be use of c-info, state data, and knowledge networking by an already existing cross-organizational team on Medicaid spending. This group was to develop specific policy initiatives to provide the same level of service at lower local cost. This initiative was recently abandoned for reasons having nothing to do with c-info or KNOC, but abandonment deprived KNOC of its first "success."

There are also challenges that had to be met at the state level. These include, for example, overcoming "data siloing"—the tendency of state agencies to treat data as treasure, rather than to share it. While early efforts to access state data taught much about the effort and nimbleness necessary to achieve a richer "silage mixture," volunteerism and serendipitous funding have not proved sufficient on this level.

An undertaking like KNOC involves risks. One such risk is, at least tangentially, implication in less desirable processes, such as further compromising recipients' privacy or participating in the criminalization of the mental health system. Personally, time on KNOC meant less time for other professional activities, and the slow pace of development and the ethical risks led me to step back in 2001. A full knowledge strategy is yet to emerge in Oneida County.

The Implications for Organization Studies of the Ethnography of Organizational Knowledge Networking

The availability of good information does not improve practice on its own. KNOC and the projects like it discussed below demonstrate that only when there are substantial structures to promote the application of knowledge generated from information can the promise of IT be realized. In the absence of substantial outside support, a sufficient mix of organizational commitment and technical innovation was not forthcoming even in this favorable local context.

I am just one of several scholars who has tried to develop an understanding of AITed work and use it to enhance knowledge networking in organizations, not only to learn about technology and sociocultural change but also to help improve social process. I now summarize what our experience suggests for organization theory.

New Perspectives in Organization Studies

Despite the analytic problems discussed below, attention to knowledge has broadened empirical organization scholarship. In particular, scholarship that recognizes the substantial relevance of nonformal forms of knowledge to organization (e.g., Nahapiet and Ghosal 1998; Brown and Duguid 1998) offers an empirical basis for a rapprochement between organization studies and ethnography. These studies help flesh out the ethnological approach to knowledge reproduction networking in organization. Of this literature on knowledge in organizations, Thomas Davenport and Lawrence Prusak's *Working Knowledge: How Organizations Manage What They Know*, published in 1997 by Harvard Business School Press, is a usefully comprehensive starting point.

Workers Have Knowledge, Too!

While implicitly accepting the capital reproductive imperative of managerialism illustrated above, Davenport and Prusak do not limit knowledge to management alone. Unlike the OECD, they acknowledge the experience and know-how of employees: "Even assembly line work, often considered merely mechanical, benefits from the experience, skill, and adaptability of human expertise" (5). Indeed, their title, *Working Knowledge*, is the same term that Kusterer used to highlight the real wisdom of supposedly unskilled employees (1978).

Theories of the Firm Must Be Knowledge Competency–Based

Oriented more to organizing than organizations, Davenport and Prusak place the new interest in organized knowledge in the context of the rise of more socially substantive, newer competency- or resource-based theories of the firm. Rather than treating the organization as an economic "black box" as in classical economics, such theories hold that firms' actions should not be ignored in accounts of economic phenomena. As Janine Nahapiet and Sumantra Ghosal express it:

> Standing in stark contrast to the more established transaction cost theory that is grounded in the assumption of human opportunism and the resulting conditions of market failure . . . , those with this perspective essentially argue that organizations have some particular capabilities for creating and sharing knowledge that give them their distinctive advantage over other institutional arrangements, such as markets. (1998, 242)

Organization matters because organizations have knowledge-based functional advantages in social formation reproduction.

Tacit Knowledge Is Relatively More Important than Formal Knowledge
Increasingly, in trying to be specific about these advantages, studies of organization focus on how they derive from knowledge embedded in routines and practices as much as, if not more than, formal documents. Much of this literature takes as given Polanyi's argument about the importance of tacit as well as explicit knowledge (1983). Thus, Davenport and Prusak critique "lean production" strategies popular in the 1990s, that meant significant shedding of employees and were therefore costly in terms of discarding knowledge. They also critique standard management practices that mean hiding, rather than learning from, errors (1997, x). All this helps them avoid the error of Nonaka and Takeushi (1995), who presume (but do not demonstrate) that the task of knowledge engineering is to turn tacit into explicit knowledge.[4]

Its Deep Sociality Makes a Complex Theory of Knowledge Necessary
A considerable literature documents the difficulties of using markets to cope with firm knowledge (e.g., Coff 1999). Davenport and Prusak frame the interest in organizational knowledge processes in relation to the failure to deliver "silver bullet" business theories. They describe "business process reengineering" as only one, quite recent instance of a long "quick fix" line. None of these have worked, because it is only through painstakingly built social forms of knowledge networking that a firm successfully transforms inputs into valuable products and services (ix). Rather than performing the trope of simplistic analysis and new nostrum (e.g., "managing knowledge"), the rhetorical ploy common to popular knowledge management texts, Davenport and Prusak recognize that the very sociality of knowledge necessitates deeper inquiry.

Davenport and Prusak take as their primary goal the initiation of a serious, admittedly still preliminary discussion of what knowledge is within organizations. Among other things, they acknowledge that not only have philosophers thought about knowledge for a long time, but what they have thought is significant. (Intellectual history and sociology of knowledge were their graduate school disciplines.) They accept the importance, for example, of recognizing how different knowledge is from information and data. They are related, and the differences are matters of degree. Still, confusion in regard to them has resulted in enormous expenditures on technology initiatives that rarely deliver what the firms spending the money need or thought they were getting. Often firms don't understand what they need until they invest heavily in systems that fail to provide it.

For information to become knowledge, humans must do considerable work. They do this through comparison, consequence assessment, connection identification, and conversation. Davenport and Prusak offer a pragmatic, working definition of knowledge, as a fluid mix of framed experience, values, contextual information, and expert insight, one that provides a framework for evaluating and incorporating new experiences and information. Knowledge originates and is applied in the minds of knowers. We can get knowledge from individuals, delivered through structured media and through person-to-person contacts, from conversations to apprenticeships. In organizations, however, we get knowledge mostly from groups of knowers, as well as through organized routines. Knowledge is often embedded not only in documents or repositories but also in organizational processes, practices, and norms.

Thus, knowledge is not neat or simple. It is fluid as well as structured. It is often intuitive and therefore hard to capture in words or understand completely in logical terms. Because it exists within people, it is part and parcel of human complexity and unpredictability. As an asset hard to pin down, it is both process and stock (Davenport and Prusak 5). There are complex processes by which knowledges develop, as in association with the authoritative, for example (B. Jordan 1998).

Scott Cook and John Seely Brown have written about such complications in "Bridging Epistemologies: The Generative Dance between Organizational Knowledge and Organizational Knowing." They begin with a comment from Sir Geoffrey Vickers:

> It's funny what's happened to this word *knowing* . . . The actual *act* of apprehending, of making sense, of putting together, from what you have, the significance of where you are—this [now] oddly lacks any really reliable, commonly used verb in our language . . . [one] meaning the *activity* of knowing . . . [Yet], every culture has not only its own set *body* of knowledge, but its own ways of [knowing] . . . (1999, 121; italics and insertions in original)

In this comment, an experienced British CEO echoes the etymological discussion of "to know" of chapter 4. In so doing, he, too, shifts attention from knowledge as a thing to knowledge networking as a process.

Unsuccessful Organizations Have Only Shallow Skills at Dealing with Knowledge

Davenport and Prusak's secondary aim is to figure out what to do with our new knowledge of knowledge. They hope to help make knowledge more effective, efficient, productive, and innovative. What key cultural and behavioral issues must we address to make use of it?

Interestingly, they trace their real interest in knowledge to a discussion of managers in successful corporations who said, "We really have no real idea how to manage value-added information and knowledge in our companies." Theirs were high-tech firms that had helped launch the so-called information revolution! Davenport and Prusak themselves admit to not having effective methods and approaches for managing and understanding how to better use the information. (Knowledge society indeed!)

Instead, what these managers wanted, and what Davenport and Prusak try to give them, was insight into best practices, creative synergies, and so on. Much of the knowledge they needed to make more effective use of knowledge already existed within their organizations, but it was not accessible or available when needed. Knowledge reproduction systems existed internally, but they were inefficient. Knowledge was unrecognized, disorganized, local, and often discouraged rather than fostered by the organizations' cultures. Davenport and Prusak acknowledge a need to study how knowledge is managed, mismanaged, and unmanaged (xii). Thus, like this chapter and chapter 3, their account of knowledge in organizations is much more a critique of current organizational practice, including its AITed versions, than a celebration of transformative change.

Organization Is about Knowledge Networking

In specifying what they mean by knowledge, Davenport and Prusak recognize that what *an organization and its employees know* is at the heart of how an organization functions. (Note that for them *this has always been the case*; it is not the result of some recent, AIT-related innovations.) In describing its knowledge as telling a firm how to do things and how they might do them better, Davenport and Prusak accept the fundamental point that knowledge is social, as much a property of collectives as individuals. A firm's ability to produce knowledge depends on what the collectivity of people currently knows as well as the knowledge that has become embedded in the routines and machinery of production. They quote Sidney Winter to the effect that firms are "organizations that know how to do things." John Seely Brown and Paul Duguid are especially strong on the sociality of knowledge:

> Assuming that knowledge is a frictionless commodity possessed by individuals makes communications technologies and social organization curious antagonists. [We argue] instead for compatible organizational and technological architectures that respond to and enhance the social production of knowledge. (1998, 90)

For such authors, in an important sense, its knowledge actually *is* the organization; at least it is the continuity that allows particular organizations to thrive over time. Similarly, Kogut and Zander propose "that a firm be understood as a social community specializing in speed and efficiency in the creation and transfer of knowledge" (1993, 392).

This perspective is indeed a different way of thinking about what an organization really is. Under Taylorism, the chief metaphor for an organization was the machine, a physical entity to be formalistically engineered. An extensive literature, to which Davenport and Prusak point, critiques the problems with trying to act as if an organization is a machine.

One possible alternative metaphor for organization is the life-form, implicit in Chandler's notion of the functional differentiation of large corporations, and more explicit in some notions of "adaptation" tightly connected to biology. Because life-forms adapt primarily through natural selection, this metaphor is less useful than that of culture—for example, "organizational culture." Organizations, like human cultures, do adapt cognitively. Cultures, however, largely develop "behind the backs" of those who practice them, while organization involves consciousness as well as cognition.

Instead, work like Davenport and Prusak's suggests an even more powerful metaphor: *Organization is knowledge networking, and, thus, at their core, organizations should be approached as knowledge networks.* Modernistically, organization makes sense as a means to foster the collection of data, its recording as information, and its reprocessing into knowledge. Postmodernistically, organizations adapt this knowledge to apply it, collect more information about what happened during application, evaluate the quality of the knowledge based on the outcomes, reapply it, and so on. Not only *is* the company its knowledge: the reproduction of knowledge is most basically what organization is *for* or, at least, that which it most sensically *should be.*

I say "should be" to highlight that this is often not the case; various aspects of an organization's functioning often get in the way of knowledge networking. Kusterer, for example, documents how the ideology that work rewards should go to formal, credentialed skills, rather than substantive ones—"know-how on the job"—forces workers and managers to relate to each other in inefficient, more or less fictional ways (1978). David Moberg's dissertation research similarly pointed out numerous ways in which the logic of capital undermines the logic of making good cars (personal communication).

Theoretically, thinking of organization as basically about knowledge networking illuminates how organization impacts the reproduction of human social formations. For an "organization" as opposed to an "organizational" studies perspective, it has the added value of highlighting the underlying similarities between employment-based and non-employment-based organizations. A global informational infrastructure makes more sense when approached as a means to foster knowledge networking than as a means to reproduce capital. The dot-com experience is an example of how valuable forms of AIT-mediated networking are undermined, not because they failed to organize but because they do not make profits for venture capitalists.

That AITs' Role Is Fundamentally Ancillary

Crucially for "cyberspace" discourse, Davenport and Prusak label as mistaken the presumption that technology could replace the skill and judgment of an experienced human worker. They call the idea that technology can replace human knowledge something proved false time and again. At most, technology can help knowledge networking in organizations, by providing new ways to exchange information and new tools (for example, Lotus Notes and the World Wide Web) that make certain forms of data easier to collect, store in repositories, and distribute. This is only of substantial value when organizations also develop appropriate approaches to the new forms of communication. This means recognizing, as Poitou says, that

> it is impossible to represent the knowledge inherent to a document or to an artifact independently of the place where they are [sic] in use . . . The computer, then, will be more helpful as assistance to documentation, rather than to knowledge representation . . . What is really needed in order to make the knowledge . . . in the corporate technical heritage more readily available to its members . . . is assistance in preparing, storing, retrieving, and processing documents (i.e., editing knowledge . . . instead of eliciting knowledge). (1996, 247–248).

Technological innovations are at most ancillary to the main aspects of knowledge reproduction in firms (Roberts 2000). Further, technology's potential can only be realized if organizations understand much more about how knowledge is actually developed and stored. These ideas, already the central conclusions of numerous earlier efforts at intelligent implementation of computing systems—such as participatory design, culture-centered computing, and computer-supported collaborative work (Allwood and Hakken 2001; Hakken 1999a)—are on their way to becoming central tenets for careful OS scholars of knowledge networking in organizations (e.g., DeSanctis and Poole 1994).

Necessary Alterations in OS Paradigms

Despite its initial commitment to Taylorism, organization studies has long had trouble with strong versions of Taylorism. This led, for example, to the early development of a "human relations school." Virtually all contemporary OS texts distance themselves to one degree or another from the "We pay you to work, not to think" rhetoric at the core of Taylorism.

In practice, however, some aspects of Taylorism have proved particularly difficult for OS to shake, such as its nonreflexive focus on only the jobbed aspects of organizations. KNOC's failure to accomplish more was due in part to its own dependence upon volunteerism. The voluntary work involved in formal social service organizations, let alone the work done through informal networks and done themselves by individuals and families in need, was not addressed in formal tools, even in c-info. Such perspectives and activities need to be incorporated into really useful knowledge networking.

Making organization formal is only one approach to organizing. Forming or joining a church, as opposed to merely entering into social relations with spiritual fellow travelers, makes sense to the extent to which a formal organization makes likely the adaptation of more of the knowledge that I and/or my fellows want. Similarly, mediating social relations via the employment nexus makes sense only when there is sufficient functional payoff at some social level, if not necessarily in each organization. Jobbed formal organizations are just one type. Thinking only about jobbed organizations, especially only about large, private sector ones, distorts in some general ways our understanding of organization.

Organization studies' overcommitment to formalism is also manifest in its preoccupation with quantification and the other aspects of the positivist program implied by the notion of a management "science." The model for management knowledge was captured explicitly in Taylor's and Gilbreth's articulation of its goal as the discovery of "THE ONE BEST WAY." This view still has echo, as in the title chosen for a newish journal, *Organization Science,* as well as in the practice of some of its writers. The commitment to formalism also follows from the fact that most organization scholars apply knowledge developed initially in social science disciplines themselves still struggling to overcome scientism. Attempts to develop the cross-organizational "knowledge teams," essential to transforming c-info-type information into social policy, are hampered by presumptions that their formation makes sense only *after* "hard, nationally normed" outcome data had been produced. I was not successful in my advocacy of an alternative approach, that such teams be built initially on clarification of their own ideas about desired outcomes, which would set the measurement agenda.

The orientations toward formal, jobbed organizations and quantified data are only a few of the many ways that Organization Studies, like other essentially engineering disciplines, (but not necessarily, I like to think, social science), limits itself to solution spaces bounded by largely unquestioned assumptions. In what follows, I address other assumptions that hamper the ability of OS scholars to explore and intervene effectively in the emerging knowledgescapes of cyberspace.

The "Management as a Necessary Role" Assumption

In essence, management is the idea that any organization beyond a small size and/or a modicum of complexity in functioning needs to be run, and this running should be done by those occupying specialized, power-concentrating roles as decision makers. Its presumption that management is "natural" is indicative of how American culture privileges management in a manner similar to the way it privileges large organizations. The management presumption is written into employment law; for example, one temporarily but effectively surrenders important rights of citizenship when entering wage or salary "slavery."

On those rare occasions when this presumption is considered consciously, management may be given one or another functional justification. For example, it might be argued that a group of specialists with powers of command are necessary if the difficult problems of coordinating large groups are to be handled quickly and their solution carried through in a sustained manner. Conversely, one could accept that other entities—a board of directors, a works council, a board of legislators, or a town meeting—could decide what to do but still argue that professional managers need to coordinate how it is carried out.

Functionalist justifications for roling management are vulnerable on several grounds. A long tradition within American political rhetoric celebrates the greater benefits of direct, participatory democracy of the town meeting variety. Just as significant is the celebration of voluntary, direct action, as initially praised by de Tocqueville. Such celebrations of democracy are often described as essential to obtaining the modicum of commitment from the citizenry necessary to the American state. Rhetorics of "power to the people" and "we can do it ourselves" are similarly corrosive of knee-jerk managerialism. There are still arenas of American life in which shared decision making and control on a substantive basis of equality is more than rhetoric: within many voluntary and/or intimate relationships, for example, or in relatively autonomous work groups.

If workers/users/investors/citizens, and so on, are to be involved in deciding what to do, aren't they just as significant in carrying it out? Virtually all studies of participatory design and participative work organization stress the stronger commitment to the organization that they produce (e.g., Ehn 1988).

The acknowledged importance of teams and other more collaborative alternatives to organized hierarchy is an additional recognition of this democratic ethic that can only weaken Taylorist command managerialism further. "Knowledge at work" talk that identifies an organization's employees' knowledge as its most significant resource has led many in OS, for example, to replace "decision making" with "facilitating groups" as the prime managerial function.[5] Thus, there is now an overt contradiction in any talk about organization that stresses maximizing knowledge networking while still accepting hierarchical managerialism. In both conceptual and behavioral terms, it makes sense to see a strong connection between general social participation and participation in knowledge reproduction at work. This is attested to by the considerable research on the conditions under which teams work and those in which they don't (e.g., A. Jordan 1998; Baba 1999; Gluesing 1998).

Thus, managerialism can no longer be simply assumed in Organization Studies. Rather, the potential risk to better knowledge networking of a hierarchy of powerful decision makers and implementers should be acknowledged and its functional necessity demonstrated empirically in each particular organizing case.

The Subordination of Organizational to Capital Reproduction

A bottom-line fixation has limited evaluation of knowledge reproduction interventions in organizations. In an influential article addressing "What Do Bosses Do?" political economist Stephen Marglin extended the critique of managerialism beyond the idea that its necessity was merely an unwarranted assumption (1974). Marglin problematized the various functional justifications for management in corporations by demonstrating how, for example, a formal managerial structure is as likely to interfere with coordination as to facilitate it. If management doesn't make formal sense, why is it ubiquitous in for-profit organizations?

Marglin offers a reproductionist alternative explanation. Managers are dominant in such organizations not to make sure that the work gets done efficiently, but to ensure that organizational activity results in the expanded reproduction of capital. We are repeatedly told, for example, that a manager's primary responsibility is to stockholders. It makes sense to see this as both the reason for her power in an organization and the best explanation, in the long run, of how she will use it. As a long tradition of labor process scholarship (e.g., Braverman 1974) has demonstrated, however, the manager's allegiance to the "accumulate, accumulate, accumulate!" imperative of capital can interfere substantially with both efficiency and effectiveness. For Marglin, far from encapsulating a functional necessity, managerialism is an ideology masking the imperatives of capital reproduction.

The point of large private organizations is to make as much profit as possible; they are formally in thrall to capital. On its face knowledge networking conflicts with this thralldom, as can been seen in conflicts over "intellectual capital." A security posture toward knowledge makes sense as a management effort to advantage capital. However, "intellectual capital" is oxymoronic from the point of view of knowledge networking, since sharing knowledge *contradicts* treating it as a scarce resource. Maximal internal knowledge flow may be incompatible with a primary loyalty to protecting the property of individuals outside the organization. To the extent that the imperatives of capital reproduction curtail sociality, this basic contradiction may actually be *heightened* in the AITed organization.

The substantive situation of the labor process itself, however, is different. The chains of capital slavery are not all the same; large aspects of organizational life, especially knowledge reproduction networking, may and perhaps must operate in relative autonomy from the reproduction of capital. Thus, the extent to which a capital reproductive imperative interferes with knowledge networking is an empirical question. A knowledge theory of organization has to recognize this possibility and develop an analysis of it.

The Misapprehension of Knowledge Adaptation Networking

Organization studies could be of even greater value to the promotion of knowledge networking were its theory of knowledge networking not in thrall to the hegemony of capital. Scholars like Davenport and Prusak acknowledge that it is not the *importance* of knowledge at work that is new, but its explicit *recognition as a corporate asset*. Whence this recognition? For them, it is driven by the need to make the most of existing assets in a world of global competition and increased consumer information. They assert that:

> Today, for the most part, it is virtually impossible to prevent competitors from copying and even improving on new products and production methods fairly quickly in an era characterized by mobility, the free flow of ideas, reverse engineering, and widely available technology . . . A global marketplace for ideas had developed and there are very few concepts and formulae that are not generally available. (15 and 16)

That its first part is a tautology is one aspect of the curiousness of this statement. For there to be such "transparency" in knowledge capital, knowledge must be being managed exceedingly well. Yet as indicated above, much of Davenport and Prusak's argument is based on its failure! Moreover, such transparency would only exist if institutions like patents and copyright had lost their importance to Microsoft, the U.S. Congress, and the World Trade Organization. Their account of new interest in knowledge also ignores the substantial increase in the power of transnational corporations—vis-à-vis governments, voluntary organizations, trade unions, small businesses, and national capital—among other things, to protect intellectual capital.

In Davenport and Prusak, mere possession of knowledge, bereft of any context, is presumed to provide the key sustainable advantage. Assumed away is the relevance of other mechanisms for pursuing the reproduction of capital, like price wars, war in general, advertising, lobbying, drug running, tax fraud, monopoly, bribery, and so on. Instead, "By the time competitors match price and quality, knowledge-rich, knowledge mining companies have moved on to a new level" (ibid.).

How comes it that, for Davenport and Prusak, "the knowledge advantage" generates such increasing returns and continuing advantages that real-world impediments can be ignored? This is because knowledge assets, unlike others, *increase* with use: Not only do ideas breed new ideas, but also shared knowledge *stays* with the giver while it *enriches* the receiver. They quote Romer: "In a world with physical limits, it is discoveries . . . that make persistent economic growth possible. Ideas are the instructions that let us combine limited physical resources in arrangements that are ever more valuable" (17).

In other words, previous worlds had physical limits, so knowledge could not then be a philosopher's stone. AITed knowledge, however, has been liberated; the mystical, ability-to-create-something-from-nothing, "cornucopia" role assigned by Adam Smith to the invisible hand in the market, and later by the neoclassical economists to capital, has now been assigned to knowledge. Capital, in the traditional sense of wealth invested to return a profit, no longer needs to be presumed mystically to be self-expanding. So how can it, and the continuing power of its representatives in social reproduction, be justified? It is conceptually reinvented as "intellectual capital."

To a substantial extent, a large number of OS studies of knowledge in organizations merely work out this or similar ideological, not empirical, imperatives. The necessity of subordinating organizational knowledge networking to capital reproduction, while simultaneously masking this subordination in the formation of value, is the contradiction at the source of much of the "hype" about knowledge and AITs. This contradiction undercuts the ability of OS to develop a really useful theory of the reproduction of knowledge in organizations. Development of an alternative political economy more applicable to organizational knowledge networking is a task returned to in Section IV.

Conclusions

Resocialing Redux

What does this journey into contemporary organization studies suggest about the extent of work resocialing? Of course, there is no simple sense in which one can extrapolate from the characteristics of contemporary workspaces to those of the future. Thus, conclusions about prospects for transformation must be tentative.

It is salutary that the most recent developments in organization studies mean knowledge is at the center of its intellectual agenda. However, there is little basis for claiming that this interest in knowledge constitutes a quantum leap in knowledge reproduction networking or general sociality at work. Illustrative of the current state of OS is the perspective articulated by Alan Belasen in *Leading the Learning Organization* (2000). This book is in many ways a useful compendium of contemporary thoughts on organization, knowledge, management, and how these are influenced by information technology. There is nonetheless an explicit paradox at the core of the book, one outlined in its theoretical chapter, titled "Integrating Management Paradigms: The Competing Values Framework" (CVF). Here, Belasen puts into a single chart the numerous roles he deems necessary for contemporary management, but performing them all simultaneously is profoundly difficult. For example, "despite the fact that in playing the role of the Director, the manager is assumed to have a task orientation, some aspects of facilitation or even mentoring (involving the two roles at the polar opposite in the CVF) must also be manifested [*sic*] in the behavior of the manager to achieve effective managerial leadership."

Belasen acknowledges the organizational structural sources of this role conflict: "A key point of the Competing Values Framework is that organizations are inherently contradictory entities, and therefore organizational effectiveness criteria are fundamentally opposing, contradictory, and may be mutually exclusive . . . .[O]rganizations operate under the burden of contradictory, competing, and conflicting expectations." Is it impossible to be a modern manager? Instead of critiquing this situation by seeking out and addressing the roots of such a prime irrationality, Belasen embraces, even celebrates it: "Responding to these expectations is vital for the success (effectiveness) of the organization." Rather than suggesting how these potential conflicts might be resolved or at least "managed," he merely names the agents who must be responsible for reconciling the irreconcilable: "Obviously [*sic*], this burden is not borne by organizations per se, but by the managerial leaders in these organizations."

In this "obviously," Belasen reveals his implicit commitment to managerialism. And how are managerial agents to do things that he well admits may be mutually exclusive? "[E]ffective managers are careful to remain within the positive zone and yet have the behavioral complexity needed to move from one set of competing values to another. To become master managers they must frame and reframe or maintain situational flexibility and possess the requisite skills to move comfortably between the various domains represented . . . " Indeed, to behave in a contradictory manner is the essential attribute of leadership, the CVF being "a universal image of the repertoire of roles manifested [*sic*] in the behaviors of effective managers . . . ."

Far from fostering the reproduction of organized knowledge, Belasen's manager is a Wizard of Oz, creating and manipulating diverse, diverting representations of the actual organized situation. She performs the master misrepresentation, that her profoundly incoherent behavior is actually coherent. One kind of knowledge is relevant to this performance, but it is the peculiar psychological knowledge of the heroically self-absorbed. Thus, "[t]he value of this framework is in allowing managers to pursue a systematic journey of self-directed learning and self-development" (all quotes from 2000, 33, 34, and 37). The actuality of managing is learning how to get good at dissembling and laying on a culture of apparent autonomy but actual control. In this regard, Belasen accepts the arguments of "antiresocialers" like Kunda. Again and again, the reader is forced to conclude that the attribute most important for the successful manager is a mollifying charisma, not the ability to extend knowledge reproduction.

We generally have sympathy for those forced into contradictory roles, but we are properly suspicious of those who embrace them! As long as it remains beholden to such managerialism, organization studies will find it difficult to develop a convincing account of knowledge adaptation networking. There is thus little evidence in such accounts that knowledge networking in jobbed organizations has escaped its unarticulated subordination to capital reproduction.

Publics and Not-for-Profits as Sites for Studying Organizational Knowledge Networking

The KNOC experience made me less sure of the unique value of publics/not-for-profits with regard to knowledge research. Such organizations are indeed less tied to the reproductive imperative of capital and therefore managerialism, but they are tied to other imperatives that interfere with knowledge networking. Formally, state bodies are accountable to the public, while nfps often combine membership with public accountability. Practically with regard to social service, public forms of accountability seem to come down to something like "electability," which means convincing individual voters that things are fine. One makes sense in terms of simplistic, stable political ideologies. In KNOC, we couldn't convince any local politician that moving from information to knowledge networking was enough of a priority to justify spending serious money on it. One state staffer explained our failure to obtain government funding by saying "politicians just don't see your issue; it's not concrete enough."

Other imperatives also distort knowledge networking in publics and not-for-profits. Oneida County shares the New York State tradition of strong ethnic identity dynamics, but these are generally channeled into political machines or their vestiges. The county lacks a tradition of strong community organizations, and thus of accountability to them; in many cases, nfp staff find themselves acting in lieu of such. For the larger ones, this is a tolerable, even comfortable situation. Ad hoc involvement in short-term grant networks allows them to find new sources of funds to replace disappearing ones, while the need for common information is not so great as to justify more serious inter-organizational knowledge networking. Because grant funding tends to be shorter term than the preceding program funding, even less attention is given to evaluation. To participate actively in real knowledge networking with staff and members takes time and opens up administrators to additional questions about their prerogatives.

In the absence of high-profile political leadership, administrators in both publics and nfps seem generally content to operate on what is locally referred to as "the mushroom principle—keep them in the dark and feed them garbage." Nor is there much reason to expect state-level elected officials or administrators to become strong advocates of real knowledge networking, a step that would require abandoning similar aspects of current practice in relation to localities. Thus, while these reproductive imperatives may not be theoretically as strong as those surrounding capital, the KNOC experience offers little reason to expect more fundamental changes in the reproduction submoment of knowledge networking in publics and nfps as opposed to in private organizations. In short, my ethnographic study of organizational knowledge networking suggests that "really useful knowledge" will be a long time in coming, and that AITs will lead only infrequently to new knowledge, let alone transformative changes in its functions.

Additional "Lessons Learned"

Nonetheless, study of how organizations reproduce knowledge is still of value to those who would use AITs to extend this process.

Building Community and Organizing Knowledge Reproduction Are Not the Same

It is now common to describe organizations as communities of practice and, therefore, to see, for example, organizational learning (Argyris and Schön 1978) as something which depends upon community. Scholars like Kogut and Zander (1992) describe firms as knowledge communities, while practitioners like Donald Norman conclude that promoting community at work automatically promotes knowledge networking (1999).

While performances of community and the networking of knowledge have substantial similarities, as argued at length in the conclusion to the previous chapter, the differences are important enough to keep them distinct. We need to study these two kinds of activities in wider varieties of their forms before we can say with much certitude what kinds of information technologies are most supportive of each. The efforts of contemporary organizations to come to terms with knowledge are thus far unfortunately largely of negative interest to this broader project of improving "real" networking of knowledge.

Reducing problems of knowledge reproduction networking in organizations to "building community" does not help much.

Identity Formation, Education, and Organized Knowledge Networking

Knowledge networking is somewhat like individual identity formation, the process through which individuals often formulate competing, even mutually exclusive notions of their selves, frequently via discourse. Individuals also act out multiple self-concepts across their interactions with intimates and broader networks of others. (How identity formation is mediated by AIT has been the focus of much research (e.g., Turkle 1995; see Hakken 1999a, 2004 [with Soeharto] for a summary).

However, while it is perfectly possible for individuals to perform multiple selves, knowledge networking requires a relatively collective stable set of approved practices, for example, shared evaluations of the quality of particular performances. As a social process in this sense unlike individual identity formation, knowledge networking permits only limited degrees of internal evaluative conflict, certainly not incoherence.

This is also an important way in which knowledge networking is similar to education. Here, it is also difficult to proceed without a modicum of coherence. This is acknowledged in terms like "a body of knowledge." A coherence presumption is at the root of "banking" theories of education, ones which presume an unknowing learner, and a knowing teacher, whose job it is to "fill the learner's mind" with an agreed-upon body of knowledge.

In both its disputatiousness and its extensive metadiscourse, however, actual knowledge networking differs from the "stable content" conception of knowledge presumed by "banking" approaches to education. (There are of course other ways to think about education, as explored in the following chapter.) The point here is that the subtle complexity of relationships like these means that the knowledge networking moment in social reproduction cannot be reduced to the community, identity formation, or education moments. Despite the decline of knowledge management in some places, the importance of knowledge networking as a distinct moment in social reproduction means knowledge will continue to be a problem at the center of sophisticated AIT practices.

For a variety of reasons, ethnography is ever more popular as a way to study knowledge and AITs. My goal is to facilitate the kinds of theoretical accounts of knowledge that will allow fieldworkers to carry out a metadiscourse about knowledge as sophisticated as the knowledges that they would construct. A carefully constructed, cross-cultural awareness of the many forms of knowledge networking is an important prerequisite to construction of appropriate roles for IT in addressing organizational knowledge problems.

Nonetheless, if the hoped-for increased sociality of existing highly AITed workspaces continues to be specious, this should give pause to those who trumpet the liberatory potential of cyberspace as a new way to be human. If in contrast a fuller resocialing really does start to emerge in organizations, the cyberfacture labor process will become a legitimate source of inspiration for advocates of new social formations. This latter will only happen generally when those developing AITed knowledge technologies adopt the kinds of approaches to knowledge outlined here.

Chapter 9

The Cyberspace Knowledge Question in Education

Introduction

Do automated information technologies in schools fundamental change the way knowledge is shared? This chapter examines the idea that there has been a "computer revolution" in knowledge sharing—the third aspect of knowledge networking. Of course schools produce some new knowledge, especially higher education (for example, research), and some higher education units also help organizations adapt existing knowledge to new situations. Nonetheless, schools are mostly about extending the network of those willing to acknowledge what is taken to be already known, about replicating more than creating or adapting knowledge.

Formal Institutions and Knowledge Sharing

If automated information technologies profoundly impact the knowledge sharing moment, it is primarily via their increasing prominence in formal educational institutions.[1] Schooling is "formal" in several senses. Schools' practices are highly organized and take place within employment institutions. Schools' purpose is to formalize knowledge sharing, as indicated by the importance of formal, credentialed knowledge in them and the many forms of credentialing (testing, grades, graduation) that they do. Being deeply implicated in constructing the fit between the number of qualified applicants and the jobs available at each level of prestige in the labor market, schools have an important, state-certified role in reproducing social privilege.

Hierarchies of various sorts are also implicated in virtually all of the school reproduction roles mentioned. Indeed, formal education is central to how contemporary social formations attain and allocate legitimacy. Schools are central to "the myth of meritocracy," the idea that those at the top are there because of their individual quality, not privilege (Young 1994). This idea is especially important to self-labeled "democratic" social formations. When maintenance of a meritocratic sense is compromised by persistent, even growing social inequalities, schools' importance increases (Bourdieu 1990).

I have had direct experience of formal educational institutions in Britain and several Nordic countries, and my career as an anthropologist has also exposed me to their operation in several other areas in the world, including Eastern Europe and Asia. However, my formal education was in U.S. schools, where I also have worked for almost forty years.

In the United States, schools have additional functions. These follow from policy directives that schools impact general social reproduction. As articulated in, for example, George Counts, *Dare the Schools Build a New Social Order?* (1969), U.S. schools are regularly charged to achieve social changes. The perhaps unique place of schools in the reproduction of the U.S. social formation follows from conservatives' generally successful avoidance of direct state forms of inequality reduction (for example, income equalization), attacked as "social engineering." *Indirect* means to achieve social change goals, like promoting new technologies and more schooling, are an acceptable lesser evil. This means schools are tasked with everything from socializing immigrants to ending racial inequality (via busing and other forms of desegregation). Through Equal Educational Opportunity, Title IX, and the like, higher education in particular has been a chief policy instrument.

Automated Information Technologies as School Transformers

The second Clinton administration followed this social formation/cultural logic when it adopted the indirect means of "wiring schools to the Internet" as its chief strategy for addressing a perceived "digital divide." Indeed, the Clinton strategy is a good example of how, whatever its pedagogical utility, any educational technology intervention will be intended to have ramifying social implications.

As in the two previous chapters, the transformative *potential* of, in this case, computers for schools is acknowledged. Again, on the Sinding-Larsen musical analogy (1991), learning may be as changed by computer mediation as music making was by the use of formal systems of notation. Moreover, a student with a good working knowledge of contemporary computing probably does have a career advantage.

However pedagogically problematic, the Internet is a demonstrable educational resource. In an interesting way, even the collapse of the dot-com bubble may improve the Internet's educational benefits. The burst may be interpreted as indicating that capital reproduction is not well served by trying to derive profits mostly from access charges, but that an accessible mass electronic space is a more value-conducive alternative in itself. If so, the Net may mature into a permanent educational resource.

AITed education is often described in transformative terms, like these of John Daniel, the former chief executive officer of the British Open University: "[K]nowledge media—[encompassing] the convergence of computing, telecommunications, and the learning sciences—change fundamentally the relationship between people and knowledge . . . " (1999, 24). Their transformo-generative capacity is typically tied to school AITs economics; for Daniel again, "computers reduce the need for face-to-face contact. This has led to a scale increase in provision, one with flexibility and capacity to adapt to demand. Expansion has not been paid for by reduced quality" (24).

On Daniel's justificatory "knowledge revolution" trope, AITs reduce radically the salience of place to education, of schools to schooling, by making formal education infinitely scalable. Daniel imagines a higher ed future in which only a few elite schools still hold archaic face-to-face classes. Most teaching is "spaced" online, provided by a small number of megauniversities teaching at a distance. He pushed the Open University—remodeled on the private sector, as indicated by his CEO title—to be one of these, until he lost his job.

AITs as Knowledge "Mythinformators"

Because of the important social functions of formal education, changes this fundamental would indeed have important social implications. The actual impact of AITs on schooling, however, is not clear. Their potential demands our attention but also needs to be assessed. The doubtful value of the dominant metaphor for school AITs, "electronic brains," is one reason why.

This chapter develops an alternative "knowledge networking" view of the AIT/schooling relationship. Evaluating AITs in schools means attending to over thirty years now of actual experience (for example, Feenberg 1999). Appreciation of the potential for change must be balanced by careful examination of hyped analogies and the lives of real students.

There are good reasons to be skeptical of the popular celebrations of computer-mediated knowledge sharing. In 1996, Diana Forsythe observed the dark side of cyber-learning in a computer-based self-diagnostic system for people who thought they might get migraines (2001). The system prompted the "user" first to describe her symptoms. It was programmed to repeat precisely the symptoms typed in, whatever they were, as the proper indicators!

This approach, one of ontological legerdemain rather than learning, was legitimated by its developers as "promoting user confidence in the system" and thus as a proper way to "increase the likelihood of patient compliance." When it came to therapy, the system was *capable* of recommending a wide range of options. However, it only *offered* those therapies that the physician or clinic deploying the system programmed into it. Selective diagnosis and therapy were both performed *without the user knowing it.*

Its proud developers boldly described the system as "increasing control by the patient," but this electronic brain is stupefying, not educative. To replace real knowledge sharing by its mere illusion would compromise ethically all knowledge networking, whether formal or informal, not just its sharing. That people can be manipulated under the guise of "self-teaching" is indicative of the deep ambiguities in and strong contradictions of cyber-learning. Of course health information was also manipulated previously. The ignorance enhanced by this particular system is more disturbing because its manipulation is hidden by the very machine that purports to be promoting the sharing of knowledge! The user is thus even less likely to be aware of being manipulated than in non-AITed health care. Forsythe's case suggests that computered learners can end up knowing *less rather than more* about where "what they know" comes from. Such ethically doubtful computing led Langdon Winner (1984) to suggest that, far from being a "knowledge society," cyberspace may well be the era of "mythinformation."

"Computopian" views (Hakken 1999a) of the higher education implications of computing like Daniel's contrast strongly with the "compputropian" ones of David Noble (2001). For several years, under the rubric "digital diploma mills," Noble has argued for a strong parallel between the practices over which Daniel enthuses and those of the correspondence arm of turn-of-the-twentieth-century U.S. higher education institutions. In Noble's account, both these developments deeply commodified knowledge sharing and, in the process, radically attenuated knowledge sharing's relationship to the other knowledge networking moments. The terms and conditions of education work deteriorated seriously, knowledge itself was cheapened and debased, and knowledge networking became directly subservient to the reproduction of capital.

The contrasts between Sinding-Larsen's and Forsythe's, and Daniel's and Noble's perspectives puts in sharp relief the radically different possible social manifestations of cyberspaced knowledge sharing. Choosing which is most accurate is further complicated by the presence of multiple, powerfully divergent educational goals and pedagogies to achieve them.

Personal Perspectives

The experiences that this chapter brings to bear on AITed schooling include personal ones, like the two previous chapters, but this time less of formal research than what Don Schön calls "reflective practice" (*The Reflective Practitioner* 1990). My answers to the cyberspace knowledge question in education are informed by this long experience with technology, both in teaching and in education-spaced formal research. For twenty-five years, I have taught anthropology and applied sociology to undergraduate and graduate students at a technology branch of the large State University in New York. Before that, I was enrolled in educational institutions most of the time from the age of four until completing my Ph.D. in anthropology. The one exception was the time, after completing course work for an M.A., that I left school, as required by the U.S. military draft, to do alternative service as a conscientious objector. Even then I split my time between doing community education in Chicago and researching/organizing in the education sector.

Indeed, it was while a freshman at Stanford in 1965 that I first confronted computers as a social phenomenon—in their "gaming" mode. An American from a small town in the Midwest, I was assigned to share a dorm room with a Scandinavian educated in elite prep schools on the Continent. In the spring, he disappeared for a week; since he had already disappeared on previous occasions, I was not worried. When he returned, he told me that he had been playing Space Invaders on the university's new IBM 360 and "had gotten lost in hyperspace." The romantic lure of such a machine—one into which one could disappear—prompted me to sign up for my first (and only) programming class.

This early experience set a pattern of ambivalence about computing that has persisted. I marvel at the relations others construct with them but carefully protect my own identity from relating too closely. Psychological "nontransference" parallels intellectual skepticism toward computer science–type theory. I first encountered this in its cybernetics form during graduate school at the University of Chicago. I could see why others found this way of thinking fascinating, but I thought it too abstract, built around the suggestive but ultimately misleading metaphors addressed in chapter 6. (My sons would probably put my ambivalence down to the fact that, by the time I was included in their computer games, they had better hand-eye coordination. I see it as part of my skepticism toward presumptive formalisms of all sorts, whether religious, art critical, physical scientific, or social scientific.)

In what follows, I interweave personal experience with others' research and practice. Those promoting AITs in education perform diverse technology metaphors informed by divergent pedagogies, each of which incline toward different social transformations. The chapter's argument is organized around the three, sequentially introduced key metaphors that have guided AITed learning. The "electronic brain" image had much to do with the failures of early, 1970s-era computerized teaching machines, as did the associated, then dominant "banking" pedagogy. In the 1980s, A second "tool for exploration" AIT metaphor was associated with an alternative pedagogical approach, "discovery." This in turn during the 1990s tended to be replaced by "virtual community" and "constructivist" views of learning. The chapter concludes that, as in the other moments of knowledge networking already addressed ethnographically here in Section III, AITed knowledge sharing must have substantially different dynamics before it makes sense to speak of a "revolution" in education.

Vision #1: An "Electronically Brained Teaching Machine" for "Banking" Knowledge

AITs are not always introduced into schools to support knowledge sharing networking. Indeed, sometimes their education purpose may only be secondary. The "e-rate," a special federal telephone tax to pay for connecting U.S. schools to the web, was introduced in the 1990s. It has been consistently justified in terms of employment—for example, that exposing poor children to computing would enable them to compete better for "new economy" jobs.

Indeed, the initial conception of what computers could do in schools, one with considerable ongoing influence, is that they are powerful mental prostheses or "electronic brains." During the early 1970s, typically as dream (Kurzweil 1999 is a contemporary version) but occasionally as nightmare (see, for example, Oettinger and Marks's *Run Computer, Run* 1969), the "teaching machine" was the dominant motif.[2]

"Teacher-Proof Teaching Machines"

I became aware of attempts to use AITs to teach early in protocyberspace. In 1971, Barbara Andrews and I were on the staff of New University Conference

(NUC), "a [U.S.] national political organization of graduate students and faculty working in, around, and in spite of institutions of higher education." NUCers were educational radicals, critical of the ways traditional, hierarchical, school-centered pedagogies contributed to educational underdevelopment, especially for women, minorities, and working-class people. We attended the annual exhibition and meeting of the Association for Educational Communications and Technology (AECT) to learn more about new computerized teaching machines. Their chief advertised advantage was "instant feedback" on drills and exercises. They were advertised as giving the learner more control over her education, but some teachers saw the machines as automation; after all, they were being promoted as "teacher proof."

The "demos" and claims of the machine makers were impressive, but, having already studied computing at Stanford and used the machines in research at the University of Chicago, I knew there was often a gap between claim and "in real life" performance. We searched out teachers actually using the machines. They told us (as we reported in College English in 1977 (Andrews and Hakken)) that computer teaching machines could be helpful in some narrow, very precise areas of learning, like medical terminology, where consistency in use was of overwhelming importance. Otherwise, the machines were of marginal value. They were repeatedly compared with language labs.

Within a few years, these early machines disappeared, and aggressive talk of "computerized learning" was replaced by more moderate talk of "computer-*assisted* learning." Yet "knowledge machine" rhetoric like that we heard at AECT has remained a regular "computer revolution" trope: "The exponential growth of knowledge necessitates the use of technology . . . " Teachers are still threatened with being left in the dust, "because the technology train is leaving the station . . . "

In retrospect, the first efforts at automated teaching set a pattern. As would be the case again and again, they were clearly "technology pushed." Interest in them was the result of attempts, often with government support, to spread use of the new "distributed," "mini" computers out of factories into new markets. To think of computers as "electronic brains" with which to replace academic "wetware" remains, as Levi-Strauss (1966) recognized about many elaborated cultural constructs, "good to think" (with, if not about). "Machined" knowledge being very like information (perhaps a bit more thoroughly processed), teaching machines fit well within the positivist knowledge project of science. These reverberate well with "banking pedagogies," the dominant teaching approach. Like a bank officer, the teacher's/brain's job was to open the "safety deposit box" of the student's mind, stuff as much into it as possible, and quickly close the door and turn the lock, lest any knowledge, spilled on the floor, be wasted.

The essentially mechanical relations of such views mean that, once knowledge is formally represented in one medium, it can be easily "re-presented" in and transferred via others. This kind of education is one-directional, transferring a stored "stock" of knowledge from one mind (a teacher)—either directly via talk or indirectly via other mediation (for example, a book or a computer)—to another, more passively receptive mind (that of the individual student). The computer/electronic brain makes an ideal teacher, not just an aid or even substitute.

"Journeyperson" AITed Learning in the "Teaching Machine" Period

My skepticism about machined learning did not prevent me from using computers when they made practical sense. As a mid-1970s Ph.D. student at American University, I volunteered to computerize the quantitative methods class. This helped me convince the faculty to treat "computer" as a "foreign" language that I had acceptably mastered.

Along the way, I became reasonably competent at using these machines in my research. This was despite being put off by the course at Stanford, which was more about showing that I was a technophilic "true believer" than about learning. I studied mathematical anthropology at Chicago and took advantage of options for computer training. We entered our data into computers and analyzed the information for David Schneider and Raymond Smith's project on kinship and sex roles, and I used sophisticated multivariate data analysis programs in an evaluation of dissertation funding while on staff at the U.S. National Science Foundation. I took with me and used a "luggable" PC in my second Sheffield field project in the mid-1980s.

Through experiences like these, I learned that, rather than trying to keep personally "up to date," I should depend on others to get "just in time" access to "bleeding edge" techniques. Of course, I only had access to such expertise because I was working in educational institutions. Still, I noticed that those with substantial knowledge of how to use computers were if anything less enthusiastic than others about the need for classes to keep students technologically "on the cutting edge."

In 1995, the Cognition and Technology Group at the Learning Technology Center of Vanderbilt University carried out a large-scale review of school-based computer-mediated instruction (CMI). They concluded that there was discouragingly little evidence that computer mediation has much impact on learning (a conclusion echoed by the British government in 2003). This was due in part to the poor educational quality of many programs. However, the report's fundamental claim concerned the extent to which CMI relies uncritically on existing pedagogies, especially banking.

An AITed Revolution in Educational Banking?

Banking pedagogies have much in common with the modernist philosophical perspectives on knowledge described in chapter 4. In these readings, knowledge is largely independent of its means of transmission. When computers are deployed under a banking pedagogy, the role of the cyber-learner remains more or less the same as it was for her nontechnologied predecessor: to "accept" readily or "absorb" a relatively stable "body" of knowledge when told to do so by an authority figure—human, computer, or cyborg. Consideration of the teaching/ educative environment is primarily psychological, on fostering individual student mental habits that promote more rapid absorption, and so on.

An "electronic brain" doesn't foster a focus on either the knowledge itself or its application; the task in CMI is merely to mediate preexisting materials in new ways. As Barbara Andrews and I had already concluded in the 1970s, a teaching machine, no matter how graphically designed, is hard put to do a better job of transmitting "facts" than an alert human. For the foreseeable future, people will also be better able to respond flexibly to the classroom exigencies of Johnny hitting Susie or Mary not having had any breakfast.

I am skeptical of the extent to which computers can "scale up" education. However, even substantial "scaling up"—for Daniel, the key advantage of cyberspace to education—could take place without having much impact on either learning style or content. Even if better "signals sent to those received" ratios were achieved, this is not in itself evidence of social transformation. Since change is concentrated primarily in the size of the knowledge body transferred, transformation would have to be further argued for, as on "quantitative leads to qualitative change," "punctuated equilibrium" (Gould 2002) grounds, or in terms of a paradigm shift of truly Kuhnian, and therefore virtually impossible, dimensions (Fuller 2000). This latter is particularly unlikely to follow from schooling otherwise "normal." In sum, it is very difficult to make an "education revolution" cyberspace transformative case, either empirically or theretically, on modernist/CMI/banking grounds.

Vision #2: A Knowledgescape to Be Discovered with New Exploratory Tools

In 1978, I began full-time teaching critical of banking pedagogies and skeptical of teaching machines. Nonetheless, I experimented extensively with several forms of computered teaching. An early 1980s interest in simulation games followed from being asked to teach some service, application-oriented courses for the SUNYIT computer science department. The department was having difficulty expanding, given a dearth of Ph.D.'s in computing. Simulations struck me as an application to teach additional to word processing, spreadsheets, and databases, as well as one with potential educational benefit.

My experience with AITed simulations in anthropology, however, was largely negative. Student interest was, like that of my Stanford roommate, clearly piqued by simulation's game dynamics. However, because the game designers didn't really understand the dynamics of the cultures they were simulating, such capture generally diverted students away from, rather than toward, the lessons I intended regarding the particularities of each culture. The culture simulations were, like SymCity™ and the artificial life forms dissected by Stephan Helmreich in *Silicon Second Nature* (1999), invariably so ethnocentric as to misrepresent any "Other."

Around this time, I was placed on SUNYIT's new "distance learning" committee. This committee arose in response to the college's receiving, unrequested, from the local New York state senator, chair of the Education Committee, a satellite uplink. This dish gave us new capabilities, but ones that we were unprepared to use. Unfortunately, the then academic vice president tried to command courseware by fiat, which faculty met largely with passive resistance. Part of their reluctance was based on fear that maintenance of the new capability would divert scarce resources from other uses (like higher salaries!), a reluctance reinforced by the administration's unwillingness to share information on the real cost of dish maintenance. Placed on the committee because of my experience, I was also concerned about the quality of the "courseware" "loaded up."

To overcome the impasse, I outlined a set of computerizable accounting simulations, tools to allocate fixed and variable costs to different functions, so that faculty and administration could consider alternative priorities. I wanted rational bases for decisions about use and criteria for evaluation, and the simulations would have allowed us to foresee the implications of each alternative. After I circulated my ideas, the committee did not meet again for ten years! I still suspect that this was because, in order to actually carry out joint decision-making implied by faculty involvement in governance, administrators would have had to reveal maintenance costs. In any case, the uplink, apparently not cost effective even for academic units allowed to ignore sunk costs, has been used only infrequently.

This experience was for me an important indication of the extent to which educational computing's use or nonuse has to do more with prestige than effectiveness or efficiency criteria. That cultural factors were more important than technical ones in determining the social correlates of AITs was the main result of Barbara's and my field research on organizational computing in Sheffield (Hakken with Andrews 1993).

Discovery as an Alternative Pedagogy

Nineteen-sixties-era educational radicals like Barbara and I had already critiqued the banking theories of learning that informed "electronic brain" machines. In our view, banking models impose an overly rigid sequentiality on the relationship of knowledge creation to its sharing. We evoked Dewey (1997) to stress the need for an active learner for knowledge to be successfully shared.

The alternative pedagogy we helped foster came to be called "discovery learning." It aimed to invert the "banking" roles by placing the teacher in the background and "putting the student in charge of her learning." Banking pedagogies and teaching machines also rigidly replicated the hierarchical differentiations among the different roles and forms of knowing. In contrast, discovery pedagogies dissolve these differentiations, minimizing or ignoring the distinctiveness of the productive/adaptive submoments of knowledge networking from the sharing submoment.

The most effective advocates of discovery learning blamed the early AIT failures on the pedagogy, not the technology. The dazzle effect was doubtless also part of why many "discoverers" advocated computing as a preferred means to "access knowledge bases." In discovery-based AITed systems, the computer was transformed from a "electronic brain" actively transferring knowledge into a tool for exploration of more passive "electronic performance spaces." Students were to travel through representational "knowledgescapes" like excited explorers in a new land.

Exploration as Transformation?

In the postmodernist claim that "there is only text," knowledge only exists when it is being transmitted. Similarly, in discovery pedagogy, active "learners" were to produce their own knowledgescapes and share them with passive "teachers," all readings presumed to be of equal value. Thus, in a postmodernist/discovery pedagogy, changes in transmission *can* produce a moderate range of changes in knowledge's character. Discovery is thus moderately more fertile ground on which to construct transformative accounts of AITed knowledge sharing networking.

The simulations I used in the early 1980s were based on discovery pedagogies. Despite claims that they were radical interventions, formal "thing" conceptions of knowledge continued to inform them and the many other software packages, and eventual "web surfing" exercises, that are the most visible representations of discovery learning. The learning efficacy of such representations still need not be addressed, because the value of mere access to "proper" knowledge was self-evidently high. The CTG applies to modalities like the self-directed "learning station" of the "discovered" elementary classroom the criticism made of the "self-constructed" undergraduate major: only the most focused, highly self-motivated, and already generally knowledgeable learner is likely to use these means to complete sustained educational tasks. When one today encounters enthusiastic "discovery"-type advocacy of the Internet or World Wide Web, one does well to remember Mitch Kapor's quip comparing the Internet to a "library where all of the books have been dumped on the floor" (Dery 1994, 4).

Discovery learning tried to reverse the sequencing of banking but replicated its formalism. While the egalitarianism of discovery pedagogies was appealing, they too often in practice justified "anything goes." Students left on their own to discover computered knowledge frequently found . . . very little. Discovery's collapsing of all knowledge networking into personal learning was an overreaction to banking theories' shortcomings, one that actually extended banking's influence. In order to be "decodable" by a student with minimal guidance, knowledge has to be "stockable" and thus has to be conceived in very formal terms. Individualized computerized discovery systems also reified the false presumption that education is a primarily individual activity.

According to the Vanderbilt CT Group, programs that rejected the "banking" theory in favor of "discovery" learning tended to fall into one of two different traps. One was that, like Forsythe's migraine program, they ended up being more about promoting the appearance of self-education than its actuality. Banking was still the objective, but the hope was that computered discovery would make it "painless," like good dentistry.

Alternatively, some more thoroughly "discovered" programs really did make accessible, often via hypertext, considerably more information on a given topic. While it was often possible to establish that a student had enjoyed her journey along the links that attracted her fancy, it proved much more difficult to establish just what she had learned. Demonstrated pleasure or entertainment value is not necessarily indicative of educational attainment, let alone of transformation.

Students, rather than discovering, tend to wander through computer-encapsulated, often radically simplified knowledge landscapes. The importance of credentialing increases the likelihood that the to-be-"cyber-discovered"-knowledge to be acknowledged to be knowledge remained essentially the same as it was under the "banking" approach. While formally "contents" rather than a single "content," the old, previously banked story maintains much of its echo.

Because knowledge is socially constructed, good educational activities share some dynamics with communities. Sharing knowledge is a fulsome activity. It is knowledge networking, not just data or information sharing. To the extent that it achieves individuation by further separating one learner from others, discovery learning tends to undermine the sociality of knowledge sharing.

Moreover, the AITed discovery approach was also justified as primarily about efficiency—that is, requiring less active teaching. In this regard, discovery learning, like the deployment of electronic brains and the employment of teacher aids, had substantial continuity with strategies to deskill teachers. Such continuities do not indicate revolution. For all these reasons, the support afforded to transformative argument by "discovery" pedagogy–based AITs is only marginally greater than that of "banking" ones.

Vision #3: Coconstruction of "Virtual" Education Communities

"Faculty Empowerment through Student Use of the Internet"

During the 1990s, I once again thought seriously about computered schooling, prompted by visions of the educative potential of the World Wide Web. About this time, however, slogans like this one, the title of a 1995 State University of New York–sponsored conference, also began to appear in "learning land." Appalled, bemused, and entertained in roughly equal parts, I searched for clues as to how as a professor of anthropology *I* was to be "empowered" by *my students'* use of computered communication. Even if the dubious rhetoric about "democratized access to knowledge" used in the promotional brochure were true for students, how would this make *me* more powerful?

Perhaps by then I should have become used to computer hype; for example, the appropriation of 1960s rhetoric, so that "empowerment" is transformed, "Big Brother"–style, into its opposite. After all, ads for the "Dodge Revolution" had already appeared by the early 1970s. With the marginalization of the social movements of the 1960s, love had come to mean hate, as in "tough love." Couldn't one expect the narrowing of academic discretion inevitably a part of putting courses "online" to become the "new liberty"?

The newspeak echoed 1970s AECT phrases like "teacher proof" and reminded me of the many machines left on the scrapheap of simplistic tech educationism. As then, I was already an experienced user of the new, Internet-based technologies. I not only sent e-mail and participated in listservs; I was doing research "online" (and would do more in the future). How naive did State University bureaucrats think we faculty were, especially given all the publicity about how millions are to be saved through technology-enabled "downsizing"? I concluded that any interest I might have for experimenting with online education would be dampened, rather than encouraged, by participation in this conference. When SUNY began to promote its new Learning Network of online courses, I remained on the sidelines. I needed better arguments for doing what was now generally called "distance learning," ones that showed me concretely how it would improve the teaching situations in which I then found myself.

"Third-Generation" Educational AITs

As in the 1970s, much ineffectuality in these approaches would eventually be exposed, the coup de gras being delivered by the dot-edu collapse that followed the dot-com and dot-telecom collapses starting in Y2K (see below). Yet in 1996, I was reconsidering my skepticism about AITs in higher education. While I was still not convinced that the new forms of AITed learning were pedagogically superior *in general*, it did seem that they might last longer than those of the early '70s or satellite courses. I suspected, however, that this had as much to do with a new significance for AITed learning in social formation reproduction. My suspicions were roused by a subtle shift in my colleagues' language: Their previous concern about how to share knowledge more effectively was being displaced by worry over protecting their "intellectual property."

The new approaches were more subtle. Like the knowledge management described in chapter 3, they deployed rhetorics of community and collaboration more than of dominance and formalism. Interest was directed to computing's communicative rather than its control affordances. Simulation displaced the search for ONE BEST WAY.

Despite the much greater computing power available by the mid-1990s, however, new AITed simulations didn't work much better than the early education games described above. Organization in the mid-1990s of Simulist, a computer list for those interested in computer simulation, prompted me to organize an adjunct discourse on applied issues in simulation. This went nowhere, at least in part, I was convinced, by an annoying lack of interest on the part of those promoting simulation to apply it to real problems. (This took place at the same time as the aborted effort with simulation technology in local social service described in chapter 8.)

In contrast, I was having good success with *non*-AITed simulations, like role playing, to create more stimulating educative environments. A good example was my graduate course on "Information and Change in Not-for-Profit and Public Organizations," in which the class project is a group simulation of a consulting firm. After self-selecting into teams, the students work with local organizations to do IT and organizational development audits, and then generate "next step" proposals for integrating IT and OD efforts. This simulation, based on Dorothy Heathcote's "mantle of expertise" approach (1991), is an example of constructivist pedagogy.

Constructivism and Third-Generation Educational AITs

In the 1990s, educational computing began to dress in the new pedagogical clothes of constructivism. This approach to learning aims to strike a more effective balance between the educational agencies of teachers and learners. It critiques the "stockable commodity easily accessed" and "individualizing" presumptions shared by "banking" and "discovery." As Fosnot describes it,

> Constructivism is not a theory about teaching. It's a theory about knowledge and learning. Drawing on a synthesis of current work in cognitive psychology, philosophy, and anthropology, the theory defines knowledge as temporary, developmental, social and culturally mediated . . . (1995, x)

The knowledge networking alternative of *The Knowledge Landscapes* . . . draws heavily on the "representational" and "social" aspects of constructivism. Both conceive of learning, the sharing moment of knowledge networking, as a process in which teachers and learners cooperatively resituate preexisting knowledges across various relevant social terrains, while at the same time coming to embody them.

A fully constructivist educational experience should be a learning process that, in its rough equality, achieves something like "invisible college" knowledge production networking. Because the learning project is arrived at by negotiation, it is difficult at the end to differentiate formal teacher from learner roles. This result is only achieved, however, after a gradual, guided expansion of students' actual abilities to participate in knowledge adaptation and creation networking, not by merely creating the illusion that this already exists. To achieve constructivism, teachers must seek out students' experience-based knowledge and legitimate it, incorporating it fully into a developing collective knowledgescape. Thus, not "anything goes," but instead a philosophical commitment to the nontranscendence of knowledge is concretely made compatible with striving after accuracy.

The pedagogical turn to constructivism, perhaps most powerfully articulated in the work of Paolo Freire (for example, 2000) and Jonathan Kozol (1992), has been a helpful counter to both passive "banking" and uncritical "discovery" theories of student leaning. Constructivism has become an influential theory of pedagogy, especially in science education, albeit not without some questionable consequences (Mathews 1998).[3] Too many applications of constructivism, perhaps because of holdovers from discovery, have given insufficient attention to the distinction between knowledge creation or adaptation networking and its later sharing with students. Process or narrative concerns too easily displace the hard effort needed to figure out what has been learned or specify what is actually new. These difficulties parallel those of participatory action research (Whyte 1991) and so-called new ethnography (Clifford and Marcus 1986).

Still, by focusing the increase in student agency on collective participation in designing the educational process, good constructivism, like knowledge networking, rejects the individuation of education institutionalized in banking and discovery. Through encouraging denser talk, constructivism supports the search for substantive bases for equalizing roles, while at the same time facilitating the necessary resocialing of educational practice. Often justified in terms of activity theory (Vygotsky 1986), constructivism identifies the construction of social relations among learners as being as important to learning as those between learners and teachers.

In ways like this, constructivism gives priority to education's communicative over its representational activities. Its network-building pedagogy undermines framings of AITed school activities focused primarily on occupational mobility. Constructivism's view of the social as a core aspect of knowledge is articulated particularly well by James Greeno of the Institute for Research on Learning. He summarizes his critique of discovery computer-mediated education in terms of research that showed a need for

> a different way of thinking about knowledge. Rather than thinking that knowledge is in the minds of individuals, we could alternatively think of knowledge as the potential for situated activity. On this view, knowledge would be understood as a relation between an individual and a social or physical situation, rather than as a property of an individual. (1988, 12)

On constructivism, knowledge is neither, as in a modernist conception, an independent "body," nor is it, as in discovery, separable from the collectivity of its knowers.

To evaluate empirically the broader education correlates of computing in schools, one must take a pedagogical stand on what schools are for. Constructivism offers a clearer view of what should be going on in education than banking or discovery and is thus the perspective most useful for careful ethnographic examination of AITs' correlates. Moreover, as opposed to electronic brain and exploratory tool images, constructivist pedagogy encourages AIT uses more in line with the way AITs actually network.

Consequently, of the three pedagogical theories, constructivism provides the conceptual terrain most amenable to the idea that AITs actually do transform education. Do third-generation AITs embody a richer constructivism than the SUNY workshop? Are they finally transforming educational practice, or are banking and discovery legacies still in the way?

"Distance Learning" and the Knowledge Sharing Question in Cyber-educational Spaces

Before evaluation of constructivist AITs, a recent discourse confusion, the misidentification of "distance learning" with AIT mediation, needs to be addressed. Part of the reason why AITed distance learning could come to stand for all distance learning is that wide acceptance of the importance of different learning styles occurred in the same time frame (the 1970s) as the first widely accessible AITs in education. Because in discovery differences in students' motivation and preparation do matter, its learning environments do tend to fit individual circumstance better and to this extent are more effective. Discovery alternatives to banking pedagogies also helped spread skepticism about congregate, and thus less self-directed, learning activities. Often promoted together, AITing and discovery became confused. A discovery "there is no one best way" conception was eagerly taken up by school cyberians, who claimed that, as knowledge could be discovered and thus re-created in many different, individuated ways, computer mediation was the "natural" way to speed and support the process.

In U.S. higher education, the congregate class, as opposed to the independent study or small tutorial, remains the generally central event. This centrality is not just mindless tradition, nor, even though classes are typically cheaper than independent studies, is it merely economics. Classes also mean that students benefit from congregately reinforced guidance in knowledging—for example, how to appropriate knowledge. The benefit follows from palpable, face-to-face exposure to each other's—both the teacher's and other students'—learning behaviors. Congregate schooling makes sense because, in Greeno (1988), knowledge is a more social than individual phenomenon. Knowledge is different from information in terms of the very social relations central to its construction, reproduction, and sharing (Hakken 1999a).

When approached constructivistically, the teacher's work is not simply "teaching" but, in terms of work sociology, "constructing educative environments." It is not appropriate for her to be assigned responsibility for students' actual learning, which in constructivism is something over which students themselves have the greatest influence. The extent to which learners are capable of taking advantage of what a given environment offers also varies considerably from group to group and subject to subject.

Still, it *does* make sense to hold a teacher responsible for constructing spaces generally conducive to learning, and experiences often have to be educative when she is not present. Are the environments she constructs ones that reasonably prepared, reasonably motivated students are likely to find conducive to learning? The craft of teaching consists in knowing which activities are likely to be conducive under which circumstances, as well as getting students to be "coconspirators" in their own education.

The standard trope in current AITed education celebrations simplistically contrasts a nonmediated, "horse and buggy" approach—for example, the "sage on a stage" in the higher education classroom—to "with it," technosavvy, innovative "distance learning." Tropes like this draw heavily on advertising to hype technology. To gloss as "distance learning" only activities that centrally involve AITs is misleading, however. Researching papers, doing projects, even reading textbooks all take place "at a distance" from the classroom. Moreover, the equation of distance learning and AIT mediation obfuscates the divergent design and evaluation implications of learning environments based on banking, discovery, or constructivist pedagogies. To take distance learning as the only site of technologied education parallels equating technology with self-controlled machines, as in the anthropologically suspect notion that some civilizations are "technological" while others are not. Characterizations like this erroneously imply the existence of nontechnological civilizations, when all human social formations are deeply dependent upon technology (Hakken 1999a).

The necessarily shared character of educational design is why I describe as "coconstructivist" the pedagogy I aim to practice. In coconstructivism, using computers for activities like simulation has substantial potential to draw attention to shared responsibility and thus to make its pursuit explicit. This means AITs have a particular potential for transforming education in congregate settings. Misguided attempts to make face-to-face learning appear arcane, however, actually divert arguing for this potentially important part of a transformationist case, which is why it is seldom invoked by transformationists like Daniel.

Distance Learning: An Alternative Concept

To evaluate the transformationist case fairly, we need to separate "distance" and "AITed" learning. I call "distanced" *any* educative environments in which students and teachers are

1. Not all in the same place,
2. Not all engaging the educative environment at the same time (asynchronous), *or*
3. Not doing it in the same way.

Differently placed, asynchronous, and complex activities have long been institutionalized in schools. The building block of U.S. higher education, the Carnegie unit, presumes three hours of student work outside of class for each hour in it. Thus, at least three-quarters of learning time is presumed *normally* to take place "at a distance" rather than "face-to-face." College students generally do read books and articles on their own and at their own pace. Further, the projects that typically determine half of my students' grades are mostly carried out at a distance in both time and space from the classroom and me. Even when in class together, we are often "distanced" by engaging the environment in different ways, as when, reporting on their work in class, they are teaching me.

Distance learning is also a standard part of primary and secondary education (for example, homework, group work, or individual work). AIT-mediated learning must therefore be evaluated in terms of its superiorities or inferiorities in comparison with non-AIT-mediated learning forms of all types, whether distance, face-to-face, or "hybrid." One must be able to distinguish those changes tied strongly to the specifics of new, AITed educational technologies from those that follow from other shifts, either among different distance modalities or between distance and face-to-face. Such clarity is necessary if we are to tease out any educational transformations actually caused by AITs, to answer the knowledge sharing question in cyberspace. The deliberate confusing of educational AITs with distance learning in general, like other common celebratory AIT rhetoric, is itself reason for skepticism about tranformative arguments.

Constructivist AITs and Transforming Both "Distant" and "Close" Education

As pointed out above, discovery was good for the most competent and self-directed students; the rest need guidance. What coconstructivist educators now see as necessary when they talk about truly new forms of learning is not substitution of internal for external direction but of shared participation for authoritarian control. The second basic finding of the Vanderbilt group was that the most positive learning outcomes from computer-mediated instruction occurred in schools *already transformed* into more effective, participative learning networks.

Such schools offer a really different education. From their pedagogical position, collaboration among students and teachers is best precisely because all agree on the point of the activity, "buying into" it because they participate in determining it. Because AITs can decouple space from place, they can provide more chances for reflexive, collaborative discourse as well as collaboratory experience anytime a question arises. Computer mediation can substantially enhance the capacities of schools when they already network knowledge sharing participatorily.

An argument can be made on constructivist grounds that AITs are transformative because they inherently generate more collaborative learning. This can be done in two ways. The more typical is negative, that learning online reduces the salience of certain social markers that normally separate teacher from students and students from each other, such as age, signs of social prestige in things like clothing, and body language and speaking style. In their absence, more attention goes to the substance of what is written or presented, which is a better basis for true collaboration.

The more positive case is that trying to learn in cyberspace makes more evident the necessity of thinking constructivistically. Participants can't rely on their habitual repertoires of learning approaches in the new media. There is greater chance that construction will be shared if it has to be done actively than if it can be avoided altogether. Moreover, AITs provide tools that can enable new degrees of, forms for, and arenas for constructivism.

Unfortunately, the CTG group found few examples where AITs led to constructivism, where computers in themselves fostered reflexive collaboration. Indeed, because most school computing also involved introduction of additional "gatekeeping" dependence—upon, for example, the technician /teacher with administrative responsibility for machine maintenance—their introduction often decreased the total control available collectively to students and teachers together. Similarly, AITs got in the way of coconstructivism when they hid pedagogical choice or were imposed and thus substituted for discussion of options.

To achieve substantial benefits, it would appear that coconstructivism must precede AITs or the two must at least be implemented together. The CTG people concluded, however, that this generally did not happen and thus that there was no reason to conclude that AITs even make coconstructivism more likely, let alone inevitable. Few schools were using information technology coconstructivistically, as a tool of collective self-governance rather than as a tool for controlling students (or faculty) by faculty (or administrators). This was the strongest evidence the CTG group offered for the nontransformativity, through the mid-1990s, of cyberspaced education.

Using AITs to Construct Virtual Communities

Since the mid-1990s, a number of new educational technologies informed by constructivism have been designed. Many of these approach construction of a "virtual community" of learners as the "missing link" in transforming education technologically. The collection *Building Virtual Communities: Learning and Change in Cyberspace* (Renninger and Shumar 2002) describes "cutting edge" projects that coconstructively model "online school communities of practice" on the "learning organizations" described in the preceding chapter. The analytic approaches of the writers, many of whom were project implementers, were at least implicitly informed by a "social informatics" (Bowker, et al, 1997) perspective, that getting the social right is just as important to good educational technology as getting the technical right.

In my afterword for the volume (Hakken 2002), I stressed its considerable practical value for those tasked with developing effective AITed learning environments. It provides an excellent point of entry into both the strategic thinking of those designing such programs and, because of the detail contained, the actualities of implementation. I was particularly pleased by the volume's implicit recognition that knowledge networking among researchers and teachers is as important to the broad sweep of education as is that among teachers and students. Another valuable dimension of the volume was the serious attention to evaluation.[4]

Design of Learner Communities

As a result, *Building Virtual Communities* offers an opportunity to gauge the quality of turn-of-the-twenty-first-century ed tech. That design strategies can "engineer" the social dimensions of learning is perhaps its basic lesson. "Community" *can* be addressed via technology, and this can be done simultaneously at a number of levels and to integrate knowledge sharing with creation and adaptation. For example, Ann Davidson and Janet Shofield (2002) describe an intervention using computers in a way that drew on shared gender identity to leverage community among students. Interesting in this case was the extent to which success with the strategy depended upon "performing" considerable teacher *non*expertise that evoked cross-status (teacher/student) gender solidarity.

Davidson and Shofield's case, involving students in a single classroom, is highly placed; indeed, it is conceivable that its success for females depended upon a palpable marginalization of the male students in the class. James Levin and Raoul Cervantes (2002) consider a case that is more typically the focus of "ed tech," creating student/teacher collaborations in an electronic space spread over dispersed places. Beverly Hunter's case (2002), using AITs to create a "community of schools," has the salutary effect of highlighting the dependence of on-line communities on "real life," geographic ones like school boards, the ones that, in the United States at least, have to fund "ed tech" in the long run. Whether knowledge is shared online depends greatly on other institutionalized forms of social reproduction.

The intervention addressed by Ann Renninger and Wesley Shumar (2002) involves fostering community among dispersed teachers who share content expertise in math, as well as among these teachers and their students. It thus focuses as much on adapting knowledge as sharing it. Alex Cuthbert, Douglas Clark, and Marcia Linn (2002) describe a parallel effort among science teachers. Mark Schlager, Judith Fusco, and Patricia Schank (2002) describe technology efforts to foster community in general teacher professional development. The initiative they describe is best seen as an example of the knowledge creation submoment, as is Amy Bruckman and Carlos Jensen's (2002) analysis of the MediaMOO initiative. MediaMOO was an effort to create and then promote new understandings of both pedagogy and technology via an online community of teachers and technology researchers. The tool called CILT described by Christopher Hoadley and Roy Pea (2002) is similarly designed to facilitate knowledge coproduction among learning technology researchers and stakeholders. Roger Burrows and Sarah Nettleton (2002) consider wired self-help networks in the relations between formal researchers and what we might call "experientialists." Experience with these networks turned out to be problematic; to some extent they even undermined knowledge production networking. Jason Nolan and Joel Weise (2002) are similarly concerned with informal, peer-to-peer activities. In tracing how Kindred Spirits became Serbia.net, they show how the attempt to build an educational activity was gradually displaced by a "purer" community-building activity, creation of a explicit identity.

These initiatives illustrate the diverse ways in which "community" and related concepts can be mobilized to address social dimensions while implementing AITed learning systems. They suggest that "virtual community" approaches to the social dimensions of other institutional forms are also plausible.

The Effectiveness of Community-Oriented Educational Technology Development

The initiatives also show that "community" can complicate such implementations and thus needs to be considered carefully in program design. (The cost of opportunistic uses of "community" as a gloss for "the social" is an issue that I return to below.) While plausible, are "virtual community" strategies effective? What are the consequences for knowledge networking, both positive and negative, of deploying them?

The interventions described in *Building Virtual Communities* were "development" or "proof of concept" projects, where resources were made available to show that something that sounds good could actually be made to work. An important benefit of demonstration projects is that they reveal unanticipated problems. Thus, efforts to foster gender community via technology to help one group (for example, girls) should be concerned about how to do this without inadvertently disadvantaging others (boys). Several of the cases (for example, the student/teacher collaborative space and MediaMOO) illustrate the importance to successful knowledge networking of very active, even activist network moderation. Similarly, several (for example, the math teacher collaboratory) highlight the severe consequences of the still-all-too-frequent technical failures. The experience of the collaboratory of schools serves as a salutary reminder that even under demonstration conditions technology is not a simple, direct, high-speed road to educational reform. To move beyond demonstration to the reality of fundamentally changed practice, one needs very high levels of careful, long-term, coordinated pressure on institutions.

Other problems in constructivist AITing also emerge. Bruckman and Jensen, for example, outline several revisions for reviving MediaMOO. One would involve a distributed architecture, thereby allowing pockets of intensive networking more or less to go off by themselves, turning MediaMOO into something of an umbrella. Another revision would promote survival of the MOO by substituting a new goal. In either case, the original goal, that of a general community of communication tech researchers and teachers, would be displaced by a narrower one, supporting community among those at an early stage in the "lifecycle" of the technology/education professional. While the first revision would have substituted the illusion of community for its reality, the second would harness the technology for a different purpose. Since neither option was argued for by the authors, I was left with the sense that the underlying goal was survival of the program.

A MediaMOO reorganized in either way was in danger of becoming one of the many technology demonstration projects whose raison d'être has become self-perpetuation. In a nation that tends to see technology as a "good thing" in itself, such "demonstration effects" are a real problem with AIT projects. To avoid this effect, we must demand that extended demonstration projects continue to generate new knowledge or reproduce existing knowledge in some substantially new way.

I had similar concerns about the intervention among science teachers. Early on, Cuthbert, Clark, and Linn (2002) assert that each of their efforts, including peer review community development, "play decisive roles in achieving the goal of improving science education." However, their article only addresses how design criteria were implemented, offering no evidence of improved science education, let alone why project activities were "decisive." Mere implementation of a design does not indicate effectiveness. In both these cases, I feel the pursuit of "lessons" engendered drawing dubious inferences that reflect celebratory "distance learning" hype.

Interestingly, demonstrations like these are the kinds of activities that pass for "research" in engineering and technology fields like informatics (Forsythe 2001). Success at showing that a system *can* be deployed is erroneously treated as evidence that it *should* be. This practice feeds into the "If we build it, they will come" syndrome characteristic of efforts like the "information superhighway."

Instead, the ultimate goal should be to decide whether a possible intervention is likely to be effective. Thus, while community can be engineered by ed tech as a means to address the social, there remain larger, more difficult questions. Under what conditions should schooling already placed be deplaced? Is there any basis here for prioritizing attention to knowledge sharing among students, among teachers, among researchers, or among the various possible permutations of these roles? Should one try to address knowledge sharing in more than one, even all of them, in each program?

With these authors, I count myself as someone who takes evaluation seriously, but conspicuous by its absence from this collection is any comparative study of deplaced versus placed or deplaced versus "hybrid" community-facilitated knowledge sharing. This absence has doubtless something to do with the "pump priming" preoccupations of funders, as well as their irresolute commitment to thorough evaluation. As indicated in the previous chapter, the absence of work that addresses these broader questions follows from the penchant of politicians for making policy without knowing what is likely to work. As long as this remains the case, AITed learning, and attempts to "engineer" community to support it, will be derivative activity, dependent upon the overflow of technology development funding. Consequently, those of us doing the work described here will have to ask such questions ourselves. The fact that they are not asked systematically by those who tout the transformative consequences of AITed learning remains, again, a strong reason for suspecting the hype. Thus, despite the obvious efforts of these writers to take demonstration evaluation seriously, the collection does not justify concluding that constructivist AITs are effective *in general*, let alone transformative.

The Theoretical Importance of "Virtual Community" to Knowledge Society

Acknowledged by the articles in *Building Virtual Communities* is a need to theorize the social dimensions of deplaced AITed knowledge sharing. How persuasive are the writers' analyses of what happens when a concerted effort is made to introduce a technology supportive of knowledge networking in a "holistic" way—that is, to try to anticipate and address the social contexts/consequences of the intervention? Is the term "virtual community" (as opposed, for example, to generic "community") of value in describing what is distinctive, or is it just more hype?

On this issue, the articles offer a widely divergent array of approaches. While several address the utility of virtual community as a concept, some refuse to use it at all. Of those that do, widely different, sometimes only lightly or untheorized conceptions are used. These disparities in approach, doubtless in part a consequence of both national and disciplinary differences, compromise the value of the volume as itself an effort at deplaced knowledge creation.

I personally find conceptual clarity with regard to community via a linguistic approach. In my experience in the United States and in Europe, when speaking English, "natives" use "community" in three often overlapping ways, to refer to:

1. Smallish geopolitical units like villages, town, and cities
2. Distinguishable-because-of-relatively-more-dense sets of social relationships or patterns of social interaction (for example, "workplace community," which can be, variously, those aspects of workplace interaction which go beyond the minimum level of necessary socializing, or, as in "a workplace community," which is a strikingly sociable worksite)
3. The existence among some bunch of people of the kinds of shared qualities implied by notions like "identity" or "solidarity" (for example, "the community of believers")

Often, claims that "community" is not descriptively relevant, as in some *Building Virtual Communities* articles, turn on the absence of one or more of these three dimensions. "Natives" might find this practice hypercritical. At the same time, the conceptual ambit of these three uses in toto is vast. If the term were to be used whenever *any* of the three were present, its use would carry little descriptive impact. Of what descriptive value then is "virtual community," a term which, in the sense of referring to phenomena that need only be "somewhat like" community, can be even more plastic?

Like the book's editors, I take my cue from Anthony Cohen (1985), who suggests we approach community from a "performance" perspective. He observes that "natives" normally evoke community when they wish to place emphasis on what is shared by a particular bunch of people (be they in the same place or the same network or "have" the same identity, irrespective of place). Such community performances stress implicitly what is *not* shared with those *not* in the bunch. Performances of community are boundary construction activities and are thus highly political.

It takes effort to ground a new activity successfully. One's willingness to put in the requisite effort depends upon many things, but clearly one important factor is being convinced that others are also willing. If a new activity is deplaced, it's harder to see if they are.

The basic point of AITs in education is—for many students, teachers, researchers, and others outside of the formal educational institutions relevant to any particular intervention but implicated in them (for example, vendors)—to deplace schooling. Thus, it makes sense that those involved in trying to promote a new activity like deplaced AITed knowledge sharing would promote performance of community. Whatever the awkwardness that might initially attend the deliberate ignoring of that which separates us, public performance of community is a good way to indicate "buy in" or "ownership." Over time, the absence of any performance of community tends to be interpreted as lack of sufficient general commitment, an absence that may be used to justify abandoning further involvement. Consequently, *it is highly likely that deliberate (engineered) community performance, which we can with justice call "virtual community," will be a necessary part of any deplaced AITed knowledge sharing effort.*

Since any effective deplaced AITed knowledge sharing intervention will probably need virtual community, it is reasonable for funders to require its promotion. However, we are most used to performances of community in face-to-face contexts. Community performance at a distance has to be imagined, which expands the range of possibly-to-be-discovered incommensurabilities. *Because people are not "naturally" comfortable with "virtual" community, it will not be easy to obtain the desired effect.*

It also makes sense to then ask, what kinds of virtual community engineering are successful, and which are not? *Building Virtual Communities* offers some hints. Several articles indicate that people perform community more willingly *after* they have seen a tangible payoff. Efforts to promote more fulsome performances of community (for example, "generalized reciprocity") from the beginning are less likely to be successful. This is also the pattern manifest in Mizuko Ito's (n.d.) dispersed "networked localities," "virtual communities" collectively constructed around online activities similar to the self-help groups or Serbia.net. Ito does see behaviors somewhat like "real-life" community but having a more tenuous, less reliable tone.

More substantive performances of community are harder to evoke online, and those performances which are evoked may be lacking in substance.

Research on dispersed AITed knowledge sharing suggests strongly that community is only as substantive as that of face-to-face or "hybrid" learning if a strong learning culture is achieved. As social beings, we find it is easier to create a strong learning culture in each other's presence. It can also be done at a distance, but only with great difficulty. In their extensive work of online education (as recounted in *The Network Nation* (1993), Star Roxanne Hiltz and Murray Turoff argue that there must be a palpable "community of learners" if online education is to work. At the New Jersey Institute of Technology, where they teach, it is accepted that this community must be engineered. They do this in part by "leveraging" certain preexisting conditions, like the fact that the student enrolling in an online course is often also enrolled in a "face-to-face" course as well. This means she is already traveling to campus, building relations with other students in her program, and so on. Since the existence of online courses allows her to do extra work and complete her program more quickly, she is willing to put the extra effort into performing community virtually, and this is easier because she already knows many of those who share her virtual classroom.

If all these findings have general validity, they indicate a real problem for those ready to jettison "bricks and mortar" educational institutionalizations for online ones or, as in my state university, heavily subsidize dispersed AITed online teaching while cutting support for other kinds. The marginal consideration often given to such social factors is why I tend to be skeptical of any dispersed AITed knowledge sharing initiative that is "technology pushed" rather than a response to a real, demonstrated need.[5]

Are Community and Other Social Dynamics Changing?

What do these efforts suggest about a potentially important change in community as new information technologies are introduced, now and in the future? Is all the talk about "virtual community" just so much hot air, or is it indicative of some more transformative change in a basic moment in social reproduction? What does the experience of dispersed AITed knowledge networking suggest about whether the character of community changes as it "goes virtual"?

Sherry Turkle has been the most outspoken transformationist among ethnographers of cyber-education (1984, 1995). For example, she argues that those learning in school how to handle multiply "windowed" online activities learn how to cope with different identities simultaneously. She argues that this experience is becoming general and that it leads to a displacement of the currently fundamental goal of identity formation. We move away from attaining an integrated personality and toward being preoccupied with effective maintenance simultaneously of multiple personalities.

However, whatever new identities her student informants think they are forging online, the process itself—an exploration of possible identities heavily mediated by the reproductive dynamics of a consumption-oriented capitalist political economy—is not all that different from what they regularly engage in already, albeit more serially, for example, at the mall. Were a multiple personality "order" largely to displace what currently takes place "in real life," to no longer constitute a stage of typical personality development, it might well correlate with a set of broader transformations. However, Turkle offers little evidence that this is in fact the case. Creating a new terrain for identity exploration, even one that promotes simultaneous multiple explorations, does not make us into a knowledge society.[6]

Howard Rheingold also makes a transformative argument on a cyber-terrain, AITed knowledge sharing in an online virtual community called the WELL. He calls the knowledge that is shared "knowledge capital" and describes it as one of the "social glues" that bind "the WELL into something resembling a community. . . . Knowledge capital is what I found in the WELL when I asked questions of the community as an online brain trust representing a highly varied accumulation of expertise" (1993, 13). The conception of knowledge Rheingold uses here is modernist, a literally "banking" one—a "brain trust" that "accumulates expertise." (How the "capital' metaphor that is inveigled into much modernist thinking about AITs further complicates evaluation of such cases is addressed in the next chapter.) Despite his rhetoric, the WELL he describes remains just an alternative means of accessing the same, preexisting expertise, itself largely unaffected by the new technology of access.

This renewed attention to community is a strong example of how the current research agenda of social science is strongly influenced, if not quite driven, by the interaction of technological developments and the reproductive dynamics of "really existing" social formations. A most remarkable feature of this interface is the yawning gap between our ability to generate information and the paltry use made of it in actual social reproduction.

When the interventions and theorizations of *Building Virtual Communities* are "metaanalyzed," they are not indicative of significant social change, especially in the way entities like communities and identities get created, reproduced, and shared. They do illustrate one clear, rather mundane but not trivial sense in which change is necessary. Humans had previously relied on technologies to support "community at a distance" before computing. In the "virtual community" urtext, *Imagined Communities* (1983), Benedict Anderson outlines one striking example, that of early-nineteenth-century national newspapers in the U.S. To sustain dispersed AITed knowledge sharing, however, even more distance performing of community is necessary. If AITed educational technologies are to displace face-to-face, minimally technology-mediated, and/or non-AITed forms, we'll have to get much better at performing community at a distance. These papers illustrate that groups of people can be cajoled into doing it, but this is hard, and it is even harder to sustain. My hybrid experience, and that of my SUNYIT colleagues who do virtually completely online courses, suggest that even with current degrees of technical reliability, the goal remains elusive.

More generally, the *Building Virtual Communities* articles do not indicate some fundamental shift in the role of community in the reproduction of contemporary social formations. A substantial shift may be, is in some sense even probably, necessary, but there is nothing here to suggest that such a shift is inevitable. If one wants to argue that such a shift is likely, one must do it on the basis of other, more general social dynamics than these cases, largely the "best" to date.

Broader Contexts of Contemporary Asynchronous Distance Learning

On my campus, AITed distance knowledge networking has remained controversial since our state senator lumbered us with the satellite uplink. Conferences like 1999 ones at Harvey Mudd College and Lehman College of CUNY brought the diversity of opinion to a broader audience. The controversy is partly due to factors like those outlined above, for example, the necessity of creating virtual community but the difficulty of doing it with current systems. The controversy follows from other issues, too: whether it is a cost-effective use of educational resources and its place in the workplace politics of higher education.

AIT mediation also has other institutional implications that scholars like Noble gloss with terms like "commodification." In the State University where I teach, the privileged vehicle for delivery of asynchronous distance (AITed) learning is the SUNY Learning Network. In the 1999–2000 academic year, SLN enrolled approximately fifty-five hundred students in one thousand asynchronous courses, an enviable teacher/student ratio of one to five and one-half. Especially if one considers the huge ramp-up costs of investment in technology, Daniel's "scaling up" is expensive, the average course being essentially a tutorial.

Enrollments per course have slowly risen since that time. The United University Professors, the SUNY professional union, now includes seven thousand part-timers, of whom at least twenty-two hundred are half-time. Thus, casualization grew substantially as SLN expanded. Can dispersed AITed knowledge sharing only be "cost effective" if it takes place in conjunction with the part-timing of workers in the academy? This is clearly the case at the University of Phoenix, often touted as the most successful online university but one more accurately seen as committed to "hybridity."

Academic Professions

Among the stakeholders with a substantial interest in the debate over AIT platforms for dispersing higher education are academic professions and the professional associations that profess to support them. In the United States, academic professions play a unique role in higher education, both in the training of professional teaching and research staff and in maintaining quality. Given the historic commitment to individualism in the American social formation, high-prestige academic professions generally perform their roles without direct control over labor markets, nor have they, unlike M.D.s and lawyers, been able to get the state to give them indirect control. This is true even in unionized higher education, as here at SUNY. Lacking the machinery of professional certification, the influence of the typical academic profession is highly mediated, via accreditation bodies, funding of research and performances, ranking graduate programs, advising publishers or editing journals, and so on.

The spread of AITed knowledge networking at a distance, especially when combined with electronic publishing and other related new forms of knowledge creation and adaptation networking, presents significant challenges to academic organizations' procedures and standards. Those in the State University of New York need only connect its Board of Trustees–mandated standardized general education program with the idea of video-linked courseware to envision potential employment implications.

Yet were AITs standard in U.S. colleges and universities, they could broaden the ambit of academic professions' collective influence, via, for example, certification of preferred educational environments. Taking advantage of such opportunities would be a historic departure for academic professions, which have, for example, stressed individual academic independence (under, for example, the doctrine of academic freedom) rather than creation of effective policy structures.

Hesitation regarding AITs has been particularly pronounced in American anthropology. The American Anthropological Association still does not have an electronic journal and has only recently made available membership lists and listservs for accessing departments. There are several reasons for this, among them the technological ambivalence of an academic profession dedicated to ennobling ways of life more often framed as "low" than "high-tech."[7]

In 1997, the AAA nonetheless formed an Advisory Group on Electronic Communication, of which I was a member. This group accepted dispersed AITed knowledge networking as one item of several with regard to which it wished to help the AAA develop policy. The Advisory Group developed several recommendations with regard to distance learning, but these were referred to a Public Policy Committee, which decided instead to create a list of experts and an array of position papers. At the time of writing, the association was preoccupied with what to do about past journal numbers.

The Bursting of the Dot-Edu Bubble

One additional recent initiative to use AITs to mediate education demands attention. Dot-edu was the idea that there was much profit to be made in AIT mediation of education. Like dot-coms, dot-edus like Unext were business ventures. They aimed to create capital by using AITs, rather than campuses, to deliver higher education, and to do it asynchronously.

Were Daniel's assertions about scale true, the profit prospects for dot-edus would have been truly outstanding. Not only could you sell more computers ("suck chips," in a memorable Intel phrase), but also you could produce online educational commodities—courses—that would vastly expand the textbook market. You could sell commodities rather than pay teachers. The willingness of the most highly prestigious institutions of higher education to partner with even disgraced financiers like Michael Milken, whose money was significant to almost all the dot-edus, gave them immediate access to more or less any segment of the higher ed market that they wished.

The dot-edu bubble burst when online education was largely abandoned by highly capitalized firms, starting in the middle of 2001. By 2002, the highly capitalized dot-edus had either declared bankruptcy or transformed themselves; Unext, for example, first tried to do traditional business training, holding weekend seminars, and so on. Now it subsists by selling its course development software and expertise to other educational ventures.

Unlike the knowledge management AITs whose failure preceded theirs, the dot-edus really did get blown away in the bursting of either the dot-com bubble or the dot-telecom collapse that quickly followed. However, like KM, the dot-edus turned rapidly from being "the biggest threat to education as usual" to being no threat at all. The dot-edu burst has not meant the disappearance of AITed higher ed, but only, at least temporarily, that of a large for-profit, proprietary sector. What is left are the public providers, like the State University of New York Learning Network, able to take advantage of initial public subsidies, an existing extension structure, and low-cost adjunct instructors to run small online sections. There is also the University of Phoenix–type private provider, whose main provision combines face-to-face and online, synchronous and asynchronous, forms of instruction.

Online higher ed has expanded the market for distance education somewhat beyond the traditional group of those strongly "geographically challenged." The several billion dollars of capital burned by the dot-edus did demonstrate two things. One is that it is possible to create online environments as compelling as the good face-to-face class. The second, however, is that when all the actual costs of organizational infrastructure, hardware, software, and "courseware" are taken into account, good online courses turn out to be expensive.

All these experiences make me skeptical of scalability claims like those of Daniel. I agree that "pure" online courses are good for some students (for example, the antisocial and extremely highly motivated, or those especially "geographically challenged") and acceptable for some teachers (for example, those whose lives/careers are helped by being away from campus). However, my observation of good online education has convinced me that there are important qualitative differences between the motivational factors in these educative environments and those of courses with a substantial face-to-face component. The skills necessary to design effective, purely asynchronous courses are not easily obtained or generalized. The 2001–2002 bursting of the dot-edu bubble was not only backwash from the dot-com debacle. Totally online distance learning has not yet been demonstrated to be, in general, a good way to teach.

Summary and Conclusions

The communications capabilities of computing have benefited my professional career, which had to be developed "at a distance"; I have been the only anthropologist at my college for all but one of my twenty-five years in Utica. I was an early adopter of the Internet for e-mail with dispersed colleagues, with whom I used it to organize meetings and panels for various scholarly organizations. Beginning with trading floppy disks back and forth, we used electronic media to facilitate shared writing. I used databases to generate letters to field contacts as well as media for storing and manipulating research data. As I began to develop "real" content courses to focus on the social dimensions of computing ("People and Systems: Cultural Perspectives on Information Practice"), I integrated listservs and other communications capabilities, such as presentation software, into student exercises. I noticed that these new media more than occasionally fostered alternative communication patterns (for example, some students silent in class participated more in online discussion). Thus, AIT supplements rapidly became an important element of all my classes, many academic committees, and other professional activities; I now develop platforms for interactive knowledge networking on the World Wide Web, like www.knowledgenet.org. Despite the fact that yet again, in the spring of 2003, SUNYIT's listserv support broke down, I promote "hybrid" approaches to distance learning, ones combining online with other forms of distance and face-to-face education.

In sum, I am a moderately enthusiastic practitioner of computered education. I am open to the idea of putting my courses completely online, but have yet to see an adequate combination of educational and personal incentives to do so. While I see positive educational benefit from the forms of computerization I have incorporated, watching my peers who have used AITs to do "asynchronous distance learning" online has been salutary. It has not convinced me that the benefits (for example, decoupling education from any common time frame) are that generally substantial. These modalities require a high level of work, both to convert good education onto or develop new courses for online formats, to reach an adequate level of interactivity, and to attain the necessarily high degree of individualization required (for example, separate direct communication with each student). They also need technological systems that are more reliable even than the ones available today.

Putting education online is hard to do profitably because it is just plain hard to do. Learning is a social process. It is hard for individuals to maintain their motivation in isolation, so dropout rates in online courses are very high. To be successful, online education seems to need small groups of students, very good (and expensive—for example, highly interactive) materials, and highly motivated instructors willing to work very hard. Perhaps most important, it needs imaginative simulations of "virtual educative communities" that foster social relations of comparable strength to those that "face-to-face" students create spontaneously.

For such reasons, I don't accept the transformationist arguments of those like Daniel. In education, computers don't substantially reduce the desirability of face-to-face contact. In actuality, provision is often not scaled up; when it is, there is often a decrease in quality, and students continue to vote with their feet.

Pedagogy and AITed Learning

It is difficult to make sense of the forty years of AITed schooling if one conceptualizes knowledge networking from a modernist perspective (and teaches via a banking pedagogy), even harder to build a case that AITs transform knowledge sharing. The postmodernist approaches associated with AITed "discovery" learning are theoretically more "revolution" friendly, but as implemented they replicated many of the patterns of banking/modernism. Nonmodernist, coconstructivist approaches to knowledge sharing provide the best grounds for making a transformative case.

The communication customization potentials of AITs highlighted by constructivism—not their "banking" delivery or "discovery" representation of formal content—also have the most to offer design of educational technologies. Networking occurs more "naturally" when students and teachers share a relationship to the learning environment. The acceleration of school interest in AITs communicative potential that occurred with growing access to the Internet in the early 1990s roughly coincided with the rise of constructivist AITed pedagogy. This could be described as the beginning of a nonmodernist, as opposed to postmodernist, knowledge praxis in education.

A developed, coconstructivist, nonmodernist approach to learning aims to achieve appropriate, significant agency in the knowledge sharing process for both teachers and learners. Their interactive collaboration is necessary not to the transmission but rather to the active co-creation of knowledge. Coconstructivist pedagogy no longer sees knowledge as separable from the process of its reconstruction in groups.[8]

Like AITs in general, and like a good book, school-based AITs can go some way toward creating simulacra of shared time and place. However, just as community computing inevitably involves more that simple replicating existing place-based neighborhoods, such face-to-face simulacra are not, nor should they be, the primary goal of AITs in distance learning.

To practice AITed coconstructivism, we need to reformulate conceptions of knowledge in the educational process. Reinhard Kiel-Slawik's presentation to the Aarhus, Denmark, "Computers in Context" Conference (1995) described his personal efforts to turn the standard "Computers and Society" course into a serious educational experience, not simple "literacy" or "exposure," while still maintaining computer science professionalism. He began by drawing attention to the tendency of computer people to want to solve social problems technically, to "let the machines think for us." Rejecting the mechanistic models so embedded in, for example, data processing, he argued that computer science needs to understand learning as involving conscious human interpretation. Thus it is necessary to program machines in ways that reduce rather than enforce sequentiality, to open space for reflection. The only way he could do this was to reconceptualize the information systems development process itself as a learning process and make this the focus of his course.

Such collaborative transformation can be substantially facilitated by AITs, yet getting the conditions right for learning is also complicated by, for example, cyberspace's decoupling of space from place. It is important to keep in mind, moreover, that while Kiel-Slawik's teaching was carried out on "high-tech" knowledgescapes, his point about knowledge applies equally to the "low-tech" organization and face-to-face, "traditional" classroom. His work is more supportive of the idea that knowledge has *always* been more social than individual than of claims that, in high-tech organizations and schools, the character of knowledge necessarily changes.

While coconstructivist approaches provide the most fertile ground for the transformationist case, evidence that AITs are being used to apply them is in itself insufficient to justify the transformationist case. Coconstructivism demands the kind of *general reconceptualization* of knowledge in *any* social formation outlined in Section II of this book. It is in relation to this broader conception of knowledge networking that any transformative case must be made. One can only do this empirically.

One way to make an empirical case is to consider in detail concrete cases, ones that at least initially are suggestive of specific transformed diffusions of knowledge in cyberspace. Such sites must be shown to be generally typical if one is to argue that knowledge sharing is indeed transformed. The ethnographic picture of actual computer-mediated learning in schools presented in this chapter is, in contrast, more like my studies of AITs in knowledge creation and in organizations. In sum, I conclude:

- That attempts to mechanize learning which accept modernist conceptions of knowledge can only with great difficulty be seen as producing transformative effects

- That nonmodernist conceptions provide a better basis for imagining transformative educational practices;
- That in order for schools (or, for that matter, enterprises) to become "learning organizations," it is necessary for them to engage in broader processes of transformation
- That to transform these processes must involve more participative forms of organizational governance, achieving a consensus over what learning is for
- That there are as yet few concrete examples of AITs that facilitate such transformed learning, certainly not enough to claim that an "educational revolution" is under way

Commodification of Knowledge

If existing approaches were to "transform the relationship of the learner to knowledge"—as they may do—the impetus would be found in social dynamics other than their demonstrated superiority. In one respect, I agree with Daniel that online higher ed does seem to change the relationship between faculty and knowledge. This is a change in the interface between academic knowledge creation and teaching about which I am particularly concerned.

Academics traffic in ideas. In many ways, our research, practice, and teaching are mixed; we chew upon, rip apart, and reassemble our own and others' thoughts and experiences, both formally and informally, in class, at our conferences, and on our listservs. These practices are at the center of how we produce, reproduce, and communicate knowledge. Before online ed, faculty generally saw their job as giving knowledge away; the more students who learned, and the more each learned, the better.

The main impact of online education to date has been a change in discourse. Professors talk about their ideas as intellectual property rather than as something to share performatively. This "successful" change has dubious long-term implications for higher education. Indeed, the gravest threat posed by dispersed AITed knowledge networking may be this alienation of our thinking from our selves.

Perhaps this alienation, a mediation of thought via the commodity form, is a consequence of copyrighting and patenting the online course. As online faculty "protect intellectual property," online education makes knowledge sharing directly open to the reproduction of capital. Even the contract clauses that increase faculty's power to protect what they know, now regularly trumpeted as major accomplishments by academic unions, are part of a new dispensation, one which implicates faculty deeply in the commodification of knowledge. Along with Jean Lave, I am astonished at the rapidity with which my colleagues seem to have accepted the property framing imposed by distance learning. We less often ask, "Is this a good idea?" but more often, "Whose intellectual property is this?"

I am not suggesting that there are no important issues of intellectual ownership in higher ed. My main point is that, just as knowledge management has yet to transform organizations, let alone society, dot-edus have only had a limited impact on knowledge sharing. To the extent that both have had an impact, it has been arguably to *reduce* the amount of knowledge shared and bend what is shared more to the dynamic of the reproduction of capital. Both developments suggest that, to date, AITs have made us *less* rather than more of a knowledge society.

The commodification of knowledge is an important development in contemporary schooling. It takes place directly when schools try to appropriate employees' working knowledge via claiming copyright over online materials, but this is only its most overt, and probably not its most important, form. The significant danger is not the loss of some potential income because universities have appropriated "knowledge capital," but the more general transformation of knowledge networking going on behind teachers' backs. As we alienate our collective thinking through turning it into "products," we hobble our critical capacities. Like indigenous peoples forced to turn their cultural resources into property, we risk undermining basic justifications for the relative social privileges (for example, academic freedom) that we enjoy. However legitimate, such new concerns for fair economic return must not be allowed to overshadow attention to fostering both previous and new forms of knowledge networking.

Political Economy and the Future of Educational Technology

The CTG study contains a large number of examples of how not to use AITs in schools. Its potential may explain the interest in dispersed AITed knowledge sharing, but its actual knowledge networking achievements do not explain why so many resources are being expended, especially on "pure" (all online) forms. The CTG correctly predicted that efforts to apply AITs to education would continue. Since this cannot be because of demonstrated effectiveness and efficiency, it must be because of some broader social reproduction dynamic, such as AITs' place at the center of economic reproduction. The search for good ways to use AITs will remain central to the agenda of educational institutions, but a substantial gap remains between commitment to their application and knowledge of how to put them to good purpose. This gap follows from resistance to examine their actual, as opposed to potential, social functions.

In the middle of 2001 when this book was first drafted, the dot-edu bubble burst. The private organizations set up to make a profit off of putting education "online," by providing educational content, not just mediating technology, returned to just building platforms (for example, Blackboard™) or left formal schooling altogether. How rapidly these turned from being hailed as the biggest threat to "education as usual" to being no threat at all.

Does this mean a halt to the creeping commodification described above? Is the dot-edu debacle an indication that the transformative potential of dispersed AITed knowledge sharing is actually quite small, or only that the preceding and bigger dot-com and dot-telecom disasters had a further unfortunate consequence?

To answer such questions, we need to go beyond ethnographic research into ongoing practices. We need to deal ethnologically with the parallel absences of evidence for transformation with regard to AITed knowledge creation, knowledge adaptation, and knowledge sharing networking described here in Section III of *The Knowledge Landscapes of Cyberspace.* The existence of such parallels suggests a deeper analytic project, one that grapples with the structurelike properties of contemporary social formation reproduction, with the political economies that foster processes like commodification. These are the concern of the following section.

Section IV

Conclusions: The Analytics of Knowledge in Cyberspace

Chapter 10

A Critique of Popular Political Economies of Cyberspace Knowledge

Introduction to Section IV

As we enter cyberspace, do the changes in knowledge's character/social functions greatly transform society? The preceding chapters, various examinations of possible or "proto" aspects of cyberspace currently amenable to field research and analysis, offer three main answers to the knowledge question in cyberspace. First, they demonstrate how the great potential of automated information technologies to change the dynamics of knowledge networking also opens the way to social transformation. If, for example, new forms of knowledge networking broadened social participation, changes could be positive, but if they obscured the process by which the criteria for redeeming knowledge claims were established, they could be negative. Second, despite this potential, the degree of transformation has not, yet, been significant. Research on the actual implementation of knowledge AITs helps explain, third, why this is so: Substantial improvement in technologies to support knowledge networking, which thus bring about positive transformation, awaits the integration into their design of social perspectives like the ethnological approach to knowledge illustrated in the foregoing chapters.

This final section of *The Knowledge Landscapes of Cyberspace* draws implications from these three answers. The very last chapter addresses their ethical and aesthetic implications for "scaping" knowledge in cyberspace, that is, for those who design, implement, or study AITs. Chapter11 offers an alternative political economy of knowledge that incorporates the perspectives of the preceding sections.

The current chapter focuses on ways in which the conclusions outlined above undermine popular analyses that *do* attribute structural agency to the AITing of knowledge. The chapter critiques three knowledge-related theories of change in basic social life patterns—that there is a "new economy," that we are moving toward a "network society," and that a "cybernetic revolution" has fundamentally changed class relations—that, like Drucker's, already have substantial popularity. These theories' structural explanations of the direction and scope of general social change in the future are critiqued by contrasting them with the actual knowledge networking systems discussed in previous sections of *The Knowledge Landscapes . . .* Their common error is to hide or marginalize their assumption of the continuing, central importance of a social relationship, capital reproduction, to social formation reproduction.

While they are not themselves persuasive, the popularity of these three theories indexes the widespread search for more compelling structural answers to a basic problem in employment social formations, the source of value. By placing them in the context of previous labor and capital answers to the "value" question, the chapter explains both why some knowledge political economies are expressed as metaphorical extensions of capital, such as "intellectual capital" and "social capital," and why this is a bad idea. In this way, intellectual ground is cleared for the truly independent, rather than these "in drag" capital dependent, knowledge theory of value and technology developed in the following chapter.

Macrostructures and Structural Explanation in Social Science

To accept any structural account, whether those described in this chapter or the alternative of the following chapter, a reader has to believe that trajectories of general social change exist and that they can be caused by adoption of technologies. However, much contemporary social thought is suspicious of general talk of this sort, especially that invoking structure. Brackette Williams is typical of those anthropologists who, in a postmodern register, are skeptical of the idea of structure, claiming that "There are only people and their practices" (personal communication). The sociologist Craig Calhoun similarly uses "post-structural" to describe his influential social theory (1995). Before social cybernauts can decide *which among* the structural accounts best accounts for the likely direction of future social reproduction, they need good reasons to accept any of the structural analyses generally referred to in social science as "political economy."

My previous book, *Cyborgs@Cyberspace?*, offered a reproductionist reading of social dynamics (see Hakken 1987). On social reproduction, since human social arrangements don't perpetuate themselves automatically (i.e., they are not carried in our genes), frequent intervention is necessary if these arrangements are to extend in time. Study of deliberate activities to promote the continuation of social arrangements from one period to the next should provide insight into how continuity is accomplished, denied, or mitigated.[1]

For example, among the articulations of the cyberspace knowledge revolution idea raised in chapter 1 was Peter Drucker's notion that because they profoundly increase/decrease the social power of particular groups (for example, technologists/manufacturing workers), changes in knowledge usher in a "postcapitalist" social formation. Drucker's notion of a postcapitalist knowledge society is "structural" in that it articulates a fundamental change in the character of social reproduction. Analyses of this sort have a compulsive quality, evoking, for example, the determining of large "systems." While some talk about cyberspace stresses its voluntary character, knowledge society talk generally presumes a new framework for social life, a set of macrosocial relations with wide ambit. The idea of a knowledge change–induced transformation of social formation type is quite structural.

Macrosocial relations are large, greater in scale than community, organizational, or even regional ones, involving "high-level" forces that precede and thus limit human volition, both individual and collective. Any connections between macrorelations and people's immediate actions or experiences are highly mediated by the very large structurations (Giddens 1991) that social scientists call "totalizing totalities." Such entities may reach beyond the nation. Macrosocial relations involve "systems" that are "general" even if they may function in open, quasi-"organic," rhizomic, and/or evolving ways. Structural rhetoric evokes forces that function, as it were, "behind our backs."

The typically totalizing slogans for the projected social formation in waiting—such as "information society"—frequently deploy the structural speech forms characteristic of economics. Despite occasional demurrals about how, for example, it is "the local" in which it is actually manifest, such talk (like many uses of the term "globalization") often has a strongly foundational or essentialist quality.

The ease with which structural terms become foundational opens them to the postmodern critique of essentialism. As discussed in chapter 4, essentialism is erroneously treating characteristics better understood as transient or highly contingent as permanent. In antiessentialism, features normally considered to be part of an object's "nature" are better understood as artifacts of particular interpretive framings.

Discomfort with "structure talk" also follows from its frequent association with discourses of mastery, whose hegemonizing concepts facilitate domination. Postmodern ideas like the knowledge regression discussed in chapter 2 build on various critiques of the structuralistics of classical social theory, critiques such as, in sociology, the symbolic interactionism of George Herbert Mead (1962), the ethnomethodology of Harold Garfinkel (1984), or the social constructivism of Berger and Luckman (1972). Anthropologists like Williams echo these critiques when they claim that there are only actors and their projects, no discernible interests, let alone structures. If its impacts only take place "behind peoples' backs," structure is irrelevant to experiential analyses that privilege human perceptions mediated by cultural constructs.

Yet it is possible, instead of concluding that social life has no discernible structure so that analysis itself might have to be abjured, to ground structure talk nonessentially. A reproductionist account accomplishes this by distinguishing between practices that merely replicate macrosocial relationships—simple reproduction—from those that for reasons of context transform them—extended reproduction. Practices do sometime have dynamic, structure-transformative implications, but these are to be accounted for in terms of conjunctions of particular circumstances, not essences.

In social reproduction, effective structural analysis of cyberspace is not only possible; it is for several reasons also necessary. Some at least loose notion of structure is implicit in the very idea that there is a legitimate analytic moment in social studies. In order to take the idea that there may be a transformation seriously enough to examine evidence relevant to it, as *The Knowledge Landscapes of Cyberspace* does, one must admit at least *the possibility* of something like structure.

Similarly, some minimum notion of structure is necessary to ethnography. To do it, one must presume general things, practices (for example, knowledge networking) present in enough social formations that their different manifestations can be compared. To communicate across languages/cultures, ethnographers must have available for use categories with substantial overlap in meaning, the kind of overlap that enables meaningful talk about what is or is not generally the case. To explain things holistically—that is, to account for specific practices in terms of broader contexts, the characteristic explanatory trope in ethnography—similarly requires a capacity for general discourse. For example, it is common to speak of "ages" or "eras," above and beyond specific places or spaces. Differences between "times like these" and "times like those" are frequently explained in terms of the dynamics indexical or at least indicative of different types of social formations. Ethnographers and social theorists are not the only ones compelled to presume the existence of things that have structurelike regularity; to construct policy, one also invariably deploys general concepts.

The cost to social analysis of labeling as essentialist all discussion of regularity in social dynamics is too high. To do so dooms one to unending ad hoc accounts of discrete events. Such knowledge can only be "local." However, this presumption itself is essentialist. It can only be rhetorical because it cannot be demonstrated to be true: to establish that every social formation's reproduction has total local autonomy, one would have to engage in structural discourse. Without identifying structures that generally support local autonomy, the idea remains mere premise. As I illustrated in chapter 2 of *Cyborgs@Cyberspace?*, most attempts to avoid theorizing structure end up merely masking it.

A fourth, still "weak" justification for talk about structure is that a large proportion of humans/cyborgs currently extend their own social reproduction by using concepts that presume structure. At a minimum, structure-based experiential relating engenders structurelike effects. For example, since the late nineteenth century, most public intellectuals in the West have used models of social formation reproduction framed in the formalisms of neoclassical economics. A notion of "productivity" developed within these models has taken on meaning outside the model's direct use—for example, in social policy. Similarly, all the talk about cyberspace "causes" certain connections to be made; "knowledge society" talk itself engenders quasi-formal, must-be-related-to-as-structural effects. Structure-presuming practices, like policies based on the "human capital" notions analyzed later in the chapter, influence experience irrespective of the concepts' analytic utility. To address the possibility that there may be a gap between talk and reality, ethnographers must generate possible alternative accounts. To open space to criticize dominant discourses, one must hypothesize alternative structuralistics rather than reject political economy a priori.

The arguments for structuralistics made thus far follow from metadiscourse over the *possibility* of structural talk. They are philosophically "weak," not derived from demonstrated structural regularities in the reproduction of actual social formations. There is a "stronger" case for thinking structurally: especially when supported by self-conscious, collective articulations, human action itself produces structure. Human interventions often produce something like inertia, and deliberate action develops momentum, in ways that have a cumulative, material impact on social formation reproduction. In actor network theories of technology (Latour and Woolgar 1979), for example, as particular technology actor networks become central to social reproduction, they incline reproduction in some directions, while making others more difficult. In Langdon Winner's phrase, "Technologies have politics" (1980).

In sum, to offer a full answer to the knowledge question in cyberspace, we must consider how knowledge change *might* result in new reproductive dynamics. Further, we need ways to talk about this that *don't presume automatically that it does so*. This means general talk, discourses capable of accounting for the notions about the structural with which people operate, the consequences of these notions, and the inertiae/momenta that potentially are manifest in "systems as wholes." Finally, we need to be clever enough that our talk does not presume that which needs examination.

Some New Structures Popularly Held to Be Induced by Knowledge Change in Cyberspace

The idea of knowledge change so important to intellectuals and scholars today takes many specific forms. To illustrate the breadth of its articulations, I examine below three diverse contemporary political economics, each of which presumes that new social reproduction dynamics are related to AITs in ways in which knowledge change is central.

The "New" Economy

Many mainstream social theorists champion the "new" economy alleged to have emerged in the 1990s, especially during their second half (Lee and Shu 1997). For example, the emergence of a new "knowledge economy" is taken by the British sociologist Anthony Giddens as structural "proof" that we are now a "knowledge society" (Giddens and Hutton 2000).

Any notion of a "new" economy implies a preexisting "old" one. Especially in the United States, the case for a new economic dynamic was the longtime, simultaneous presence of several positive economic phenomena: continuing expansion, fast growth, low inflation, and low unemployment. According to the "old" neoclassical economics academically regnant in the United States, these factors couldn't co-occur for long periods of time. A related, apparently also outmoded, "old" law was that of the inevitability of business cycles, of alternating growth and decline. Since the copresence of the first four phenomena, and the absence of the last, indicated that the old "laws" of economics no longer applied, the new economy demanded a new economics.

As in Giddens (Giddens and Hutton 2000), AITs are generally treated as one, if not the most important, factor responsible for the new economic dynamic. For example, with AITs raising productivity faster than income, profits could continue to increase amid continuous expansion. However, attempts to justify empirically such connections between the new economy and AITs were stymied for a long time by a problem that came to be called the "IT productivity paradox." From at least the 1960s, increased investment in AITs was associated with *declining* rather than increasing productivity statistics (Attewell 1994). In the words of the Nobel laureate Robert Solow, "You can see the computer age everywhere but in the productivity statistics" (1987, 12).

The embarrassing absence of the expected, AIT-induced increase in productivity was explained, unhappily, by various ad hoc means, typical of which is the "old" idea of "convergence." In convergence theory, to be the application leader is risky because it costs a lot and others can quickly take advantage of your efforts at much reduced cost. Such economic calculations mean that the advantages of applying new technologies thus tend to dissipate. It is easier, and much cheaper, for most producers to wait to adopt an innovation until a few have worked out the bugs.

Convergence is not a very satisfactory explanation of the IT productivity paradox. Convergence would predict a slow pace of technology deployment, but firms deployed AITs quickly *in the face of* declining productivity statistics. Moreover, convergence predicts declining profits, but these were generally increasing. From the convergence perspective, the actual correlates of AITs look even more paradoxical.

In any case, about 1995, U.S. productivity statistics started up, and the embarrassing "IT productivity paradox" could be (temporarily, it turns out) put to rest. In particular, "new economy" advocates seized upon the argument that the latest corporate knowledge technologies—inventory control, demand forecasting, flexible scheduling of production, Computer-Supported Collaborative (or Cooperative) Work, intranet knowledge bases, interorganizational data sharing—had narrowed the gap between supply and demand so much that a truly epochal productivity surge had finally overcome whatever, such as a convergence effect, had caused the problem (Lee and Shu 1997).

Even skeptics like Federal Reserve Chairman Alan Greenspan began to speak of an AIT-induced productivity increase. The idea of a "new economy" need no longer be treated as hype, because the increase in productivity really did indicate dynamics quite different from the old ones. These dynamics were closely connected to new characteristics/functions of knowledge brought about by use of AITs. In this way, a positive answer to the cyberspace knowledge question became central to the new economics.

The Network Society

A month's review of any large-circulation Western newspaper with a business section during the late 1990s would establish the importance of AITs in "new economy" popular thought. Arguably the articulation of the links between knowledge and social change most influential in both scholarly and politically liberal policy circles, through his influence on both Tony Blair's and Bill Clinton's policies, was that of the geographer/urban sociologist Manuel Castells.

In his notion of the "network society" (2000a), Castells strives both to name and to account for the general dynamics for a new type of social formation, one that he believes now dominates social reproduction.[2] In substantial part for Castells, the new dynamics derive from a profound shift in the locus of social process. A "space of flows" displaces the grounding of human activity in "particular places"—or, in the phrasing I prefer, space is "decoupled" from place. With globalization, the salience of units like "cities" and nations to social reproduction substantially decreases. If geography is no longer a particularly meaningful framework against which to organize accounts of social relation and interaction flows, what alternative framings should replace it?

"Networks," Glaser and Straussian "grounding points" (1967) replacing geographic ones, are Castells's alternative theoretical register. His justification for calling the new social form "the network society" is not that networks themselves are new. Rather, networks *re*emerge in new forms and displace the hierarchical forms so characteristic of organization and governance in the industrial society. The new networks can do this because of AITs, which, even under conditions of capitalism, disperse activity, distribute intelligence, and unhinge knowledge from place.

"Network enterprises"—intra- but especially interorganizational networks—replace firms as the chief unit of capital accumulation and states as the chief units of governance, creating a new, globally operating economy. Network society has very different dynamics from Industrial Society. Electronic networks facilitate a more individuated identity formation and replace the collective units of organic solidarity so important to Marx, Weber, and Durkheim. The result is consummately Blair- and Clintonite, a capitalism with neither a capitalist nor a working class:

> In the last analysis, the networking of relationships of production leads to the blurring of class relationships. This does not preclude exploitation, social differentiation, and social resistance. But production-based, social classes, as constituted, and enacted in the Industrial Age, cease to exist in the network society." (2000b, 18)

Castells has only recently substituted "network society" for "information society" as his rubric for the new social formation type. On the anthropological ground that "knowledge and information were central in all societies" (2000b, 10), he now feels that the "information society" label is misleading.[3] Generally deploying network theory in a contemporary sociological, Barry Wellman (Wellman and Haythornthwaite 2003) mode, Castells holds networking to have been the most typical form of human interaction until displaced by the historically recent rise of hierarchies like states and corporations. However, by undermining these latter forms, AITs compel networking's reemergence: "But for the first time, new information/communication technologies allows [*sic*] networks to keep their flexibility and adaptability, thus asserting their evolutionary nature . . . Networks de-centre performance and share decision-making."

A Castellian network is an oddly autonomous, even self-determining entity:

> It works on a binary logic: inclusion/exclusion. All there is in the network is useful and necessary for the existence of the network. What is not in the network does not exist from the network's perspective, and thus must be either ignored . . . or eliminated. If a node in the network ceases to perform a useful function it is phased out from the network, and the network rearranges itself – as cells do in biological processes.

Despite the last biological metaphor, Castells's networks are essentially informational, not organic, entities (see chapter 6 or, for that matter, the dialogue from "The Matrix"): "A network is a set of interconnected nodes. A node is the point where the curve intersects itself" (all quotes from 2000b: 15).

Here, as at many other points, imprecision in language, especially about the causes of these dynamics, impedes understanding. Nonetheless, these quotations capture the "foundational" quality of Castells's account of the implications of AIT for social reproduction. Liberated from the inefficiency and ineffectiveness of hierarchy, on the one hand, and place-boundedness, on the other, AIT-compelled networks manifest their underlying potential to evolve and remake social reproduction in their own image. The resulting social formation is driven by a "flow, flow, flow!" imperative, replacing the "accumulate, accumulate, accumulate—this is Moses and the Prophets!" (Marx 1871) dynamic to which employment social formations were heretofore bent.

Like other cyber-enthusiasts, Castells views these changes in epic terms: "[The] . . . new set of information technologies represent a greater change in the history of technology than the technologies associated with the Industrial Revolution . . ." (2000b, 10). Most important for the purposes of this chapter, Castells assigns a key place in the ascension of the network society to change in the social functioning of knowledge:

> [Characteristic of] this new technological paradigm is the use of knowledge-based, information technologies to enhance and accelerate the production of knowledge and information, in a self-expanding, virtuous circle. Because information processing is at the source of life, and of social action, every domain of our eco-social system is thereby transformed." (10)

Knowledge changes (predictably fudged to include informational ones as well) generalize their social impact via the network enterprises described above. These new forms (replacements of the firm?) are found "[a]t the heart of the connectivity of the global economy and of the flexibility of informational production" (10).

Change in Worker Power?

Knowledge has also recently attained a privileged place in radical as well as mainstream and liberal political economies. In 1994, the once New Leftists Carl Davidson, Ivan Handler, and Jerry Harris launched *cy.Rev: A Journal of Cybernetic Revolution, Sustainable Socialism & Radical Democracy*. In contrast to leftists critical of AITs-related knowledge changes (for example, Noble, Stoll, Aronowitz, and DiFazio), *cy.Rev* celebrates the computer revolution. Indeed, for it, the key to the revival of an American left is not to critique cyber-knowledge rhetoric but to embrace it:

> [A]n important revolution going on in the world today . . . [is] being driven by new developments in information technology . . . Digitalized knowledge has now become the major component in the production of new wealth. The information society is supplanting industrial society as surely as industrial society replaced agrarian society. The depth of these changes, however, has been largely ignored by much of the left community." (1994, 1)

Once their importance is recognized, previous Marxist notions must be revised in light of changes in knowledge:

> [N]ew insights into the nature of changes in the economic base [occur because] knowledge has become the most important tool of production...[in] what we'll call 'information capitalism" . . .
>
> The changes here are having a dramatic impact on both the relations of production and the nature of work. There are new social divisions being created along with a realignment of classes and strata around many critical issues.

> The ground for organizing the class struggle is shifting; there are new dangers of prolonged joblessness, repression, chauvinism and war. But there are also new opportunities creating new possibilities for a democratic and ecologically sustainable socialism. (1994, 4 and 1)

Like Castells and so many others (for example, the NSF), here Davidson, Handler, and Harris elide the information/knowledge distinction. They go on to add Alvin and Heidi Toffler to the list of important contemporary political economists, taking from them the idea that

> the main reason for today's ongoing revolution in the productive forces was the invention of the microchip. This revolution began in the 1950s with the merging of transistors, themselves the first major practical application of quantum mechanics, with the mass replication of miniaturized integrated circuits...The microchip's impact is changing everything about our world and the way we live. Civilization is undergoing a quantum leap on the order of the agricultural revolution launched 6000 years ago and the industrial revolution launched 200 years ago. We have now entered a third period of human history . . .
>
> Intellectual capital, developed and held by knowledge workers and encoded in software and smart machines, is the key element of wealth in today's information capitalism. Physical labor and industrial machinery are now secondary to the value added by information." (29 and 6)

"[N]ew challenges for Marxism and radical theory" follow from changes in basic class structure:

> Knowledge workers today are in the position of the old industrial proletariat. They are key to the enhanced production of surplus value. Just as blue collar workers contained two sides—the conservative labor aristocracy as well as the most progressive sector of labor supportive of democracy and socialism—knowledge workers will divide into two as well. One sector will form the social base for the defense of information capitalism regardless of its excesses. Others will deeply understand the potential the new technology has for creating and sustaining a new social order. This progressive side also is born from the conditions of its own labor, which are enmeshed [*sic*] in the most advanced forms of capital. (10 and 11)

As a final jibe at those unable to appreciate how radical the knowledge-induced changes are, *cy.Rev* warns:

> [W]hat is worse than the dangers posed by the third wave is the attempt to ignore or stifle the information technologies fueling it. This was a deep flaw in the structure of the "command economies" of the Soviet block . . . The growth of the new technology requires open, accessible, and decentralized sources and outlets for the flow of information. (1994, 31)[4]

Why the "New Economy" Became "The Economy Formerly Known as 'New'": The Weaknesses of "Knowledge Society" Political Economies

Just as I am not a professional philosopher or full-time informatician, I am not an economist. My intention here is to point out rather obvious empirical weaknesses manifest across this broad range of economic discourses on knowledge. This critique is a necessary preliminary to consideration of the alternative political economy of knowledge to follow in chapter 11.

In a new economy account of the old economy, any tightening of the labor market would tend to produce inflation and "overheating" of the economy, requiring higher interest rates. With enough increase in productivity, however, employment and wages could rise without setting off inflation. The fact that wages and employment rose while prices didn't in the late 1990s "proved" that productivity could increase so much that convergence was no longer a problem.

The second draft of this chapter was completed in early 2002. This was several years after the "Asian economic flu," two years after the bursting of the dot-com and roughly one after the tele-com bubbles, and just as the last of the (first wave of the?) for-profit online universities or dot-edus were being bought out or declaring bankruptcy. In a time of continuing economic retrenchment, talk of a "new economy" had more or less disappeared, replaced by a nervous "looking over one's shoulder," as in the Wall Street Journal article on February 14, 2002, in which I first encountered "the economy formerly known as 'new'" phrasing. Yet even during final drafting in 2003, as in this article, economists' talk is still largely Panglossian—focused on why the continuing recession wasn't an "old style" one, why AITs weren't the real reason for it, and how in fact they would rescue us from it.

Two additional dynamics characteristic of the time in which writing took place should also be mentioned. One was that the power of both corporate and individually held capital to promote its reproduction was still growing, as manifest in The Second Bush's energy policies and proposed tax cuts. The second was the increasingly anarchic quality of the world's economy, especially the gap between economic developments and the ability of corporations, the

U.S. state, or the World Trade Organization to influence them. Rather than a stable economy in which markets clear and reach equilibrium more quickly, "economic diseases" continued to spread chaotically. Disastrous ecological change was generally (but not universally) acknowledged to be a direct consequence of economic activity. Even computer use had been recognized as a substantial contributor to energy shortages. A social movement against corporate globalization, one highlighting the inability of remaining political structures to influence events, had emerged, one more ambivalent than either Castells or Davidson et al., let alone mainstream economics, about AITs.

The increased ambit of capital reproduction, anarchy, and economic distress may all be linked. In the age of Enron, the impression one had was of a world increasingly beyond control. The shrinking ambit of both "old" nation-based and "new" multinational tools to influence shifts in economic dynamics, let alone cushion their effects, engendered a sense of narrowed rather than extended prospects for influencing social reproduction. The connotations of the term "mythinformation," seem a more appropriate characterization of the new millennium than "knowledge society."

New Economy?

Such conditions (perhaps only temporary but still empirically observable) are incompatible with by the three political economies examined in the last section. Why were they not predicted by "new economy" structuralistics?

The beginnings of an answer emerge through reconsidering productivity. Its changed dynamics in the mid-'90s were taken, as argued above, as the decisive explanation for the new economy. The basic idea was that increased knowledge increased worker productivity so much that, sometime about then, the pent-up but yet unrealized potential for increased productivity in AITs broke through. Because the new AITed knowledge technologies were being first deployed about this time, the correlation was taken as a causation, the increase in productivity seen to follow from their applications: The "information to knowledge barrier" was finally breached, and Chairman Greenspan need no longer be an IT skeptic.

Beginning in the summer of 2000, however, productivity statistics in the United States began to fall again. One is tempted to attribute this to "knowledge management fatigue syndrome," but this view, like the argument described in the previous paragraph, assigns too much influence to knowledge technologies. Rather, like the increases after the mid-'90s—and, indeed, the declines from the '60s to the mid-'90s—the 2000 decline is more likely an artifact of the bizarre ways that productivity statistics are calculated than a "real" phenomenon. The measurement problems are most obvious in the service sector. There being no service sector equivalent to the "widget," the countable, generalized unit of the manufacturing sector that is set against hours worked, measuring productivity in service remains a fundamental problem for formalists. Productivity economists have therefore generally treated salary as a proxy index of productivity in this sector. Salaries in the service sector, adjusted for inflation, declined through much of the '60s to the '90s. Consequently, the decline in general productivity statistics of the late-1960s–1995 era may be an artifact of the pronounced shift from what to economists counts as goods to services production. This alternative explanation makes even more sense when one recognizes that this was an era of higher unemployment and declining trade union power, leading to stagnant/falling wages in both goods and service sectors. Wages in service finally rose only with the general economic expansion of the mid-1990s. After 1995, but especially in the "Y2K" run-up of 1999, expansion even slowed the rate of corporate downsizing. The subsequent statistical decline in 2000 productivity makes sense in relation to falling manufacturing employment, especially in the computer industry, and the consequent increase in the proportion of service employment, where salaries returned to stagnation.

In August 2001, second-quarter U.S. economy productivity statistics ticked up again. Some Panglossians interpreted this as a sign that the economic decline was "bottoming out" and predicted that convergence effect would again disappear. Other economists pointed out, however, that if, as was the case in 2001, employment declines while output remains stable, "productivity" statistics always rise. This happens, as it did in the period in question, when massive corporate downsizings reappear.

In short, changes in productivity statistics reflect shifts in employment and social, and therefore economic, power. They do not necessarily directly reflect changes in production technology, including knowledge technology. At the time of writing, Greenspan had not yet reinvented himself as a productivity skeptic. Perhaps were he a mere economist he would. Because, however, his slightest hesitation can cause a market decline, Greenspan, like other mainstream economists, tends to "get stuck" in celebratory mode. This is particularly true with regard to productivity, even though analyses like that immediately above suggest that what productivity statistics actually measure is not at all clear.

Irrespective of their analytic shortcomings, their ideological importance to the legitimation of existing social reproduction patterns means new economy rhetorics give momentum to the status quo. Once performed, the rhetoric of productivity's alleged AIT-induced increase came to play a role in economic discourses, and in the broader social arrangements they justify, one too important to be easily abandoned. That AITs increase productivity is just too good a story to be deflected by mere statistics. Mainstream structural accounts continue to echo new economy thought, even if the slogan is abandoned.

Instead of the really different dynamics of a new economy, however, we got knowledge management fatigue. To be able also to see around rather than only in new economy structuralistics, one needs heightened critical sensitivities. In particular, alternative conceptualizations are needed if the actual role of knowledge change is to be evaluated empirically.

Network Society?

Talk of a "network society," like that about a new economy, had drastically fallen off by mid-2001. The rise of the former is explicable in terms of the struggles of disciplines—and "schools" within disciplines—for space in the "marketplace of ideas" long an adjunct of capitalism. (See Abbott 2001and also chapter 6.) The sudden silences in regard to them are interpreted more parsimoniously as "rhetoric fatigue syndromes" than as reflective of important subsequent changes in general reproductive dynamics.

Besides, "network society" is not an empirically useful notion. In essence, Castells confuses the increasing *ideological* value of computing's knowledge relationships, an admittedly significant cognitive terrain, with structural change in social reproduction. "Flow, flow, flow" is good rhetoric but not an analytically justified replacement for "Accumulate, accumulate, accumulate."

This trope is not the only echo of Marx in Castells. In his Parsonianized but still recognizable "stages" account of cultural evolution, hierarchies displaced networks, the "natural" forms of social expression of early social formations. "Rationalized, vertical chains of command and control" "outperformed" networks "as tools of instrumentality," (Castells 2000a, 12), only themselves to be displaced in turn by newly energized-because-AITed networks. Through the dictatorship of the proletariat, the mature communism envisioned in *The Communist Manifesto* replicates "primitive communism," but on a higher level. This is structurally parallel to the relationship that AITed network societies are supposed to have to "premodern" ones.

Two generations ago, we Marxist anthropologists were arguing for attention to "really existing" different social formations, like gathering/hunting, as a remedy for rigidly essentialist Marxist accounts of social evolution (Hakken and Lessinger 1987). We were critical of deterministic, arguably teleological, cultural evolutionary formulations then, and similar criticsm apply to Castells. We can acknowledge transformative possibilities without assuming, as Peter Pan does with Tinkerbelle, that believing in them makes them so. To argue that society really *is* profoundly transformed via the new focus on knowledge, one must ignore the embarrassment of knowledge management fatigue. If knowledge technologies were responsible for the new economy, shouldn't knowledge management's failure, and therefore the "unnetworking" of organization, be held responsible for the world economic slowdown? It seems most reasonable, however, to remain skeptical about the strength of the link between knowledge technology and world economic dynamics.

Chapter 8 presented an alternative, less foundational and more descriptively accurate way to conceptualize the changes in production at which Castells points. This is the possibiliy of an emerging "cyberfacture" stage in the history of the labor process under capitalism, one potentially as distinct as factory-based manufacture was from putting out, or later Fordist machinofacture was from manufacture. Theorizing a new stage within the same social formation type, rather than a new type altogether, means focusing on shifting arrangements within the same basic underlying institutional pattern. It is more parsimonious, albeit of less rhetorical power, than "network society" hype.

Technicist Political Economism

Postmodern social theory properly alerts us to be suspicious of facile transformative determinisms of overly structuralist theory like Castells's. To develop effective alternative structuralistics to the dominant neoclassical ones, one must be equally cautions of the political economism of Davidson and his colleagues.[5] While to my knowledge *cy.Rev* is no longer being published, its structuralistics influenced debate in, for example, the antiglobalization movement.

As argued in *Cyborgs@Cyberspace?,* political economistic structuralistics interfere with being empirical about computing and social change, whether anti- or pro-capitalist in their foundationalist technodeterminism. In the "lite," Davidson version, knowledge change–inducing AITs cause a revolution in the forces of production which in turn moves social dynamics onto new terrain. *cy.Rev* adopts the same knowledge theory of value as the procapitalist OECD: "Digitalized knowledge has now become the major component in the production of new wealth." Its naive positivism about AIT echoes Bernal and the other interwar socialists committed to a scientific-technical revolution, for whom this "way forward" substituted policy for politics (Hakken with Andrews 1993).

While more pessimistic about technoscientifically induced changes, the political economic "dark siders" Stanley Aronowitz and William DiFazio (1995) are equally presumptive about the determining force of new technology. They see an "ineluctable" tendency in AIT toward the destruction of jobs, especially good ones.

Involvement in the neo-Marxist battles over political economy of the 1970s taught me two lessons. One, from the Althusser wars (Althusser and Balibar 1970), can be stated, if overly simplified, as the priority, in the long run, of social relations over technical relations of production. A second was to emphasize the extended over the simple moment in social reproduction, to stress the recurrently transformative, richly dialectical character of social dynamics, as well as the relative autonomy of multiple moments within them (Hakken 1987). These lessons are equally lost on Davidson et al. *and* Aronowitz-DiFazio.

There are occasions when it makes good explicatory sense to abstract the mechanical elements out of an economic congeries, and it may be appropriate to describe the consequent momenta (and inertiae) of reproduction in structural terms. This is only justified, however, as long as one keeps in mind that structural abstraction means simplifying the reproductive complexity of actual social formations. Abstraction is thus a legitimate moment in social analysis, but it should not be taken for the totality of social analysis.

In actor network theory (for example, Callon 1986), to give any abstraction analytic permanence, as when one identifies a social property as a part of a machine rather than the broader technology actor network (TAN) of which it is a part, invites essentialist distortion. The technical capability of a TAN is only a potential that must be concretely actualized, not an easily separable "factor." Further, technical capability, like knowledge, is contested, constantly requiring reproduction, which, in the process, is extended and reconstructed differently. Since TANs vary greatly in their degree of stability, it makes little sense to speak of anything, whether disemployment or free flow of information, as an "ineluctable" implication of AITs. It is better to concentrate on the various ways in which social groups differentially appropriate artifactual potential and, in the process, actively transform the relevant TANs.

To treat social dynamics as technologically determined while ignoring the processes through which some technologies are rejected and others implemented is an example of what sociologists call "hypostatization." There are strong disemploying potentials in contemporary employment-based social formations. However, accounts of these tendencies that trace them largely or fundamentally to something inherent in knowledge technology are facile (Hakken 1999a).

The Value Theoretical Context of Popular Knowledge-Based Structuralistics

While there are good reasons to be skeptical of the popular political economies of cyberspace knowledge critiqued above, their popularity is indicative of a need for better structural accounts of contemporary social change. Is there a more satisfactory way to incorporate knowledge AITs' potential into models of contemporary political economy?

Employment Economies and Political Economy

In the West, structural accounts tend to begin with the economic. The social science that invented the idea of an economic moment in social reproduction was itself created, about two hundred and fifty years ago, as a metadiscourse on the rise to reproductive dominance of a particular kind of activity, that variously associated with "markets," "commodity production and distribution," "industry," and "employment." In social formations of this new sort, the employer/employee relationship tended to displace older ones, like that between the serf and the lord or the believer and the church. To label the new social formation in terms of the social relationship most salient to it, it is the "employment social formation."

The new social formation fostered a new discourse that recognized and celebrated this new "economic" activity. Employment's replacement of other relationships was justified because it was a new, felt-to-be-better source of "value," or the wealth of nations (Smith 1991 [1776]). The task of the new social science, the foundational project to distinguish it from social philosophy, was to account empirically for value's creation (Toulmin 2001) and thus its centrality to social formation reproduction. Its accounts were to avoid the moralistic approach (B. Williams 1985) of older discourses, to replace "ought" arguments with "is" descriptions of new "laws of value" determining human events.

The knowledge produced by this project came to be known as "political economy." Adam Smith and his Scottish moral philosophy colleagues, its chief advocates, believed they were constructing foundational accounts of social value, accounts that broadly paralleled Newtonian understandings of matter.

The Value Question and Labor Answers

In Smith's eighteenth century, figuring out where value came from was also a pressing public policy issue. Such knowledge would determine the legitimate activities of the state in a social formation bent to an employment dynamic. Since it was initially posed, three basic answers have been given to the question of where value comes from: labor, capital, and, most recently, knowledge. Until the late nineteenth century, political economists in general, from Smith and Ricardo to Marx and Mill, adhered to a labor theory of value. This was the idea that the increased value in an employment social formation came from a new productive factor, labor power, analyzed by Marx as a commoditized form of the capacity to do work.

Constructing labor socially as labor power enabled comparison of a wide range of diverse activities. Labor power was perceived as generally displacing land, raw materials, or rent as the most dynamic element of value creation. This new approach to labor was in the common view the factor most crucial to the capacity to accumulate value, now taking the form of profit. Enabling more buying and selling, employment institutions could foster more rapid accumulation of a social surplus than mercantilism or rent-producing agrarianism. Employment allowed commoditization of new markets and exploitation of new productive instrumentalities (for example, technologies).

In addition to a labor theory of value, these scholars also generally accepted some corollaries about the dynamics of employment social formations. One was that the capacity of the new arrangements to expand value was not permanent. In the long term, employment-based profits had a tendency to fall. As long as there was competition, employers would tend to bid up wages until wages approached the selling price of the commodities produced (an earlier form of the convergence problem already discussed in this chapter).

Via monopoly, accumulation could be extended into the medium term. In the end, however, the pace of commoditization would inevitably slow, and surpluses would tend to shrink. Smith was enthusiastic about how innovating new technologies of production could postpone the slowdowns, but like the other classical political economists, he accepted the long-term tendency of the rate of profit to fall. Thus, in classical political economics, the new society would enjoy only periods of accumulation; it was doomed to both periodic crises of profitability and ultimate decline. This political economy was dismal science.

Capital Answers to the Value Question

An alternative theory of value traced it not to labor but to investing profits, or "capital." A minority of political economists (for example, Marx's target Senior) argued that when wielded knowledgeably, capital was a value-creating factor independent of labor. What one got via a bank loan to buy, say, newer machines or raw material was control of an entity that had an independent, inherent tendency to expand.

In the late nineteenth century, political economic revisionists like Walras and Marshall rose to prominence with a new "economics" based on capital theories of value. "Modern" in its use of formal models, their neoclassical "revolution" provided the foundationalism that has dominated Western academic economics ever since. The models presume that the structure of all societies is similar, because they all tend to respond to the universal condition of limited resources or scarcity with the same allocation mechanism, the market. Neoclassical models also presume psychologistically that individual exchange of commodities is the prototype human action and that individual exchanger's actions are predictable (in terms of what are today called "preference curves").

In neoclassical economics, the reason economies are not all the same is the existence of complicating mediators "external" to the core market relations. Some externalities impede market dynamics, while others improve them. Their stress on the inherent tendency of markets to achieve equilibriums in supply and demand means neoclassical economists tend to regard collective human intervention as an externality likely to distort the "natural" market. They thus discourage state policy in principle.

In Smithian political economics, economies are analyzable in terms of laws and can be treated as being not abstractions but "really existing" deep structures. Unlike the classical political economists, however, neoclassicals asserted that the inherent capabilities of capital to expand would free employment economies from the tendency of profit to fall. This is "good time" economics.

AITs and Capital Theories of Value

Predictably given their dominance, neoclassical models were the ones initially mobilized by accounts of the AIT–macrosocial change relationship. Because conditions of scarcity still obtain, the arrival of cyberspace did not mean revising the basic economic model. Like state intervention, new technological developments are market externalities. However, because they create unprecedented opportunities for entrepreneurial virtuosity (new chances for capital to work its value-generative magic), new technologies are generally applauded, as they were in Smith's political economy.

Consider, for example, the structuralistics of the Organization for Economic Cooperation and Development. The OECD is a kind of "think tank" for leading capitalist economies. Its *Jobs Study* (1994), conceived as a strategy document for the world's twenty-five most powerful economies, was the focus of the spring 1996 meeting of the G-7 nations. "Apply new technologies to create new jobs . . . " is how the U.S. Chamber of Commerce summarizes the *Jobs Study*, application of new technologies being primary among the "strategies recommended to overcome rigidities that cause unemployment" (1996). New technologies create jobs because economic growth is attributable to the development of technology and industrial research and development (R&D): "Research and development–and protection of the intellectual property R&D produces–raises living standards, thus boosting demand for labor and generating high-wage jobs" (ibid.).

This is only one example of how capital theories of value privilege enthusiastic performance of computer revolution rhetoric. Cheery optimism about cyberspace is possible because technology is "black boxed"—that is, technologies feed real economies' dynamics but do not independently affect formal economic laws. While technologies like AITs change the content of actual economies, their structure remains unchanged. As an externality, technological change does not demand structural explanation.

If capital is more responsible than labor power for extending a society's reproductive scale, the new social formation should be called "capitalism." Moreover, such capital theories of value privilege the moment of capital's reproduction over that of labor. Indeed, all other reproductive moments (work, knowledge networking, social interaction) should be subordinated to those social arrangements that facilitate the expanded reproduction of capital. This is usually accomplished by commoditizing these other moments, increasing the proportion of the range of activities under their ambit that is mediated by the employment relationship. In this way, more capital is created. Privileging capital's reproduction also privileges those who own it. In capital theories of value, general social formation reproduction is mortgaged to the reproduction of capital.

Critiques of Capital Value Theories

Capital theory was the theory of value that Marx critiqued in *Capital*. In his view, capital should not be seen as value generative in itself because it was really just congealed surplus labor, ripped off from workers. They were forced to give it up because of the vulnerability consequent to not having independent access to means of production.

It was the "something from nothing," magical quality of capital theories that led Marx to coin an anthro-talk term, "commodity fetishism," to caricature them. "Fetishization" is the attribution of independent agency to things humans have made, like goods, or made up, like spirits and "capital." The fundamental critique of capital theories of value is that they treat capital as an independent thing capable of generating its own consequences. This essentialist attribution obscures capital reproduction's dependence upon an underlying social contradiction, the unequal social relationship between worker and owner, which prevents any ultimate social stability.

The radical political economists of the nineteenth Century heard considerable class propaganda in capital theories of value. Rejecting the idea that capital has any essence, they saw capital theories of value as mythic, even mystical.

Recently, the interest on the capital theory mortgage has risen. Consider the renewed influence during in the economic turmoil of the 1990s of Joseph Schumpeter's notion of "creative destruction" (1976). In a Schumpeterian reading, a capitalistic social formation can avoid implosion only through periodically destroying the technical basis of the regnant regime of capital accumulation. Innovation is the necessary vehicle for accomplishing survival through this necessary destruction.

Schumpeterianism is a capital theory of value. Like Senior's, it postpones indefinitely the secular profit decline predicted by labor theories of value. "Creative" destruction, however, means massive institutional dislocation, which in turn undermines the reproduction of many groups and social forms, including important forms of capital. The imposition of an automobile economy marginalizes the foundations of a horse one, taking down the makers of buggies as well as buggy whips. Online shopping promotes "disintermediation" and endangers fundamental aspects of existing commercial business.

Moreover, by linking theoretically the extended reproduction of capital to technological upheaval, Schumpeterianism compromises the neoclassical presumption that the economy is autonomous from other aspects of social formation reproduction. Schumpeterian capital theories of value lead back to substantive, institutional economics.[6] The value that capital was alleged by the neoclassicals to produce on its own instead looks quite similar to the admittedly nonproductive social relationship of rent.

According to the nineteenth-century critique of political economy, rent was dependent upon the differential power of social groups. A sharecropper pays part of her crop to a landlord because the collective landlord has the power to force starvation, not because the land produces something on its own. Similarly, an entrepreneur pays interest on a loan because she has no better way to finance her business, not because the loan qua loan adds value. (This of course is the root of the Muslim conviction that all interest is usury.) Similarly, workers accept less in wage than the value of what they produce because they have no preferable choice. Profit comes from this surplus value, not because of any magical value-generative powers of capital.

Expanded Capital Power and Critiques of Capital Theories of Value

The structural links between technology and capital are not just highly complex. As indicated above, changes stimulate new accounts of value. One recent empirical development fostering much theorizing (for example, globalization) is the growing power of capitalist institutions like transnational corporations to influence their own reproduction.

In a capital theory of value, an increase in the power of capital over social formations is interpreted as additional evidence of capital's contribution to value, and therefore of the validity of capital value theories. However, an alternative interpretation is also possible, that the increased influence of capital over social reproduction is a response to the greater reproductive *difficulties* capital now encounters. In this view, capital's increased power follows from its vulnerabilities rather than its value generativity, that capital exerts more influence on the dynamic of social formation reproduction because otherwise it could not reproduce itself.[7]

Continuing corporate downsizing and disemployment strengthen capital vis-à-vis labor. Selection of technology continues to be, as Braverman argued (1974), regularly filtered through a class sieve. These are only two of several ways in which capital appears to be of continuing, even more not less, relevance to social reproduction. Such economic and technical phenomena are on their face more indicative of changes in the reproductive imperatives of capital than of a decline in its importance.

An approach that focuses on capital is not compatible with "postcapitalist" notions, which imply a reduction in capital's influence on social formation reproduction. In the alternative, instead of indicating the demise of capitalism, the resurgence of Shumpeterianism indicates a discourse problem, a crisis on the legitimating power of capital theories of value. The emergence of alternative knowledge value discourses like those critiqued in this chapter, ones that trace value to things other than capital as historically understood, make sense as indicators of this crisis, as indirect acknowledgments of capital's reproductive troubles. The critiqued political economies of knowledge only appear to be alternatives to capital theories of value. They do not arise because capital reproduction is less important but in response to a need for new accounts of value that overcome the problems in twentieth-century value accounts.

If AIT-induced changes in knowledge are not the chief causes of turn-of-the-twenty-first-century social changes, how are we to account for the prominence of knowledge in popular structuralistics? In locating the reproductive dynamic inside of technology, the theories critiqued above divert attention from capital's problems of reproduction. The alternative political economy of knowledge presented in the next chapter does not locate a dynamic internal to and inherent in AITs. Unlike the structuralistics examined above, the alternative attends to change in the dynamic of capital reproduction.

Knowledge Theories of Value as Capital Theories "In Drag"

An indication that defense of capital may be entering a manic phase is the burgeoning set of metaphorical extensions fostered by the new political economies, for example, "knowledge capital." The invention of this notion, like the knowledge society idea itself, may be part of the broad ideological search for more compelling justifications for the role of capital in employment social formations.

A critique of the basic premises of those approaches that presume a capital theory of value is a necessary prerequisite to constructing a valid political economy of knowledge in cyberspace. Terms less subservient for talking about value are needed. To clear the way for new terms, the remainder of this chapter critiques ideologically the knowledge terms like "intellectual capital" in which the popular political economies dress underlying capital theories of value.

Intellectual/Knowledge Capital

"Knowledge capital" is only one of several metaphoric extensions of capital, but of course it is the one in which a capital theory of value is most directly glossed by a veneer of a knowledge theory. In conceptualizing knowledge as a kind of capital, "intellectual/knowledge capital" frames knowledge as a thing. It encourages thinking of "getting a return to knowledge" in the same way one might "get a return from investing money." Also implied is that, if benefits can arise from knowledge as a factor of production, it like capital merely needs to be added. Construed along capital theory lines, knowledge, too, "magically" creates value.

Unfortunately, such terminology entails all the impoverishments of "thing" conceptions of knowledge discussed in chapter 5. For example, it suppresses recognition of knowledge's dependence on the collaborative activity of the people having and using it, of the centrality of knowledge networking to knowledge use.

Knowledge capital is only one of several forms of capital that appear to have been "discovered" in the "new" economy. To highlight the absurdity of this astonishing terminological effusion, I now refer to resources invested with the intent of making a profit as "capital capital"!

Human Capital

"Human capital," for example, is a concept coined by neoclassical economists to help explain why women, people of color, those with disabilities, working-class people, and so on, receive lower wages. In human capital theory, the wage I command is primarily a consequence of my investment in myself, so those who are badly paid are in this situation because they did not take advantage of opportunities for education and training. Were they to forgo the gratification of consumption and enroll in higher education, they would be trading small immediate for greater long-term benefits, "maximizing their human capital." Those who didn't do so wouldn't have maximized their personal potential to work value magic in the way "capital capital" is supposed to. In this respect, human capital theory is an extension of the capital theory of value.

On human capital, a person's capacity to network knowledge is also "thinged" and individuated. There is of course a point of view from which one can metaphorically see going back to school as "investing in one's self," but this point of view is limited. Reducing a self to "human capital" has broad identity project implications. It tends to create important silences by diverting attention from a broad range of other considerations equally relevant to such a decision. By placing the onus of responsibility on individual choice, human capital framings marginalize awareness of structural impediments to opportunity, like discriminatory structurations (Giddens 1991). By individuating value discourse, human capital perspectives generally ignore the social institutions that determine why some skills—as well as some people's skills—are valued more highly than others by labor markets.

To judge the extent to which human capital constitutes a reasonable model of how value is produced, consider the situation of white male heterosexuals from at least a middle-class background. Even such individuals don't behave in the manner described by human capital theory, because, as with other microeconomic presumptions, to do so would require possession of perfect information of labor markets. To behave in this way also means to ignore other important information. Even the most career-fixated student at SUNY Tech considers other factors, such as his family situation or work schedule, when making his schooling decisions.

Human capital conceptions of value distort perception of social dynamics. Its framings tend to encourage workers to blame themselves for their unpleasant experiences at work. Performances of metaphors like "investing in one's potential human capital" do, however, facilitate capital reproduction. They do this by making capital (and therefore capital theories of value) appear more "natural," just doing what people do. In addition, they indirectly help prevent questioning of capital capital's magic.

As described some time ago by Stephen Marglin (1974), in a corporation, bosses' chief function is to facilitate the reproduction of capital. It is not to facilitate an individual's redemption of her human capital, and it often interferes with production maximization (see Kusterer 1978). Harry Braverman's primary message about technology—that corporate decisions about investment are more a function of the long-term reproductive needs of capital than of their technical impact on production—remains as suggestive in the new millennium as it was in the old (1974). Channeling talk about the decisions of individuals or the practices of institutions via "human capital" illuminates little and obscures much.

Cultural Capital

To my chagrin, a social scientist, Pierre Bourdieu, contributed directly to the metaphorical metastasis of "capital"; he might even be described as its urpractitioner. He did so in his effort to explain the presence of "distinction" as an important dynamic in modern society (1990). Briefly, Bourdieu asked how it is that, in social formations (like France) formally dedicated to inhibiting (through, for example, estate taxes) the inter-generational transmission of privilege, considerable privilege gets so transmitted anyway. That is, the children of high-status parents tend themselves to be high status.

Such privilege, Bourdieu answers, is now reproduced indirectly, via distinction. Through taking them to museums, reading them books, and in general preparing them for entry tests and prestige activities, privileged parents "invest" in their children, providing them with additional means to access advantage. The privilege potential is redeemed through apparently egalitarian, meritocratic institutions like schools, universities, bureaucracies, and corporations. The term Bourdieu coined to label the value thus given, dinstinction's means of reproducing privilege, was, unfortunately, "cultural capital."

A confirmed radical and progressive, Bourdieu intended to critique the inegalitarian results of this process and its institutional forms, especially in the academy. In its malleability, its capacity to be latent and even disappear as a consequence of institutional change (say, as a consequence of change in elite taste), class privilege does have some mythic properties similar to capital capital. The metaphor Bourdieu chose might be acceptable were it not so easily co-opted into the general legitimating project of capital value theory. Some schools now regularly refer to their superior "cultural capital" in recruiting students. In the context of "knowledge capital," "human capital," and so on, "cultural capital" ends up reinforcing that which it would critique.

Bourdieu makes a strong case that the social support of museums, recital halls, and colleges disproportionately afford means to already socially advantaged individuals to privilege their children. These means supplement other institutions—networking introductions, socialization into facile performance of social graces and artful exercises of taste, and admissions into private universities of "legacies"—that already support the generational reproduction of privilege. Both sets of institutions have other social reproduction implications, and determining how to support them collectively without reproducing privilege is a significant social policy issue, beyond that of an individual choosing whether to be seen at a concert or stay home and watch TV.

Once again the simile—seeing the sending of your kids to college as somewhat like investing capital in a firm on the expectation of profit—may have some descriptive value. As a parent I would urge others not to bet their future welfare on this kind of analogy, however. Focusing functional explanation of support for educational institutions in terms of a similarity to capital also feeds into the cynical, anti-intellectual discourses through which the supporters of unfettered capital like Rush Limbaugh critique "cultural elites." Moreover, it diverts attention from more direct forms through which such institutions support the reproduction of capital, such as the use of university endowments of stocks to concentrate capital and make it more mobilizable. In conjunction with the other extensions of capital analyzed above, "cultural capital" is an unfortunate concept.

Social Capital

In his famous article and recent book, *Bowling Alone* (Putnam and Feldstein 2001) Robert Putnam metastasizes capital further, tracing much contemporary social malaise to "a decline in social capital." Even though more frames are bowled that ever before in the United States, a much smaller proportion of them are bowled in league competition. This empirical pattern is presented as a synecdoche for a broader decline in sociality. Americans increasingly spend their time outside of the organized social relations that were previously an important support for collaboratory activity, whether aid in an emergency or "garage" development of new commodities. Celebrating "self" more than group or community, our networks are now less dense.

As we individually spend our time commuting to work at a distance or surfing the Internet, our places of residence lose resilience. Much of this used to come through collective experiences, like working with those who live in our neighborhood or spending time in Oldenburgian "third places" (2000). Our ability to handle difficulty, individual and collective, is indeed generally reduced.

By framing this decline in sociality as "a decrease in social capital," however, Putnam commodifies its solution. What we need is not more social capital (read, "contributions to charity") but social relationships of a different, more multiplex quality. Like the other capital metaphors, Putnam's extends the reproductive ambit of the value myths of capital and its attendant distortions. The concept "social capital" too closely associates sociality with capital. In doing so, it obscures the relative autonomy of other aspects of social formation reproduction from the reproduction of capital. It shares this property with "cultural capital."

I understand and sympathize with Putnam's, Giddens's, and Amatai Etzioni's desires for more community to moderate the dynamics of both state and market (or more accurately, the reproductive imperative of capital). As a parent, I depend upon other parents paying attention to their kids; when they don't, my children, too, are at greater risk. To capture the attention of these parents, I might even try the rhetorical ploy of comparing their actions to those of a company that fails to buy new equipment.

However, I would be very unlikely to choose the alienating activity of capital reproduction as a general model of how to approach the problems of raising children. Treating such moderately useful metaphors as core constructs benefits capital reproduction by making capital appear transcendent. Places of residence need community, and community comes from voluntary extensions of sociality. Your time and your self are as important as your wealth, and much more important than that portion of your wealth ripped off from others in amounts sufficient to be invested.

Deconstructing "Capital"

The foregoing has critiqued metaphoric extensions of "capital" to other construct realms. In their "thingness," "cultural capital," "social capital," "human capital," "personal capital," and doubtless other similar terms narrow thought and, like knowledge capital, tend to mislead. The thing about a metaphor, as Ulf Hannerz argues, is that, like a horse, one needs to get off before it is too late (personal communication). With a hammer in one's hand, one sees nails everywhere.

Metaphorically extending the ambit of capital might be defensible if this had analytic value, promoting something more than mere awareness of similarity. There is worth in the notion "capital." Indeed, understood as investment for profit, "capital" is a construct essential to understanding contemporary social reproduction. "Capital" is not a cultural construct like "ghost," whose conceptual existence is clear but whose actual impacts are harder to detect. There is no doubt that capital matters. In most current social formations, if I wish to bring a commodity to market on any but a modest scale, I really do need access to, in a quite legitimate use of the notion, "venture capital."

Still, it is not easy to state a "vanilla" notion of capital in general. For example, "money" and capital are often used interchangeably, but they are not the same thing. Nor is it easy to identify the point at which it makes even metaphoric sense to think of capital as "productive." Not only can one invest one's capital badly; one can do so deliberately, as in a tax dodge. But is it then still capital? If it is, then what other forms of "unproductive" capital are there, and how does one separate unproductive capital from any valued thing? Is "capital" just another term for entities that have worth?

The intent here is not merely lexical, to straighten out definitional conundrums. Rather, it is to illuminate how, just as "content" approaches obscure the social dimensions of knowledge, an important range of social phenomena are obscured by "knowledge thing" representations of capital. Consider, for example, what capital has in common with "authority," or "charisma." Under the appropriate conditions, the wielder of each of these forms of power can compel the activity of other humans. Like Marx, I think it important to note that for capital, in contrast to these other forms of power, this capacity depends on its fetishization, upon a collective "forgetting" of from whence it comes. In simple terms, the medieval ruling class became a capitalist ruling class, converting its relative monopoly over land and raw materials into a relative monopoly over access to machinery and markets. Workers accept less for wages than the value of what they produce, because this history means that they have no real alternative. Their relative powerlessness is the reason a substantial portion of the value they produce is alienable from them.

It is true that individuals ripped off in this way tend to become annoyed or "alienated" psychologically. The point of critiquing "thing" capital constructs is not psychological but sociological, to show how they institutionalize the forgetting of indignity (Sennett 1999). Capital theories of value induce worship of capital as a magical thing and thus obscure how capital is based on institutionalized alienation. To frame capital as a "thing" of any sort is to be complicitous in this alienation. Without this alienation, capital would cease to be: If workers in general had independent access to markets and the means of production, capital would not be necessary to put production in motion, and it would "disappear." As wealth, of course, money would still have value.

The situation of capital is in some ways similar to the promise carried for years on each U.S. dollar bill, that it was "redeemable for silver." For many years, this promise was no longer valid—it was in fact illegal for private citizens to hold "specie"—but most of the U.S. citizenry "forgot" this fact, and a myth served a useful circulatory purpose.

As with money, we perform capital via a collective Wittgensteinian language game. The capital game requires us to ignore alienation and accept its claimed self-generative properties. This game is performed, for example, each time we accept the notion that underdeveloping nations require outside capital, that without it, they have nothing with an inherent tendency to grow. The attempts of Cuba, Brazil, and so on, to operate on an alternative view, that "more freedom" for capital means less freedom for peoples, showed how, unless they accept slavery to capital, nations are frozen out of the world economy.

The authority of a police officer depends upon the sovereignty of a state, and the wealth of a TV preacher depends upon his ability to project certain personal qualities. So, too, the power of capital rests upon certain social arrangements themselves dependent upon acceptance of some myths. Its reproduction is best served when the applicability of its fetishized self-image is accepted unquestioningly, when "Accumulate, accumulate, accumulate!" is indeed treated as the message of "Moses and the Prophets."

Of course, those individuals and groups who depend upon the reproduction of capital for their well-being tend to advocate social arrangements favorable to the reproduction of their privilege. To the extent that power needs to be exercised culturally ("behind the back" in social formations committed rhetorically to democracy), such social inequality cannot normally be argued for directly. It is in this sense that those who extend the metaphor of capital to other realms contribute to the reproduction of its social hegemony and therefore the dominance of groups highly dependent upon it.[8]

Instead of giving analytic value, however, the metaphors examined here obscure. While indexing important social issues and legitimacy problems, these concepts as a group have a negative analytic impact. Given the extensive complications of this social history, this negativity is perhaps inescapable. Like vanilla capital, its metaphorical extensions' overuse and underthorization are suggestive of a rear guard defense of a mythic pattern of thought, of capital value theorizing under stress.

The overenchantment of contemporary social science with capital metaphors undermines critical faculties: The more they are used, the harder it is to see their limitations. When all resources are presented as alternate forms of capital, social science becomes social apologetics. These new, noxious weeds in social science's conceptual garden are indexes of the stresses on capital reproduction. We should acknowledge their value in this regard, but to clear ground for a real alternative knowledge theory of value, not a capital theory "in knowledge drag," they need to be uprooted.

Summary: Toward "Straight" Knowledge Theories of Value

Just as social science originally congealed around a new answer (labor power) to the value question, talk of a new economy has often pointed at potentially new characteristics and roles, including in value creation, for knowledge in cyberspace. Searches for a new knowledge structuralistics are also responses to the shortcomings of the dominant theories of value, such as their failure to account for important, Shumpeterian, institutional phenomena.

Unfortunately, popular cyberspace knowledge talk holds over discourse conventions from the regency of capital theories of value. Instead of offering truly new political economies, they merely place a knowledge gloss on what remain basically capital theories of value. Just as skepticism was warranted with regard to the new economy, it is proper with regard to theories that merely dress capital theories of value in "knowledge drag."

At the same time, although knowledge may be *labeled* a form of capital ("knowledge" or "intellectual capital") in accounts like these, one can also perceive in them a strong impulse to make knowledge a *replacement for rather than a form of* capital at the center of production. As argued in chapter 7, a quite liberating resocialing of work, one facilitated by expanded use of knowledge technology, is indeed possible. This possibility is an important reason behind the refocusing of the value debate on knowledge. The switch in the focus of value discourse to knowledge is further facilitated, perhaps even compelled, by all sorts of ideas about teams, dispersed work, virtual organizations, participatory design, collaborative work, and so on.

A clearer, thorough perspective on knowledge and value would have other consequences. It would force a new discourse on management, one in which the necessity of management was no longer presumed a priori. Management's place in production would become narrower and more contingent, dependent upon its success at mobilizing expertise in particular forms of labor. With management reduced to the labor of coordination thorough development of more comparable notions of management and workers, knowledge would change class dynamics as well as our understandings of them.

Such accounts, however, would also put at risk current legitimations of management that associate it with the self-generative magic of capital. Instead of risking a thorough rethinking of management in knowledge terms, some may wish to retain the idea that management possesses privileged knowledge about how to unlock the magic of capital. They might then deploy notions like "knowledge capital," either overtly or metaphorically, in ways that presume the inevitability of the social relations of (capital) capital. As long as management is tied to a capital theory of value, the liberatory potential of organizational knowledge technologies will be severely limited.

Approaches that link knowledge to capital, including those that construct knowledge *as* capital, obscure rather than illuminate the potential of knowledge in the transition to cyberspace. To take advantage of the potentials of AITs to facilitate knowledge networking, as well as to foster the broader social development that this would make possible, we need truly independent knowledge theories of value. These in turn could generate political economies more appropriate to the extension of contemporary social formation reproduction.

To create such knowledge theories, it is necessary to liberate knowledge constructs from enslavement to capital reproduction. Freedom will not come by restating capital theories of value in terms of knowledge. Once "knowledge capital" has been deconstructed, knowledge structuralistics like those presented in this chapter are recognizable as first steps toward articulating a third, distinct, neither labor nor capital but knowledge, theory of value.

New knowledge theories should be evaluated in terms of whether they offer a more satisfactory discourse on where value comes from. Indicating what a genuine knowledge theory of value would address is the task of the following chapter.

Chapter 11

An Alternative Political Economy of Knowledge in Cyberspace

The evidence of a knowledge-driven transformation is not yet sufficient to justify calling ours a knowledge society. However, there are good indications that technologized support systems, if adequately informed by social design, *could* change knowledge networking substantially enough to change social reproduction considerably. Consequently, it makes sense to be thinking about what kind of knowledge society we *want.*

A sense of what is possible is limited, however, by popular political economies like the three critiqued in the previous chapter. At the same time, their very existence is indicative of weaknesses in the persuasive power of current political economic myths, including capital-based ideas about where value comes from. In the last chapter, popular ideas about societies' structural properties were analyzed in the context of previous value theorizations. Their shortcomings are manifest in knowledge-related metaphorical extensions of capital such as "knowledge capital."

To specify the kind of knowledge society worth striving for, we need truly independent, rather than "capital"-dependent, structural perspectives on knowledge The current chapter aims to articulate a political economic perspective that theorizes knowledge in cyberspace independently of capital theories. The first analytic step is to account for the most recent social changes not in terms of knowledge but in terms of a vibrant but vulnerable "turbo-capitalism" (Will Hutton). The second step is to indicate the really different political economy of knowledge whose realization is, for the moment, blocked by turbo-capitalism, as well as the knowledge theory of value distorted by capitalist value mythologies (Nick Dyer-Witheford). The chapter concludes with a summary of what these alternative structuralistics suggest about the reproductive preoccupations of social formations in the foreseeable future.

Current Capital Reproduction

The Place of a Capital Theory in a Knowledge Theory of Value

Several of the theoretical critiques of capital theories of value outlined in the previous chapter have been around for a long time, yet capital theories remain dominant. While the chapter raised the possibility that interest in knowledge political economies is a manifestation of problems in the reproduction of contemporary social formations, it also acknowledged an increase in the influence of capital. The notion that power of capital over social reproduction is increasing seems to contradict the idea that capital-based economics should be replaced by knowledge ones.

The last chapter also suggested that capital's increasing ambit may be necessitated by new weaknesses in its ability to reproduce itself. The rise of new knowledge theories of value, even if they turn out to be ultimately based on capital theories, is nonetheless an indirect recognition of problems in capital value theorization. But doesn't the expanded centrality of capital in contemporary social formations empirically justify capital theories of value?

Were this so, the search for new, knowledge-based alternative theories of value would make no sense. Moreover, the influence of capital on the marketplace of ideas may itself have compromised discussion of value. The failure of critiques of capital theories of value to become economic orthodoxy may have less to do with their analytic quality than with economics' ideological service to the reproduction of capital. The inability to recognize directly how momentous capital's contemporary problems are would also explain the contradictions in the knowledge theories of value considered in the previous chapter.

In short, to specify what a knowledge society would really be like, and thus what a knowledge theory of value must account for, we first need an adequate account of the contemporary role of capital in general social formation reproduction. This account must explain capital's current power at the same time as it avoids being dazzled by, for example, metastasizing capital metaphors.

The Recent Expansion of Capital's Reproductive Ambit

Throughout the history of employment social formations, capital's influence on general social formation reproduction has tended to grow. It is arguably greater now than at any other time. The increased centrality of transnational, corporate capital to most social formations today is arguably the most distinctive aspect of what is called "globalization."

Computing Myths, Class Realities, Barbara Andrews and my 1993 study of Sheffield new technology, examined various predictors of the social correlates of computing initiatives. The best predictor of outcomes was the workspace groups that an initiative mobilized and whose interest it served. In the second decade of the 1980s, even in "Labour's Home" in the north of England, the group most able to influence the technology/employment nexus remained the one already in this position, the private owners/controllers of means of production. It was workers who most strongly felt their effects.

Thatcherism and Reaganism were two very visible examples of a general 1980s tendency, the use of state power to accommodate the expanding reproductive ambit of transnational capital. In *Cyborgs@Cyberspace?* (1999a), I described a prodigious expansion of the influence of capital over general social formation reproduction in the Nordic countries. This expansion was as an important reason for the declining influence of Nordic working life legislation on the way AITs were actually used. The books made similar points about unemployment, especially that the alleged disemploying/job creating tendencies in new information technologies were so highly mediated by the reproductive dynamics of capital as to have little independent effect. Similarly, chapter 8 of *The Knowledge Landscapes* . . . described local government projects in the Upper Mohawk Valley of the United States unable to use public means to influence how technologies get institutionalized.

The current situation is illustrated clearly by 1990s changes in Wallenberg family-controlled—which is to say virtually all—capital in Sweden, to consider one example. In essence, the Wallenberg corporations abandoned the national-level bargaining that had purchased social peace since the 1920s. Instead, they reoriented directly to a world market. This development was indexed vividly by the emergence of the transnational firm ABB, particularly it's chameleonlike attempt to be, in the words of its slogan, "the best corporate citizen wherever we happen to be." This was a very different face for what some anachronistically insist on still calling "Swedish" capital. Danish and Finnish social formations similarly accommodated the more globally exercised ambit of capital. Despite the potential for relative autonomy provided by nationally owned oil, even Norwegian enterprises and state institutions increasingly adapted themselves to the global demands of capitalist institutions.

As loci of decision making have shifted to increasingly assertive supranational corporations, the influence of nation-state structures, including state-sponsored participatory institutions to promote economic democracy, has contracted. Capital's increasing influence contrasts with the decline of trade union power and the narrowing of the range of options available to previously influential working people's (for example Labour, Social Democratic, or, in the United States, northeast Democratic Party) politics. Ideologies inhibiting working-class influence also gained wider ambit.

Turbo-Capitalism, not Knowledge, as Dominant

The structural theories of cyberspace examined in the last chapter asserted that knowledge was the generative source of recent change in social reproductive dynamics. Is it reasonable to trace developments like those described above to new knowledge technologies? This is the view of the knowledge revolutionary Anthony Giddens, who exercises a substantial theoretical influence over British "new" Labour. Giddens highlights "the new role of knowledge as a factor of production . . ." (Giddens and Hutton 2000, 4). He speaks of the "new knowledge economy that almost certainly operates according to different principles from the industrial economy . . ." (1), one that is "changing the very character of how we live and work . . . " (5). Like Davenport and Prusak, Giddens accounts for revolutionary change in terms of something more broadly spread: "[M]ost companies know pretty quickly what other companies are planning, because of the general profusion of information. Secrecy is much more difficult. Given the global nature of contemporary communications, there is no geographical isolation any longer" (26).

Here Giddens, like so many of the writers already examined, blurs the difference between information and knowledge and invokes popular but simplistic space/place contrasts. More substantively problematic is his ignoring of how, as described in chapter 8, the chief ostensible task of corporate knowledge technologies was to *prevent* general dissemination of company knowledge!

In Giddens, world-transformative changes are undermined by the abject failure of knowledge technologies to accomplish their intended goals. Giddens's interlocutor Will Hutton offers a different structuralistics. For Hutton, knowledge's influence is not causative but instead is mediated through its role in what he, following Edward Luttwak, calls "turbo-capitalism." This "very particular kind of capitalism" is one that "has emerged victorious from its competition with communism." It is a triumphant form,

> a capitalism that is much harder, more mobile, more ruthless and more certain about what it needs to make it tick. . . . It's overriding objective is to serve the

> interests of property owners and shareholders, and it has a firm belief...that all obstacles to its capacity to do that – regulation, controls, trade unions, taxation, public ownership, etc. – are unjustified and should be removed . . . (Giddens and Hutton, 9–10)

Hutton regrets the eclipse of forms of capital alternative to this share- (stock-) oriented turbo form:

> I would say that communism, although it failed, did have one good impact; it kept capitalism on its guard—in a sense it kept it aware that it had to have a human face . . . . (9)
>
> [T]he alternative tradition of Catholic capitalism, social market capitalism, or stakeholder capitalism . . . is [also] retreating . . . (10)

Hutton does acknowledge a connection between this resurgent capitalism and AITs. Unlike Giddens, he stresses that turbo-capitalism drives technology rather than being driven by it. Steroidal capital takes advantage of the opportunities to extend its reproductive ambit that are opened by technological change. In a Schumpeterian register, Hutton comments that turbo-capitalism

> is particularly powerful at a time of great technological change because not only does it encourage new entrants into markets, it also shakes up the sometimes powerful but sleepy companies who currently hold a lot of market power . . . (13)
>
> Technological change sometimes has the effect of producing a sort of quantum leap, forcing a sort of restructuring of the whole of the capitalist economy. A quantum leap of this kind is happening through the impact of the information revolution at the moment . . . although . . . it has as much to do with the spread, character, and ambition of capitalism as the march of science . . . (20)

Thus, while there is a connection between change in knowledge technology and turbo-capitalism, the connection is not the simple, one-directional, "cause-effect" one described by Giddens. Indeed, to present knowledge as if it commanded capital is to obscure what is taking place:

> Of course I agree that there is a dynamic sector of the economy where knowledge is very important, and all firms can access and use the new processes to some degree. But I am also not sure that the inference we are meant to draw—that everything is cleverer and more knowledge-based and therefore that the fundamentals of capitalism have wholly changed, is right[,] . . . that the rules of the capitalist game have changed. (23–24)

Hutton's analytic point is that knowledge-related phenomena are bent to the reproduction of capital, rather than that capital is being bent to knowledge networking. He goes on to comment that "[A]lthough commoditization is an ugly word, . . . it does capture the process by which capitalism tries to turn every relationship into a commercial exchange . . . " (17). Intellectual capital is not a new form of capital. While "Intellectual property rights are increasingly what makes capitalism tick," it is "control of the idea rather than what the idea gives to production" that counts. "[A]ll the difficulties about exploitation, private ownership, and instability remain remarkably the same" (25).

Instead of a knowledge revolution,

> what really took place in the 1990s was a great power play: Asian capitalism versus American capitalism. US capitalism wins, with the Asia crisis of 97/98 actually being the flashpoint and the financial markets working in a way that furthers US interests . . . I think it puts an important question mark over globalisation. There is a dimension of globalisation that is about opening up the world to American interest in particular and Western capitalism in general. ...[U]nderneath the glitz there remains the exercise of raw power . . . (41)

Hutton rejects the idea that new technologies are the primary force for change. For him, this remains capital, a still nation/region-linkable but newly active form of it. Capital's increasing active role has developed because, contra neoclassical economics, capitalist systems *don't* tend toward neoclassical "equilibriums":

> The rationality of capitalism doesn't lie in any . . . tendency to produce a stable equilibrium. Its rationality lies in its inherent capacity to accommodate risk, to experiment over investment for the future, and to be creative about new forms of production and consumption . . . (19)

In the relentless pursuit of its reproduction, turbo-capital especially is generative of *in*stability, the reason for extending social formation reproduction:

> The notion that capitalism should be seen as a creative process rather than tending to unimprovable equilibria is one of the great strengths of the [second, late-twentieth-century] Austrian school of economists' championing of capitalism. Friedrich Hayek says that markets are brilliant means of capturing the collective judgments of individual intelligence because they allow decentralized decision-making, but we should not think of them as stable . . . (20)

Rather than "markets working to produce a self-correcting equilibrium, what you have watched is a wild process of experimentation and overshot involving some crazy and avoidable risks and economic pain. Heaven knows what will happen next and to whom . . . "(40).

Casualties of the accelerated instability of turbo-capitalism include the institutions of social democracy, such as the "welfare state": "[I]t was more or less inevitable that the whole policy nexus would become unsustainable as soon as the financial deregulation caused asset price booms—bubble economies really—property booms and the rest of it . . . " (40). Hutton consequently is critical of those like Giddens, whose knowledge theory encourages a "naive trust in markets" that provides ideological cover for greater capital power. Rather,

> [T]he injustices you [Giddens] want to correct are not independent of the capitalism you admire – they result directly from its operation . . . (45)
>
> [B]neath the technological change some rough and tough old capitalists truths are being reasserted . . . [and] . . . beneath the glitz of modernity a lot of people are as exposed as ever to some hard brutalities . . . " (30).
>
> [C]alling for structural reform of labour markets and the welfare system as stand-alone recommendations . . . really mean . . . that non-wage costs should be lowered, work made more insecure, and the . . . system of social protection weakened. (Giddens and Hutton, 35)

In sum, in Hutton, phenomena like globalization are not caused by an emergent political economy of knowledge before which all must fall, but by contingent changes in the dynamic of capital reproduction. Technology change, the increasingly global reach of the corporation, and increased competition are all real. However, they do not follow from any particular inevitable dynamic, "laws" endogenous to knowledge technology. Rather, they follow from deliberate policy interventions, including the weakening of nation-based trade unions (the only effective trade unions there are, yet) to control access to labor. These interventions have also weakened the capacity of geography-tied capital to enhance the conditions of its reproduction, for example, through tariffs. Forceful performance of knowledge "mantras" do impact social reproduction, but not because they reflect structural "truths." Rather, they are an ideological influence in policy discussions, one that diverts attention from the increasing ambit of capital and therefore of any attempt to mitigate its undesirable consequences.

In Marxist terminology, readings like Hutton's stress "social relations" rather than the "technical relations of production" of central interest to

Giddens and the theorists discussed in the previous chapter. The contemporary era is one of renewed, very great if not unprecedented capital dominance and hegemony, certainly comparable to the 1920s in the United States, Britain, and even the Nordic countries. This centrality is associated with several phenomena, including defeat of the Soviet Union, new limitations on states' actions, and assertion of new capitalist cultural legitimations (for example, intellectual property) in the face of the challenges of the '60s.

Turbo-capitalism does take advantage of the Kunda-type AIT-enabled reorganizations of the labor process explored in chapter 7. Yet while AITs add options for reorganizing the labor processes, they do not compel it. It is only ideologically, via the *notion* that such reorganization is necessitated "ineluctably" by technology, that AITs influence comes to appear structural.

While AITs clearly *can* be used (as they have) to legitimate reimposition of a strong capital regime, they do not *have* to be used in this way. Their obvious ideological value alone should prompt doubt about "knowledge as technological imperative" lines of argument.

A Cultural Theory of Contemporary Value Contradictions

Hutton more parsimoniously analyzes the dynamics of contemporary social formations than social revolutionaries like Castells. That is, he properly attends to capital's expanded power without extending it mythically, by giving capital knowledge clothing or by deploying metaphors that terminologically exaggerate its influence while also diverting critique.

Hutton's kind of analysis can be restated in more anthropological terms. Doing so allows greater specification of the responses that distinguish the reproductive dynamics of contemporary employment social formations from previous ones and thus the actually new challenges to capital accumulation.

Value and Culture in General

On a general social reproductionist account (Hakken 1987), human social reproduction depends on cultural reproduction. That is, what differentiates the dynamics of the reproduction of human from other types of social formations (whether species-specific plant or animal, or general ecological) is the extent to which human ones depend upon culture. Human social formations only last if existing humans convince new ones (whether "recruited" through sexual reproduction or immigration) to adopt compatible sets of cultural constructs. That is, the "newbies" are convinced to accept, or at least to act as if they accept, that the cultural constructs of their elders accurately describe actual social reproduction. This is one example of the kinds of deliberate interventions necessary to promote social formation reproduction

Moreover, for any *particular* human social formation (what anthropologists call "a culture," as opposed to the general human *type* of social formation) to perpetuate itself, it must withstand both natural and cultural "selection." That is, it must meet the (culturally structured) biological needs of its adherents as well as the threats to its reproduction in its cultural environment, the other cultures with which it is in contact. This is equivalent to saying that new types of social formations arise by displacing older ones.

Anthropologists use "myth" to describe the stories that humans tell that account for cultural dynamics. Because, as argued above, human social formations must reproduce socially, myth development is a necessary component of cultural and therefore of social formation reproduction. To an anthropologist, the political economics developed to account for the rise of the employment social formation constituted the early mythologies of capitalism. As described in chapter 10, it was neither money, markets, production of goods for sale, nor even forms of mass production that were the distinctive feature of the new "employment" type of social formation that came to prominence in the eighteenth century; all existed in previous social formations. What was new was the extension of the commodity form (mediation by markets) into two new arenas of social practice:

1. Actual human labor became labor power (the capacity to do work).
2. The difference between the value of what workers produced and the value of their wages and other costs of production, or profit, became capital.

That is, once the institutions of labor and capital markets came into existence, one could buy and sell work in the form of labor power, and surplus value (profit) in the form of capital.

The Cultural Contradictions of Capital Mythologies

To become widespread, mythologies must provide convincing, if not necessarily accurate, accounts of the dynamics of social reproduction, accounts convincing enough that they themselves are also reproduced. A contradiction at the heart of the reproduction of capital has limited the cultural reproductive ambit of employment social formations' myths, particularly the labor theory of value. On the one hand, sellers had to convince potential buyers that the things one wanted them to buy were worth the price assigned to them. On the other, producers had to convince workers to produce these things at pay rates lower than the selling price.

In the social formations in which the institutions of labor and capital first developed, the markets were largely in luxury goods, and deserfed uncommoned new workers had little alternative but to accept the wages offered. As the commodity form penetrated more aspects of social reproduction, however, workers became important as consumers as well as as workers. They extended their collective ability to influence general social reproduction, specific labor markets, and states.

Capital markets require stability. This was initially provided by states that, for example, promoted sufficiently transparent banking and meaningful exchange rates (the necessity of which is vividly illustrated by the experience of the ex-Soviet Union). States also periodically served as crucial sources of investment, first in canals, later in Internets.

To the extent that workers were able to exert influence, the contradiction at the heart of the labor of value became more blatant, and its value as a justificatory myth correspondingly declined. The capital theories of value that displaced labor theories mythically resolved this problem. Under them, value arose not from ripping off workers, but from value-generative qualities inherent in capital. Moreover, freed from having to be moored in the real worth of things produced—that is, as its reproduction became mediated by ever more dense narratives and thus decreasingly corresponded to events in the real world—the mythically powerful entity, capital, also becomes more malleable.

However, as illustrated in the capital metaphors critiques in the previous chapter, this mythic malleability regularly engenders new contradictions. If public entities can lower interest rates to stave off recession, why not keep rates low so that small businesses stay afloat? If public moneys can be used to guarantee the profit level of military contractors, why can't they also fund worker cooperatives? If they can rescue savings and loans, why not communities?

Such questions indicate how vastly extended myths of capital reproduction are more difficult to control. Its continuing actual dependence upon labor to produce the value turned into profit makes capital increasingly difficult to reproduce in the real world. It can only do so by bringing more and more domains of existence within its ambit, as is happening now to education (see chapter 9). The gap between the cultural reproductive potential of "capital" in its latest mythic forms and the reproductive demands of so-called late capitalism as a social form increases, threatening the reproduction of entire social formations.[1]

Why Reconstructing Capital as Knowledge Doesn't Work

We now can see why knowledge has recently been theorized both as capital *and* as value. Encouraging and feeding off twenty years of actively "metaphorizing" capital, its theorists developed "knowledge capital" as a way to help organizations address a serious problem, one that arises as soon as one acknowledges a place for knowledge in production.

Once one has analyzed the "knowledge resources" of one's organization and acknowledged them to be significant, it makes cultural sense to think of them as "capital" and therefore as something to be protected from the competition. However, knowledge is hard to secure. For example, given that it can be transferred without being lost, one's security department can't even rely on its presence to indicate that it has not been stolen. One can't prevent leaving employees from taking it, either. Indeed, if knowledge really were the chief form of capital, the capitalist system would probably be doomed.

However much cultural sense it makes, theorizing knowledge as capital is a conceptual trap. Knowledge theories in an intellectual capital register merely further extend the ambit of an apparently infinitely malleable, and therefore increasingly mythic, substance—witness, for example, the "value" of dot-coms and the "vaporware" on which many were based. Saturating the world with capital metaphors only increases the difficulty of reproducing actual capital.

Steps toward a Knowledge Theory of Value

If knowledge is to be recognized as having a central role in cyberspace, it will not be by treating knowledge as capital. The promise of a knowledge theory of value is to resolve the contradictions of, rather than further complicate, capital theories of value.

A first step in constructing a knowledge theory of value is to acknowledge the important contribution of labor to value, as theorists like Davenport and Prusak (1997) do. An important additional intellectual source of the shift of value attention to knowledge is the "turn to the social" of the institutional "neopolitical economics" of the 1960s. One important aspect of this development, the anthropologies and sociologies of work described in chapter 8, underlined the knowledge similarities between what workers and managers give to production; each, for example, depends on "know-how," albeit of different sorts—how to coordinate versus how to habituate (Kusterer 1978).

Recognition of the interdependence of capital and labor would inhibit the metaphoric effusion of increasingly empty capital forms, but this is not enough. This section develops a knowledge theory of value in cyberspace alternative to both labor and capital theories.

Summary of Elements Already Presented

Many parts of this "real" knowledge theory of value have already been presented in the previous chapters of *The Knowledge Landscapes . . .* Chapter 3 pointed out the negative practical consequences of knowledge management's efforts to treat knowledge as fungible, as composed of discrete, easily equatable, and transformable bits. This tendency follows from analogizing knowledge too closely to capital. Instead, a practice approach to knowledge was proposed, a process one built on knowledge networking.

Chapters 4 to 6 situated this practice approach to knowledge in the multiple intellectual contexts that any knowledge AIT structuralistics must take into account. Chapters 7 to 9 demonstrated how it was possible to apply this complex theorization of knowledge in multiple research and practice domains. Finally, by indicating the major drawbacks of trying to fit contemporary social formation reproduction dynamics into a "capital" straitjacket, Section IV presents the theoretical benefits of an alternative value account.

Just as the output of individual workers varies with their competence, so the group output depends upon how well work is coordinated. As neoinstitutional work social science showed, both labor's and capital's reproduction depends on what individuals and groups know and their ability to put this knowledge to use. Especially if, under genuine competition, productive units have access to similar labor powers and comparable machines and raw materials, "know-how" could easily be the main factor differentiating one firm from others.

Framed as "know how," knowledge is a substantial factor in production. Awareness of the potential of new automated information technologies to "leverage" deployment of know-how, in part a consequence of the publicity surrounding the preoccupation with knowledge in informatics examined at length in chapter 6, certainly contributed to the resurgence of general interest in knowledge. This interest was also a consequence of the entry into markets of knowledge products overtly based in informatics.

Another source of knowledge value interest was the recent organization theory described in chapter 8, especially the increasing acknowledgment of the dense sociality of organizations and of organization itself as a process. Abandoning the effort to identify a rational, positivist management science based on discovery of ONE BEST WAY, organization theory moved beyond mere grudging recognition of the modicum of informal organization that inevitably accompanies formal organization. Instead, theorists came to view not the knowledge that an organization holds but its capacity to learn new knowledge as its chief asset. With recognition of the profound sociality of this capacity to learn comes acknowledgment that organization knowledge is not merely a nominal collection of knowledges actually bounded by the heads of individual organizational members. Knowledge has an important locus in individuals, but its locus in organization is perhaps even more profound. Moreover, because the nonformal knowledge of individuals, work groups, and the organization is substantial and often decisive, it becomes difficult if not impossible to separate organization knowledge from organization itself.

Organization as Knowledge Networking

In other words, at base, organization is knowledge networking. This is the key point in a political economy of knowledge. Further, to the extent that organization dynamics are bent to some other imperative—whether the reproduction of capital or labor—knowledge networking is "distorted."

In the new phase in the employment social formation that I refer to as "cyberfacture," organized knowledge networking is bent both by computerization and turbo-capitalism. At the same time, like any new stage in an evolving social formation type, cyberfacture may lead to a more profound transformation in organizational knowledge networking. As described in chapter 8, some of the knowledge networking strategies responding to the contradiction between "deplacing" work, on the one hand, and relying on more collaborative (for example, "team") forms of coordination, on the other, have the potential to compel a "resocialing" of work. Loosening the ties of know-how to current worksite politics, these strategies open the way to overt recognition of the substantive skills of all workers, including the unskilled. This logically leads to pay schemes that compensate individual workers for all they actually contribute to value, rather than schemes that primarily reward organization members (disproportionately managers) for their contribution to profit. Were such schemes to be broadly applied, they might well be indicators of a "postcapitalist" social formation (to borrow Drucker's phrase but not his argument).

Such developments are not out of the question. The "call" to the OECD conference discussed in chapter 10 (1994) acknowledged concerns about the new economy. These included how "innovation destroys some jobs . . . " and how the "technology equals jobs" formula has a downside, such as the social psychological costs to workers of lost workplace identity. These are identified as reasons for wanting alternative narratives: "There is a need for a debate on alternative ways of organizing labor and the use of technology."[2]

At least some participants saw the conference as a breakthrough in the introduction of alternative perspectives in the jobs/technology debate. Keith Smith, head of an important policy group funded by the Norwegian Research Council and chief conference rapporteur, summarized the conference (1997) as:

- Presenting innovation as a learning process, one cumulative over time, which leads to the idea of spatially differing technology paradigms;
- Viewing technology as flexible—for example, much of it is tacit, not easily constrained, so there are questions to be asked about how or even if it can be codified;
- Seeing knowledge as not individual—rather, its creation is collaborative, inhering in organizations as much as people; and therefore
- Recognizing how the use of knowledge rests on specific, even cultural, infrastructures, on concretely different systems of innovation.

With their increasingly strong economy and oil wealth, the so-called Sheikdom of the North was in a position to think very differently about jobs and technology. In his conference paper (1997), the Norwegian economic historian Francis Sjersted argued for a radical experimentation with ways to conceptualize social participation in which the job was much less central, beginning the process of decoupling access to social wealth from the particular job one finds oneself with (or without).

The explanatory strategies of the institutional economists in Oslo were not oriented toward identification of the presumed formal, machinelike processes "built in" to all economies. Rather, the search was for new capacities for and exercise of alternative social power based on different national/cultural dynamics. It is for such projects that knowledge theories of value hold most promise.

A Classical Knowledge Theory of Value

"Knowledge as the key productive force" perspectives like those outlined above should be alternative to, rather than subordinate forms of, capital theories of value. One example of an attempt to theorize such notions explicitly is Nick Dyer-Witheford's *Cyber-Marx: Cycles and Circuits of Struggle in High-Technology Capitalism* (1999). The book's chief argument for its relevance to the current era is what Marxism offers to answering the knowledge question in cyberspace.

Dyer-Witheford begins with a footnote in *Capital* on the work of the early informatician Charles Babbage:

> Commenting on capital's ever-increasing use of machines, [Marx] notes that "mechanical and chemical discoveries" are actually the result of a social cooperative process that [Marx] calls "universal labour . . . all scientific work, all discovery and invention. It is brought about partly by the cooperation of men now living, but partly by building on earlier work." The fruits of this collective project are, Marx argues, generally appropriated by the "most worthless and wretched kind of money-capitalists." But the ultimate source of their profit is the "new development of the universal labour of the human spirit and their social application by combined labour. (3 and 4)

One can see here the germ of a theory of value that gives substantial weight to knowledge, albeit one grounded on the collective and social dimensions of labor. Dyer-Witheford describes how, in other comments in the *Grundrisse*, Marx

> foretells the future technological trajectory of capitalism . . . At a certain point, Marx predicts, capital's drive to dominate living labor through machinery will mean that "the creation of real wealth comes to depend less on labour time and on the amount of labour employed" than on "the general state of science and on the progress of technology." The key factor in production will become the social knowledge necessary for techno-scientific innovation—"general intellect." (4)

Contrasting Marx's attention to universal labor to Babbage's allegiance to the reproduction of capital, Dyer-Witheford poses a "contest for general intellect" between Marx and Babbage. The contest was, in essence, over how a theory of knowledge was to be inclined—toward capital, or toward labor.

In concluding his general defense of the relevance of Marxism in a high-tech world, Dyer-Witheford glosses Marx's view of intellect as an evolutionary account of employment social formations. That is,

> at a certain point in the development of capital, the creation of real wealth will come to depend not on the direct expenditure of labor time in production but on two interrelated factors: technological expertise, that is, "scientific labor [*sic*]," and organization, or "social combination." The crucial factor in production will become the "development of the general powers of the human head";

> "general social knowledge"; "social intellect"; or . . . "the general productive forces of the social brain." (219 and 220)

Thus, in the mid-nineteenth century, Marx began to develop a knowledge theory of value. "What Marx describes is eminently recognizable as a portrait of what is now commonly termed an 'information society' or 'knowledge economy' . . . " (Dyer-Witheford, 221). As both labor and capital declined in importance as society developed, this knowledge theory would not only supersede the labor theory of value; it would also obviate any need for a capital one.

A Contemporary Knowledge Theory of Value

However, as articulated, this Marxian knowledge theory of value is ambiguous, having very different implications depending on which of the "two interrelated factors" is stressed. Modernistically, in the vein of the "scientific" Marxism of Engels, the theory could emphasize the content of "the general state of science and the progress of technology . . . " Indeed, "the power of knowledge" could even be "objectified" against labor and for capital. On first reading one might indeed see Marx as stressing how "the accumulation of knowledge" gets "absorbed into capital."

Alternatively, nonmodernistically, stress could be given to the social side, neopragmatically critiquing scientism and emphasizing "social application by combined labor." Dyer-Witheford prefers this latter reading, an "optimistic" Marx:

> However—and this is the whole point of Marx's analysis—such a level of technological advance . . . contains within itself the seeds of a capitalist nightmare. By setting in motion the powers of scientific knowledge and social cooperation, capital ultimately undermines itself . . . First . . . as advances . . . reduce the requirement for direct labor, . . . the very basis of capitalism's social order . . . is eroded . . .
>
> This is reinforced by a second tendency, the increasingly social nature of activity required for techno-scientific development, which unfolds not on the basis of individual effort but as a vast cooperative effort . . . [B]oth private ownership and payment for isolated quanta of work time appear increasingly as irrelevant impediments to the full use of social resources. (Dyer-Witheford, 220)

An Extended Contemporary Knowledge Theory

Dyer-Witheford argues that the contemporary case for transformative optimism is most fully developed in the theoretical work of the largely European journal group *Futur Anterieur*. For them, it is true that " . . . the revolutionary tendencies Marx identified . . . are occurring, but [still] in forms prescribed by an order that continues to organize itself on the basis of the wage and private ownership . . . ," the reproduction of capital. They go on to critique Marx:

> In this situation, it is not enough to focus, as Marx did, on the objectification of social knowledge in new technologies. Rather, the critical issue is that of the nature of the human activity required to create, support, and enable this technoscientific apparatus . . . [H]ere . . . we encounter [a] paradox. While capital has developed machines to subordinate and reduce labor at the point of production, this development itself demands the emergence of a new range of social competencies and co-operations—the cultivation of 'general social knowledge' . . . [or] 'mass intellectuality.'
>
> Mass intellectuality' is the ensemble of "know-hows" that supports the operation of the high-tech economy. It is "the social body" as a "repository of knowledges indivisible from living subjects and from their linguistic cooperation . . . , immaterial labor." (Dyer-Witheford, 221)

In Dyer-Witheford's reading, for social formation reproduction today,

> the crucial question thus becomes how far capital can contain . . . "this plural, multiform, constantly mutating intelligence" of mass intellect within its structures . . . [I]t 'appears to domesticate general intellect without too much difficulty.' But this absorption demands an extraordinary exercise of "supervision and surveillance," involving "complex procedures of attributing rights to know and/or rights of access to knowledge which are at the same time procedures of exclusion . . . "

As opposed to the debilitating dialectical idealist views of Belasen (2000), whose contradictory views on management were described in chapter 8, *Anterieur* offers the following:

> "Good 'management' of the processes of knowledge consists of polarizing them, of producing success and failure, of integrating legitimating knowledges and disqualifying illegitimate knowledges, that is ones contrary to the reproduction of capital. It needs individuals who know what they are doing, but only up to a certain point. Capitalist 'management' and a whole series of institutions (particularly of education) are trying to limit the usage of knowledges produced and transmitted. In the name of profitability and immediate results, they are pro-

> hibiting connections and relationships that could profoundly modify the structure of the field of knowledge." (Dyer-Witheford, 222 and 223)

Interestingly, the writers of *Anterieur* go on to analyze "teams" and "participative management" as sites in which these contradictions are particularly manifest. Beginning in a Kundaesque vein, they speak of how sometimes

> new team organization is even more totalitarian than the old assembly line . . .
>
> However . . . [i]n delegating . . . certain managerial responsibilities to workers, capital is partially relinquishing its claim to act as the mediator and coordinator of production. There is a potential tension between capital control of enterprises and the increasingly self-directed nature of work . . .
>
> [A] massive contradiction arises for capital: it has to stimulate and harness subjectivity by encouraging increasing worker responsibilization, even creativity, in order to grasp a social and communicational surplus value in the workplace . . . This . . . comes to constitute a competitive edge in the global fight for shrinking . . . markets. But in doing so, capital has to be careful in depriving worker subjectivity of any implication in terms of power and control . . . In this way, capital silences subjectivity just at the same time it calls it into life. Capital has not found, yet, the ways to deal with this contradiction. (Dyer-Witheford, 224)

Knowledge Workers of the World, Unite!

The *Anterieur* writers go on to argue, like good Marxists, that some workers have been able to mobilize "cooperative" aspects of the new work organization to create counterpower. In my view, the open source movement may develop into such a social movement (chapter 6). In Dyer-Witheford's reading, the *Anterieur* group also provides a theory of transformative networking. This they present through analysis of the contradictions of the "general intellect" being worked out in

> media and communication. General intellect is "a labour of networks and communicative discourse; it is not possible to have a 'general intellect' without a great variety of polymorphous communications . . . communications to use in a creative fashion the knowledges already accumulated, communication to elaborate and record new knowledges."
>
> Capital has developed technologies of information—mass media, telecommunications, and computer networks—to consolidate markets and ideological control. But here too it has been unable to develop the objective, fixed, machine side of "general intellect" without also involving the subjective, vari-

> able, human aspect . . . [*Anterieur* writers] reject media critiques framed only in terms of "manipulation."
>
> Nowhere has [the need for such rejection] been more apparent than in the field of computer-mediated communications . . . [I]n the development of this extraordinarily powerful technology capital has depended on a mass of informal, innovative, intellectual activity—"hacking"—on whose creative commerce [it; *sic*] constantly draws even as it criminalizes it. It was out of capital's inability to contain such activity that there emerged the astounding growth of the Internet. This is surely the quintessential institution of "general intellect . . . [or] . . . collective intelligence." (227 and 228).

Dyer-Witheford finds substantial grounds for optimism about "the capacities of mass intellect to reclaim advanced capital's means of communication." A potentially explosive

> volatility arises not only from a dynamic of immiseration [as in classical Marxism]—with more and more people being expelled from production by automation—but also from a reappropriative process in which "mass intellect" begins to fold back into itself the organizational and technological knowledge necessary for the running of society . . . [Such a] "constituent power" . . . [means] . . . the task of radical politics [is] the creation of a "republic" that dissolves both capitalist command and state authority. (230)

Dyer-Witheford concludes:

> In the era of mass intellect, a purely Luddite stance is not enough. To grasp the tactical and strategic changes present by capital's failure to control the technological dynamics it has set in motion, activists must be . . . Luddites on Monday and Friday, cyberpunks the rest of the week. (236)

Conclusions

The purpose of this chapter has been to sketch out a knowledge theory of value that could appropriately analyze the structural dimensions of cyberspace. While not generally seen as a core political economic discipline, anthropology is a practice that in general aims to recognize the importance of both emics and etics, the cultural elements with which humans collectively construct their world and the multiple physical, biological, and material conditions that limit what is culturally constructible. As such, anthropology shares more with institutional/political economic perspectives than with neoclassical economics. A group of self-identified anthropological "substantivists" arose in the 1970s (for example, Sahlins 1972) to counter the simplistic adoption of neoclassical terminology by ethnographers. These scholars, for example, critiqued the presumption of a universal "social surplus" whose allocation was the scarcity-driven, necessary preoccupation of economic activity (Hakken 1987).

Because a satisfactory ethnology of cyberspace has to account for both dynamic change and the form that change takes, it, too, is more properly grounded in such substantivist political economics. Sociocultural economy approaches that acknowledge a plurality of capitalisms (Blim 1996) engage cyberspace more promisingly than does Castells's theoretical project. Anthropologists should relate to Castells's ideas as suggestive hypotheses demanding critical evaluation, not as ethnologically demonstrated propositions. Such evaluation may support some of Castells's arguments. For example, the deplacing affordances of AITs-in-use do in my view justify Castells's developing disenchantment with analytic categories, like "cities," that privilege geography of the old style.[3] Mimi Ito's (no date) efforts to theorize "networked localities" is suggestive of one strategy for coming to terms with the "decouplings of spaces from places" that are new in social relations.

In use, AITs can support diverse tendencies, including that of capital markets to "go global" and new forms of workplace deskilling. AIT-based technologies of surveillance at work can tilt power even further toward capital. At the same time, AITs are technically just as compatible with expanded work humanization, expanded state intervention (for example, computerized monitoring of the environmental effects of production), and expanded worker control, as demonstrated by, for example, Nordic systems development projects. Chapter 5 of *Cyborgs@Cyberspace?* discussed data suggesting that analyses of the wellspring of value added are shifting toward the collective performance of the workforce. Chapter 8 of *The Knowledge Landscapes* . . . similarly argues that already the successful organization is held to be the one able to realize capital by getting its workers to participate most actively while at the same time convincing customers of the genuineness of workers' performance. Such organizations eventually confront the conflict inherent in all attempts to promote worker control while still keeping work subjugated to the reproduction of capital. A unionism less tied to collective bargaining would find here terrain on which a social activism for the contemporary era might be built. Social activism on these grounds, combined with social experiments which decouple income from labor, would be indicative of a truly different cyberspace political economy.

The Knowledge Theory of Value and the Future of Social Formation Reproduction

In short, the theory of knowledge developed throughout the book can provide the basis for a viable knowledge theory of value. If it were applied to policy and in organizations, what would be the result? Could such knowledge theories of value extend the reproductive ambit of employment social formations into the future? The obstacles to be overcome are formidable. As described in Section III, the commodity form continues to expand its long march through the institutions of social formation reproduction, colonizing new arenas like education. Turbo-capitalism eliminates or severely weakens institutions with some independent ability to influence social formation reproduction (educational institutions, governments, families, voluntary organizations/not-for-profits).

Because it continues to foster anarchic practices, capital's continuing dominance does not bode well for humans. Our capacities to extend social formation reproduction via AITs depend upon reversing the dominance of one social relationship, that of capital, and ultimately displacing it by a process, that of knowledge networking. In Sheffield, Barbara Andrews and I saw the beginnings of something like this (1993). When the computers came, the most important determinant of what happened was not the technology qua machines, social relations in the abstract, nor the iron laws of the market. What was important was how the technology was perceived and which potentials were actually appropriated by the people in actual social relations. While the dominant social relations clearly marginalized some constructions, and economics and mechanics certain others, there was still a broad range of interpretive flexibility in the actual performance of AITed actor networks.

How momentous is the task of replacing a capital with a knowledge political economy? Does it necessarily mean ending capitalism? Like Meghnad Desai (2002), Hutton doesn't think so:

> Obviously globalisation favours shareholder [U.S.: stockholder]-value-driven capitalism and . . . is being driven by it, so it's hardly surprising that variants of capitalism that try to balance the other interests in the enterprise, like those of the workers, and to behave more ethically—stakeholder capitalisms—are under pressure. But that doesn't mean that the principle of stakeholder capitalism is wrong; it means rather that some of the means of achieving it have to be updated and modernised . . . (Giddens and Hutton, 31)

For Hutton, "stakeholder capitalism" is a form of capitalism in which capital reproduction doesn't run rampant. Instead, it is designed and disciplined in a manner that equally benefits all social stakeholders. Because turbo-capitalism is neither a technologically driven inevitability nor an unstoppable structural imperative, there remains considerable opportunity, as well as pressing need, for the "greater governance of the global economy . . . " The question remains

> to what extent we can modify capitalism so that it can live with other values like quality and social justice . . . (19)
>
> Every form of capitalism must possess a legal framework in which to do business . . . [C]orporate, banking, pension fund, employment, trustee, contract and commercial law reflect conscious choices about what kind of capitalism any particular society wants—and my contention is that it can be biased significantly to favour interest other than property owners and private shareholders. (34–35)

In Hutton, is it is possible to redomesticate capitalism. Such a project could use knowledge technologies to construct substantial counters to the reproductive influence of capital.

One need not share Hutton's optimism about pushing the turbo-capitalist genie back into the bottle. Nor, in theoretical terms, does acceptance of interpretive flexibility with regard to the political economics of AITs mean "anything goes." Just as turbo-capitalism and "postcapitalism" are not the only possible social formations of the future, so cyberspace structuralistics are not forced to choose among neoclassical capital mythology, political economic technological determinism, or chaos. One can be subtle about causation without abandoning it altogether.

AITs are better viewed as terrains of contestation than as ineluctable, independent forces. Technologies do have politics, but like all politics, they manifest multiple, contradictory tendencies. Their role in particular situations depends upon how multiple constructions play out, and contexts influence, through conflict. Capitalism is an inherently anarchic political economy, the "new economy" a mirage, but contradictory forces and conflicting constructions mean the future is yet to be determined. Such moments of underdetermination can be moments of opportunity. In the next and final chapter, I examine the implications of the answers to the cyberspace knowledge question presented so far for those who would intervene in AITs creation, whether as ethnographers, designers, or activists.

Chapter 12

An Ethico-aesthetics and Politics for Changing Knowledge in Cyberspace

This section's preceding chapters examined the structural connections between automated knowledge technologies and knowledge change. AITs designed with appropriate consideration for the social have substantial potential to broaden democratic participation, resocial work, and accelerate knowledge creation. Yet the chapters concluded that, so far, this substantial potential for change has been displaced by other dynamics, especially that of capital's reproduction. In the absence of deliberate interventions to expand the scope of a knowledge political economy, social change through change in knowledge networking will remain either insignificant or negative.

The triple collapse of the dot-com, dot-telecom, and dot-edu "bubbles" may result in some additional space for knowledge technology initiatives. Lower expectations regarding return to capital create more "elbowroom" for pursuit of cyberspace knowledge projects to build the alternative political economy of knowledge described in the last chapter. We cannot know how extensive this autonomy will be or for how long it will last, but networking to support knowledge rather than money is at least for a time more realizable. Realizing this potential should be at the core of today's cyberspace agenda.

This concluding chapter outlines a "knowledgistics" for this time of opportunity, a socially informed, collective, professionally relevant program to change knowledge dynamics, one that moves changing knowledge to the center of informatics practice.

These knowledgistics are first oriented ethically. Because cyberspace is designed—it does not just "happen"—special ethical responsibilities fall on those who, like cyberspace ethnographers,[1] study it, as well as those who create and inhabit it. Also, because constructing cyberspace is so largely a matter of design, these ethical imperatives have substantial aesthetic implications. Thus, the second, aesthetic grounding of the chapter's knowledgistics is in the form of a standard trope in artificial science, a statement of the new knowledgistics' design principles, the goals that should inform both technically informed social interventions and socially informed technical ones.[2]

Practices do not become ethical and aesthetic merely because humans desire to behave ethically and aesthetically, however. Nor, no matter how ethical and aesthetic their design practice, will ethnographers and even socially inclined informaticians be able to change the dynamics of knowledge in cyberspace on their own; without allies, Cyberia is indeed a cold space.

Because ethical and aesthetic practices are only truly fostered by judicious intervention into the possibilities created by actual social formation reproduction, the chapter presents an additional grounding, in politics. It summarizes the preceding argument in terms of the political challenges that follow from the limits on political action in cyberspace, as well as the opportunities suggested by the answers to the knowledge question visible in *The Knowledge Landscapes of Cyberspace.*

Ethics in Cyberspace and Cyberspace Ethnography

Can One Act Ethically in Cyberspace?

Since Plato and Socrates first posed it, answers to the question "How should we live?" have constituted the domain of ethics. Western notions about ethical effectiveness have been predicated on the possibility of meaningful ethical action, or what philosophers call "ethical agency." Greek ideas of agency presumed the social privilege of the young male citizens who studied at Plato's Academy. These young men already had considerable agency because of their social standing. The elitist social formation that benefited them was justified in terms of their ability to act ethically.

Greek conceptions of ethics leave us with a quandary regarding the scope of agency: Is it only those with the superior knowledge and sensibility that come with class privilege who can create a better society? In an important sense, the French Revolution opened the door to a broadening of this same sense of agency rather than offering an alternative to it. In the kind of modernist approach described in chapter 4, ethical agency is still largely framed in Greek terms. That is, one gets one's agency first from mastering the universal rules for proper behavior, the derivation of which is the proper activity of philosophical ethics. Without this knowledge one cannot act meaningfully, but with it, one can to apply the rules in one's life. Modernist ethical agency is individualistic, and it is about effort, not necessarily impact.

Some writers, like the humanist Mark Slouka in *War of the Worlds* (1995), feel cyberspace generally inhibits ethical practice. On modernism, to be ethical, we must be able to project ourselves into the situation of the other. In this way, ethical behavior depends on referencing interacting in a more or less stable and shared "real world." To the extent that we spend our time in some self-generated, artificial, or 'virtual' cyberworld, our capacity for ethical action is impaired.

The broader implications of cyberspace for especially Kantian ("treat others as ends, not means") modernist conceptions of ethical agency emerge in Sherry Turkle's *Life on the Screen* (1995; blurbed as "the first Internet ethnography"). Turkle argues that in cyberspace individual identity is pluralized. As current operating systems (for example, Windows™) allow users to work in several different software programs simultaneously, the accomplished netizen typically projects several different identities at the same time (one in a chat group, another on a listserv, a third in her MOO (multiuser domain, object oriented), and so on. Turkle argues that the capacity to maintain multiple distinct identi*ties*, rather than to construct an underlying coherent identi*ty*, is the treasured personality skill in cyberspace. "Multiple personality" becomes the norm rather than a recognized psychic disorder.

In a Slouka reading of Turkle, the integrated, coherent self necessary for agency, and thus a precondition for ethical action, becomes in cyberspace a handicap. There are additional reasons to be worried about ethics in cyberspace. The profound flexibility of the computer as a medium carries with it the dangers, articulated by Jean Baudrillard, of hyperabstraction (1995). At least for some, his famous epigram about the Gulf War taking place primarily on television has suggested that the expanded potential for simultaneously transforming both signifier and signified may ultimately produce mass confusion. The consolidation of capital reproduction on a global level increases the scope for permanent mystification—for example, the kind of commodity fetishism that permanently "forgets" that exchange value ultimately needs use value. To the extent that, via the complex manipulations of programming, computers hide the "raw" data from which information is obtained, separating valid from invalid analyses becomes more difficult. The famous operator maxim "garbage in, garbage out" becomes a common phrase to justify denying any responsibility.

Richard Sennett's *The Corrosion of Character* (1999) further illustrates the difficulty of acting morally under such conditions. Perhaps his ethnographically most striking example is the experience of an older woman who has successfully entrepreneured a bar in Manhattan. She gets bored and quits to work with an advertising agency that sells supplies to people in her former position, but she leaves in frustration after a short time. She discovers not only that as an older person she is presumed to have nothing new to contribute but that in the new, performative world, no one cares for knowledge of the actual work process, only its representation.

Actor Network Theory and Nonmodernist Ethics

One could of course construct a more optimistic account of ethics in cyberspace, say, by stressing the way the Internet extends knowledge of "the other." Yet on most Modernist accounts of ethics, like those presented above, cyberspace undermines agency, meaningful individual action of any kind, including moral action. There may be more room for characters but less for character and skill, let alone the thick ethical constructs that often arise from community or work culture.

Modernist conceptions of ethical agency, however, have lost considerable ground in professional philosophy, as outlined in chapter 4. Moreover, they have also been undermined by science, technology, and society's broadening of the notion of agent to include nonhumans (Callon 1986; Fuller 1992). This is especially the case under actor network theory, arguably the most prominent (albeit contested) theoretical basis for STS (Hakken 1999a).

Briefly, Bruno Latour and Michel Callon developed ANT as an alternative account of how specific technoscience practices achieve prominence.[3] In the older, modernist view, science sees the world in the way it does because this sight most accurately reflects how the world really is. Similarly, those technologies that become strongly institutionalized do so because they reflect most effectively the real laws that govern the universe that science sees. In such strongly realist ontologies, there is little room for culture or a social science of technoscience. Ethics can effectively be reduced to "do good science."

On an ANT reading, however, accepted theories derive their hegemony from social process, not necessarily because they fit reality best: A given technology actor network has come into palpable existence because identifiable "actors" recruited sufficient allies and fostered enough practices to achieve stability. Always provisional, this stability involves complex "negotiations" among people, organizations, practices, and artifacts. Any of these, not just the humans, can hold "agency," that which distinguishes an actor from an "actant."

ANT's redistribution of agency calls for a rethinking of what it means to be ethical. STS debates over the similarities between human and nonhuman agency parallel public discussion over whether one can rape another in an online chat group, or when it becomes reasonable to hold a cyborg criminally responsible. While often sensationalistic, such discourse can deal seriously with, for example, the conflict between Langdon Winner's dictum that "technologies have politics" and AITs' potential for social transformation as opposed to merely individual action. Any rethinking of the tight Kantian connection of agency to transcendent individualist ethics must acknowledge the potential for nonhuman agency of actor network theory.

Resolving Problems of Agency in Cyberspace Ethics

Personally, I guard against the problems of reduced agency in cyberspace by making sure I engage in a wide array of both basic and applied research. I also complement research with active practice in systems design and implementation. Fortunately, with the decline of scientism, there is increasing scope for projects that no longer search after the ONE BEST WAY. How ethically I can act in cyberspace clearly depends upon the general ethical agency afforded by cyberspace.

Nonmodernists invoke a collective ethics of consequence. In nonmodern readings like those offered in chapters 8 and 11, the nontranscendent potentials of cyberspace create opportunities for renewed, especially more collective ethical practice, like that based in work cultures. The "resocialing" readings of chapter 8, for example, are suggestive of renewed ethical sensibility, for which the knowledge political economy of chapter 11 could provide a basis.

At the same time, nonmodernism also makes attainment of agency more complex. Its "relative relativization" of rules of conduct means that an individual trying to figure out what to strive for ethically—in order to be an agent toward something—cannot rely on universals. She should of course pay attention to the notions about proper conduct of the relevant culture, and to a certain extent her agency will reflect them. However, she still cannot "derive an 'ought' from their 'is.'" Further, such notions are insufficient guides in situations of culture conflict, when she will need to figure out some defensible notion of "greater good" without transcendent rules on which to rely.

Contemporary ethics seeks answers not in modernist identification of abstract, universal truths but by juxtaposing possible values against diverse empirical reflections. On nonmodernist philosophy, answers to the question "How should we live in cyberspace?" should be informed by study of its likely dynamics. Out of such study one can construct "thick ethical constructs" (Williams 1985).

Those who would be ethical agents must therefore also be committed to rigorous study of the world (Young 2000). As discussed at length in *Cyborgs@Cyberspace?*, ethnographic study of protocyberspace activities can discipline our thinking about this potential new social formation. A descriptive ethnography of ethics in cyberspace could help us sort out whether decline or renewal of ethical agency seems in the offing. The early cybernaut, however, is appropriately cautious about the permanence of what she experiences.

Decisions about what approach to take are an important part of all research. If we can articulate a defensible ethics for ethnographic study *of* cyberspace, we shall have a good beginning point for a general ethics of knowledge work *in* cyberspace. Under what conditions can cyberians expect to do ethical research?

The Ethics of Ethnography

The good ethnographer prepares for her fieldwork by learning as much as she can about what others have to say about her site and how they have approached her research problem.[4] Hakken (1999a) and Sections I and II of *The Knowledge Landscapes* . . . simulate in somewhat exaggerated form this preparatory labor. Once in the field, however, the ethnographer learns by doing and watching ("participant observation"), connecting to others in the field though multiple forms and to varying degrees of active engagement (as in Section III). For example, one tries to create a dense relationship with at least one "good informant"—a person who has thought a lot about the culture and who is good at expressing her thoughts.

In a very specific sense, the ethic of anthropological research has been somewhat passive. That is, we were arguably behaving ethically if we were able to:

1. Ground our practice in an appreciation of and support for, and produced generally accurate representation of, the specific "really existing" cultures we studied.

For the anthropologist whose field was one of the simpler social formations, the conjunction of these ethics with construction of the active relationships required by the ethnographic way of knowing raised concerns of the "Hawthorne effect"[5] variety: How do I make sure that the actual social realities influence my thinking more than my thinking structures my perceptions of them? These concerns are palpable toward the end of Colin Turnbull's monograph on *The Forest People*, the BaMbuti of the Congo (1972), when he ponders the research ethics of making a crutch for an injured child. Turnbull is concerned that such an intervention will "change the culture" and thus distort his results. He tries to balance the benefit for one child and his group against a less-"Hawthorned" broader scientific good. (With my students I like to compare and contrast Turnbull's thoughts with those of Hortense Powdermaker when in 1935 she ponders whether to try to stop a lynching in Indianola, Mississippi [1966].)

Such concerns led to articulation of a second ethical imperative, to:

2. Pay particular attention to whether what we think we have learned reflects what's really going on, rather than being an artifact of our presence.

"Hawthorne" concerns were very relevant in the colonial (or neocolonial) context of much earlier fieldwork. (That they remain relevant today is indicated by the controversy surrounding Patrick Tierney's recent *Darkness in El Dorado: How Scientists and Journalists Devastated the Amazon* [2000].) However, most ethnography today is carried out in complex social formations where one field project is unlikely to have much impact on general social dynamics. Much ethnography today is also applied, carried out in the context of interventions whose purpose is to change practices. Under these conditions, ethical research must supplement its anti-"Hawthorne" steps by also taking up how to:

3. Make sure that the intended change and the way it was to be accomplished were ethically defensible.

But even this was not the whole ethical story for precyberspace ethnography. As Richard Fox (2000) points out, ethnographers have always confronted ethical conundrums of communication beyond mere accurate description and defensible intervention. As Tierney's work makes clear, mere description—indeed, any other kind of ethical "coasting" that ignored the impact of our work on societies whose actions affected our informants—was never a viable ethical posture. Thus, in order to be ethical, our practice also had to strive to:

4. Anticipate effectively the rhetorical consequences of our representations, both for those represented and those represented to, particularly what politics (for example, their ethnocentrisms) our anticipated audiences were likely to associate with our representations.

The Ethics of Ethnography in Cyberspace

These four imperatives are still relevant to ethnography in the artificial world of cyberspace, but they are not sufficient. In my experience, many of the "practical" methodological problems faced by the cyberspace ethnographer are analogous to those of other ethnographers (Hakken 1999b). However, her ethical position is different in that it must be expanded, especially in comparison to that of other, "basic research" colleagues.

This is partly because the "basic" cyberspace researcher's position is extended from that of the previous "applied" researcher. Since cyberspace is an artificial world in the making, being active in it means that the ethnographer is inevitably a cocreator *of* it. Thus, even research intended to be "basic" has, like applied interventions, to be evaluated in terms of whether it is likely to promote defensible outcomes.

There is also another complexity, derived from contemporary ethics. The social formations cyberspace ethnographers study are as much potential as "really existing." What we study often seems like the proverbial webpage always "under construction." Reproductive dynamics are not yet stabilized; the manifestations we experience may even turn out to be, like so many promised programs, mere "vaporware." If the key move is to relate notions of proper behavior to a rich empirical background, how are values to be empirically grounded if the social formation doesn't yet really exist?

Moreover, cyberspace ethnographers' acts of representation may well become part of an eventual stabilization process. More than other ethnographers, we are directly involved in the cocreation of the culture whose protoforms we study. For this reason, we cannot depend just on the practices we study and the notions of our audience to set the parameters of the representations at which we aim. What does it mean, ethically, to study a culture, participation in which, since it only exists "in embryo," inevitably means we are implicated in its construction?

Even the appropriate means of representation are unclear. We find ourselves asking, because every representation is an act of intervention, what is an appropriate representational strategy in the current moment of possible transition to cyberspace?

Because of all these factors, the ethical posture of cyberspace ethnographers must be more "active." This is not to suggest that we can abandon the ethical imperatives to understand the dynamics to which we are exposed as clearly as possible and find ways to characterize them accurately. Nor can we ignore the political implications of our descriptions. However, any notion that the current dynamics of this social formation's reproduction are its "mature," stable forms is likely to be a chimera for a long time to come. We thus cannot even rely on "probable consequences" to guide our representational performances. How, then, are we to decide what to present, how to represent it, and to whom?

Under rule number 4, ethnographers should consider the politics of their representational acts in terms of their audience and thus should anticipate the likely impacts of different rhetorical phrasings. We have considerable information about the likely impact of "optative mode" cyber-performances like those described in chapter 1. Many of the rhetorical choices that I have made, for example, have been informed by a conscious decision to emphasize the continuities of cyberspace with other social formations. This has also been done because research reveals substantial continuities that the dominant rhetorics consistently underemphasize.

The Contradictions of Cyberspace Ethnography Ethics

However, if fieldworkers in cyberspace are to take seriously our role as its inevitable coconstructors, we need more than negative advice, for example, about what to avoid. To compensate for our inability to presume anything like stability of reproductive dynamics, we need to articulate what we want, and to do it even more thoroughly. I thus propose a fifth rule for the ethical practice of ethnography in cyberspace; to be ethical it must:

5. Involve substantial cultural advocacy; that is, significant effort to articulate and communicate the kinds of practices that we believe should *be characteristic of cyberspace and how to foster them.*

The articulation of such a rule is of course very far from the so-called value-free methodological individualism characteristic of the moment of "high science" in the social science of the 1950s and early 1960s (Mills 1959). This fifth rule also calls for something beyond the '60s-type ritualistic articulation of the ethnographer's personal values.[6] The cyber-ethnographer needs to work out her techno- as well as fieldwork ethics. These former need to be worked out systematically enough to be embodied in practice, in the "participation" moment of participant observation. The very active form of action research (Whyte 1991) that this implies of course carries substantial risks. It increases potential for "Hawthorne effects," situations where our research presence has hard-to-discern impacts on the dynamics we observe.

As long as only humans were presumed to be capable of it, agency could be associated with the quality of consciousness—"I think, therefore I can be an ethical agent." The "cyborgic" critique of consciousness (Hakken 1999a, chapter 3) is corrosive of such "consciousness-bound" theories of agency. When in the early 1980s I began my personal efforts to make more space in anthropology for attention to AITs, I encountered a particularly rigid form of this boundary. Anthropology implicitly (and modernistically) privileges consciousness by location of our object of study, the carrier of culture, as the individual "human being," conceived of as a life-form bounded by the skin. By following Durkheim too closely in identifying "culture" with the sacred within the skin, and technology with the profane outside it, we open our discipline to an intellectual, and therefore ethical, schizophrenia. On the one hand, we operate with a general anthropology based on evolutionary conceptions of technology ("gathering and hunting," succeeded by "horticulture," and so on); on the other, technology tends to be treated as exogenous to culture.

To overcome this limitation (and as a way to take seriously the actor network theory–induced dispersal of agency described above), I find it useful to conceive of the carriers of culture as cyborgs, not human beings. In homage to Bruno Latour's *We Have Never Been Modern* (1993), I thought about using the title *We Have (Almost) Always Been Cyborgs* instead of *Cyborgs@Cyberspace?* for my previous book. I have even argued on occasion for renaming of our field "cyborgology."

It is to be hoped that anthropologists, freed from the tyrannies of modernism in cyberspace, will cease trying to ground disciplinary professional ethics on pursuit of invariant rules for governing the agency of individual life-forms. Rather, we may be able to see our "basic" tasks to include the identification of potentially thick ethical constructs located in the spaces we share and the other resources for agency of our "extrasomatic modes of adaptation." A cyborgized physical anthropology, one freed from an essentialist commitment to the biological, would play a substantial part in creating such an anthro-ethics. Perhaps this will come through rethinking agency in and in relation to cyborgs like transgenic organisms.[7] In any case, a Cartesian conception of agency is no longer adequate professionally.

The practice of ethnography is increasing "privatized," through its growing popularity in modes of social reproduction more directly implicated in the reproduction of capital (product development and marketing as well as systems development). This phenomenon needs to be assessed in conjunction with the possibly decreasing academic anthropological interest in ethnography (as manifest in the decline in sociocultural applications to NSF anthropology, although not to other programs). Together, these suggest that our ability to monitor the ethics of ethnographic practice will depend upon dense dialogues with other professionals.

In the largely artificial world of cyberspace, brought into existence through acts of design and use of artifacts, a broader approach to agency is necessary. The cyberspace ethnographer will have to try to embody her ethics in practical projects. To do this, she will have to find "informants" also willing to be "coagents," to collaborate in putting similar ethics into practice, in this very participatory form of action research (Whyte 1991). In such a world, designers act ethically when they design in accordance with ethically defensible goals. At least part of what makes a goal defensible is that it makes sense in relation to the actually emerging dynamics of cyberspace, dynamics that must be demonstrable, not just appealing and/or possible. Nonetheless, because of as well as in spite of these special conditions, ethnographers and other cyberspace professionals both continue to strive after ethical agency in cyberspace while acknowledging continuing changes in agency.

Aesthetics and Design Criteria

Among the implications of the ethics articulated above is the notion that the design of cyberspace artifacts should include substantial advocacy. Even knowing what we want artifacts to do doesn't tell us which artifiacts to build, however. Ethnographers and other cyberspace workers trying to address this problem ethically can gain assistance from another quarter, that branch of philosophy called "aesthetics." Just as nonmodern ethics abjures the search for transcendent statements of "the good," nonmodern aesthetics no longer focuses on articulating abstract standards to, for example, differentiate "the beautiful" from "the ugly." Rather, nonmodern aesthetics strives to understand the choices made by, for example, an artist, the frameworks in relation to which she makes her choices and their relationship to frameworks we would advocate. Thus, the comparative aesthetician develops accounts for the ways judgments of form, balance, taste (Bourdieu 1990), and so on, are actually made by people in different cultural contexts. The intent is to understand the range of aesthetic practices, their contradictions, and their connections.

Applied ethics involves helping individuals and groups achieve more coherent embodiments of their notions about how people like them should live. Similarly, applied aesthetics is about understanding why we as individuals and groups respond to cultural representations and performances as we do, finding coherent ways to evaluate these responses, and identifying those initiatives that foster the responses we wish to encourage. Applied aesthetic practice is based on the presumption that there is a strong connection between having greater awareness of our reactions, influencing them, and being able to pursue those goals that we think of as really important. In this way, a good aesthetic sense may well be a prerequisite to effective ethical action, just as it is to constructing good ethnographic representations.

Recognition of the importance of aesthetics to informatics has grown recently, as manifest in recent program developments at universities on the Nordic cities of Aarhus and Malmö. It came into its own with recognition of the important visual and graphic design aspects of, first, user interfaces and, more recently, webpages. The emergence of the Internet as a potential center for realizing capital had the salutary effect of broadening cyberspace design, from the relatively narrow question of information system design to the broader one of fostering generally effective channels for automated communication.

This emerging cyber-aesthetic sensibility has to be given an ethically strong reading, to develop into, in the chapter title's words, an "ethico-aesthetics." From this perspective, a webpage is well designed not just when it has well-chosen colors, understandable words, and the right number of interactive buttons. It also is well designed only if it effectively promotes knowledge networking. What design ideals, then, should be foremost in our minds as we create the future? The following list is an initial effort to state AIT design criteria that are compatible with this ethico-aesthetic sensibility.

Support Collaboration, Not Individuation: Cyborgize the Interface!

The technologies we develop to support knowledge networking must encourage collectivity and group agency. In the applied educational program in which I worked until recently at the State University of New York, I tried to prepare students for the teamwork increasingly a part of careers in social service organizations. Because of the orientation of recruitment literature, however, students entered the program focusing primarily on what it could do for them *as individuals.* Similarly, an aspect of most approaches to "distance learning" that undercuts their transformative potential is their perpetuation of the romance with individuation via so-called individualized learning. In both the classroom and the asynchronous online "web class," individuation undermines both learning from each other and creating new levels of shared knowledge.

For Marx, alienation was more than a mere matter of losing property; losing control of what they had made meant that workers lost control of a part of their selves. The collectivizing of this alienation was an important features of the manufacture stage of the labor processes that he described as replacing the craft stage. Still, as extensions of their hands, the tools used in manufacture tended to remain in workers' control. Thus their labor *process,* if not its product, remained compatible with their "species being" or "nature." In the next, machinofacture stage of the labor process, the worker becomes an extension of the tool, the image captured so well by Charlie Chaplin in *Modern Times.*

In informatics, the difference between manufacture and machinofacture is that the latter subverts the collaborative interface between the human, machine, and cyborg moments of a work process. There is a psychologism at the base of much of the work in human-computer interaction, one that frames the basic interface problem as the need to fit *a* machine to *a* human being. This pyschologism also individuates AIT design. In contrast, design based on Computer Supported Collaborative (or Cooperative) Work (CSCW) or the "tool perspective" of Participatory Design builds tools to be controlled by the workers as a collective—not as individuals but as a group.

A cyberspace work ethic would no longer strive to control workers' bodies. It would systematically reverse Taylorist individuation, which impedes the elemental sociality that arises as workers articulate their shared experience through talking to each other on the job. The potential for enhancing sociality at a distance is the productivity advantage of knowledge AITs. Thus, reversing individuation is not only a good thing ethically; it is a necessary part of furthering the relative independence of the cyberspace moment in the evolution of work.

Support Knowledge, Not Information, Networking

Confusing information with knowledge is one of the main current impediments to knowledge networking. If knowledge support systems do not include tools to validate knowledge that supplement face-to-face activities, they remain mere information technologies.

Recent developments have broadened the range of activities that count toward validating. While they still have great prestige, the positivist protocols of Western "natural" science that privileged experimentation in laboratories have begun to accommodate to other ways of knowing. One reason for this is recognition of situations, as in corporations, in which the knowledge one wishes to access or share for success in the market is embodied in a person—say, a retiring district sales manager. To become a collective resource, such knowledge need not be abstracted into data, transformed from tacit into explicit. It can be kept deeply situated in workspace culture, communicated in stories more existentially compelling because they retain experiential relevance.

STS research like that of Rabinow (1992) suggests that embodiment of knowledge is common in technoscience, which is not the domain of purely positivist pursuit of universals described to schoolchildren. Scientists and technologists arrive at shared constructions of knowledge socially. For example, standards for judging a claim as valid are arrived at through the construction of social relationships best understood as alliances. Technologies become standard not because they conform best to the properties of the physical world but because the most effective advocates are allied behind them.

Since knowledge is culturally constructed, the structures through which, and the processes by which, information becomes knowledge in one culture need not necessarily reflect—indeed, often will not reflect—those in others. Attempts in general to divorce the social context from the content of knowledge are likely to be misleading and are probably doomed in the long run.

For these reasons, organized systems to support knowledge networking need to be able to present knowledge in registers different from the encyclopedic, classroom talk of scientism. They may often need to foster the sharing of these different kinds of knowledge, from evaluations of the latest lab tests of a drug to the drinking habits of a potential customer. This may mean that an organization needs to conceptualize multiple knowledge networks, one for each type of knowledge. Equally, they need to plan for and provide mechanisms to manage conflict over what is known. Since conflict is as likely to be based on different notions of what makes something "known" as on simple differences of content, there must be good notional spaces for reflexive metadiscourse. Organizations also need to develop electronic analogues of the social conventions that humans have developed to deal with such conflicts in face-to-face situations and build them into knowledge technologies. Their incorporation would indicate a new stage of maturity in automated knowledge technologies.

(Re)socialize Learning and Extend the Reproduction of Culture

One of the more celebrated spillovers from cyberfacture into broader thinking about organizations is Argyris and Schön's notion (1978) of "the learning organization." Too often, however, this notion has been misapprehended in terms of an organic metaphor, that of adaptation. Organizations don't just need, like life-forms, to adapt to change in their environment; they need, like cultures (as Argyris and Schön make clear), to learn how to adapt better and more quickly. Thus, we need to design knowledge support infrastructures that extend organizations' adaptive capacities, that promote the anticipatory action afforded by learning, not just adaptive reaction.

Tamoko Hamada (1998) nicely captures the ethnographic critique of popular business appropriations of the anthropological notion of culture. Rather than organizational culture being something planned, imposed from the top and manipulated, it, like other cultures, evolves from the bottom. While its development can be influenced by the deft, it is beyond control.

Rather than Goebbels's propaganda, I again suggest Marx as the provider of the appropriate metaphor for the organizational culture most conducive to transformed knowledge networking. Marx distinguished between simple reproduction and reproduction of an expanded scale. He contrasted the former to social formations where outputs from one period never equaled initial inputs. A tendency to foster transformation of the social and the technical made reproduction on a greatly more extended scale characteristic of employment societies.

Notions of truly knowledged organizations can be built on Marx's metaphor. The learning organization uses AITs to reproduce its culture but aims to do so on an extended scale. It not only uses AITs to broaden access to available information; it also uses them to provide powerful symbolic tools to validate this information. It promotes networking of knowledge rather than its commodification. Further, because it understands the basis of ethnogenesis in autonomous collectives, this organization rewards those groups that use the information to position the organization to anticipate change.

A Prolegomenon to an Ethico-aesthetic Politics of Cyberspace

A central point of *The Knowledge Landscapes of Cyberspace* has been that the possibility of a both distinctly different and substantial cyberculture needs to be taken seriously. The preceding section offers one way to do this, by turning desirable social correlates into design criteria. Anthropological fieldworkers have already had another kind of impact; Gary Downey (1998), David Hess (2001a), Henry Lundsgaarde (1992), Jean Lave (Lave and Wenger 1994), Julian Orr (1996), Bryan Pfaffenberger (1990), and Lucy Suchman (1987) are just a few of those who have had diverse but substantial influence on our sense of what is possible in cyberspace.

Ethnographic description and design can have an impact because cyberspace is at most a protoculture; its eventual dynamics are "open." That is, cyberspace is a social formation that may come into prominence but will not necessarily do so. Since it has no necessary form, one can't be sure that its existing, protomanifestations will be the ones that come to dominance in the future.

Since openness also applies to the dynamics we would promote through design, an additional obligation is placed on cybernauts, both its ethnographers and developers. Not only do we need to develop a collective sense of what we want cyberspace to be, a cyberspace ethics, and an aesthetics that helps us embody preferred values as we choose among the design options we face. We also need projects that build our ability to influence social formation reproduction, which of course means we need to collaborate with each other and our informants/partners. In a word, we need a politics of design, a practice of intervention that maximizes opportunities to participate effectively as well as ethico-aesthetically in the construction of cyberspace.

In what follows, the implications of the answers to the knowledge question in cyberspace presented in this book are summarized in terms of the challenges that such a cyber-politics will face. Some implications follow from important features of cyberspace's early or "proto" forms; others are likely to increase in significance when and if cyberspace becomes the primary mode of adaptation.[8]

Challenge #1: A Politically Useful Political Economy

To date, AITs have supported diverse tendencies. On the one hand, they have been used to extend, even "globalize" the reproduction of capital and new forms of workplace deskilling—for example, existing AIT-based technologies of surveillance at work tend to tilt power further toward capital. On the other hand, knowledge technologies have also expanded democratization (through cyber-referenda), enabled more effective collective intervention (for example, computerized monitoring of the environmental effects of production), and expanded the potential forms of worker control, as demonstrated by, for example, Nordic systems development projects.

Moreover, organizational analyses of the wellspring of value added increasingly focus on the collective performance of the workforce. The switch in the focus of value discourse to knowledge is further facilitated, perhaps even compelled, by all sorts of ideas about teams, dispersed collaborative work, virtual organizations, participatory design, and so on. The organization able to realize value by getting its workers to participate most actively while at the same time convincing customers of the genuineness of workers' performance is increasingly the one labeled successful.

As charted in the earlier chapters of this Section, such developments have refocused the value debate on knowledge. Once notions like "knowledge capital" are deconstructed, popular knowledge structuralistics are recognizable as preliminary, inadequate efforts to articulate a third distinct, neither labor nor capital, theory of value. A real knowledge political economics, a fully developed theory that makes knowledge a *replacement for rather than a form of* capital as the chief factor of production, would offer a more satisfactory account of where value comes from. It also initiates a truly independent politics of knowledge.

The theory of knowledge developed in this book as an initial stab at a knowledge theory of value. Just as social science originally congealed around a new answer (labor power) to the value question, a resituating of knowledge as knowledge networking is necessary if the use of knowledge technology is to expand.

To take advantage of the potentials of AITs to facilitate knowledge networking, we need broader social change. In a social formation whose reproduction is still dominated by capital, such politics will inevitably sharpen the conflict inherent in all attempts to promote worker control while still subordinating work to capital. However, when the computers come, their correlates are determined by which of the possible potentials are appropriated by the people in the relevant social relations. Social activism grounded in a true knowledge theory of value could, for example, foster a unionism less tied to bargaining. This new knowledge praxis can aid development of a really distinct cyberspace social formation by providing a terrain on which a social activism for the future can be built.

Challenge #2: A Knowledge Politics of Organization

Knowledge Management Fatigue Syndrome

To network knowledge, workers need to create social relations. This is the point where work process is open to effective cyber-intervention. The "thin" knowledge management practices of 1990s organizations, however, tended to be merely slightly more complex, even merely renamed, forms of information or even data sharing. General confusions about knowledge were consequently a regular manifestation of first-generation knowledge systems' talk and the software sold to implement them. Their central failure was not coming to terms explicitly with the social, their difficulties essentially recapitulations of previous attempts to conceptualize knowledge in purely technical ways. Hence, these systems had greater continuity with pre-AITs than with a new social dynamic.

Some organizational systems, like Japanese CSCW, collaboratories, and virtual educational communities like MediaMOO, did attempt to simulate social life. Even these systems conceptualized knowledge networking too narrowly, however. Instead of trying to create virtual face-to-face practices, systems need to create new, AIT-based social practices. These need to support collective production and effective dissemination of knowledge in distinct ways that take advantage of new, AITed capabilities.

Certainly, knowledge management did not fail because the problem of knowledge at work evaporated. Contemporary social formations are now doubly dependent upon solving it. Organizations are increasingly teamed, and they will not in general decrease their scale of operation. Moreover, the world's economies took a deliberate Schumpeterian leap of "destructive destruction," hitching the future of social formation reproduction to the gossamer but disconcertingly diaphanous fabric of a "high-tech" economy.

In this "globalized" world of dispersed activity, coordination is the dominant economic problem. If only as a consequence of their wide distribution, AITs will certainly be a part of any general solution to our daunting coordination problem. Yet better coordination requires a broader understanding of the more general process of knowledge networking. A better understanding of knowledge's cultural dimensions must be built into knowledge technologies before we can use AITs to solve the specific coordination problems of organizations. With such understandings, it may be possible to to design more appropriate roles for AITs, to "get right" the role of AIT in organizations.

Resocialing Work

What counts as work frequently has ethical implications—legitimating, for example, access to the social product (Hakken 1987). Such connections are frequent enough perhaps to constitute a cultural universal. Because it could transform the labor process by "resocialing" work, cyberspace may give work ethics a radical twist (chapter 8). Resocialing would mean that work would be once again substantially more social (deindividuated) and that this greater sociality would be celebrated, neither ignored (as in human relations) nor suppressed (Taylorism).

The analytic case for resocialing is based on the idea that, once again, a series of contradictions has increased the relative autonomy of the labor process. Some autonomy results, for example, from the contradiction between the "globalization" of production and the increased dependence upon the social at work for production of value. Current changes in organization increase labor processes' dependence upon workers' initiative. "Know-how" is applied best when applied spontaneously, as in cooperative, self-managed groups, rather than on individual command. In the past, such groups formed through body-to-body, face-to-face interaction. Innovation is more frequent and the labor process is evolving along a relatively more independent trajectory. Hence, the delicious contradiction of the cyber-workspace: Work becomes more dependent upon workers' abilities to create close social relations, at the same time as globalization inhibits their construction.

A second, cultural process comes to parallel the physical one in the computered labor process. Staff participate in endless workshops, training, and other performances intended to convince all participants to "buy into" the commodities produced. As communicating symbolers, computers are vital to this parallel process; they are thus the appropriate symbol of cyberfacture.

In some organizations, cyberfacture also reverses individuation. Elimination of individual offices means that work cannot help but be more social, and the group takes on many of the traditional roles of the boss. This is a "resocialed" socially revived workplace, one where innovative practices and groups grow new forms of socio- and ethnogenesis.

This increased scope for the social in cyber-work is a potentially significant locus of (relatively) autonomous work. It thus provides at least a vision of the separation of work from conditions of labor. The declining power of Taylorist narratives in workspaces parallels the rise of multiple discourses that contest the privileging of "hard" science so much a part of modernism. The multiculturalism of complex knowledge networking at work and at home could become a pluralism tolerantly reflective of the multiple communities to which each human/cyborg actually belongs. Promotion of such possibilities would be an important part of the ethico-aesthetic politics of cyberspace ethnography and practice.

Managerialism

Recent, salutary developments in organization studies mean knowledge is at the center of its intellectual agenda. Nonetheless, there remains a paradox at this discipline's core, one with structural sources. As Belasen acknowledges, "organizations are inherently contradictory entities, and . . . organizational effectiveness criteria are fundamentally opposing, contradictory, and may be mutually exclusive . . . ." (2000, 33).

To manage this situation may well be impossible. As long as it maintains a unquestioned commitment to managerialism, Organization studies can only celebrate this paradox, to pretend that leadership lies in the ability to behave in a contradictory manner but not be perceived as doing so. As long as it allows itself to be colonized by managerialism, the movement will remain a coconspirator in its own deception.

A thorough switch to knowledge as the chief source of value would require a reassessment of OS managerialism, a consideration of the possibility that management may not be an a priori necessity. Management's place in production would probably be reduced—narrower, more contingent, and dependent upon its success at mobilizing the expertise of particular forms of labor. It may be reduced to, for example, the labor of coordination. Thoroughgoing development of more comparable notions of management and worker knowledge would also be likely to change class dynamics as well as our understandings of them.

Such a knowledge account of value would undermine those legitimations of management power that presume the self-generative magic of capital. Not-for-profit and public organizations are less tied to the dynamic of capital reproduction and therefore to these forms of managerialism. Their experience should be a better indicator of what the eventual political economy of knowledge will be like than that of for-profit organizations. However, in my ethnographic experience, nfps and publics remain subject to other imperatives, like "electability," that also interfere with knowledge networking.

A tradition of strong accountability to community organizations might make a difference. To participate actively in real knowledge networking with staff, members, and publics takes time and opens up administrators to additional questions about their prerogatives. In the absence of high-profile political leadership, administrators in both publics and nfps seem generally content to perform a public version of managerialism.

In sum, ethnographic study of AITed organizational knowledge networking suggests that, under managerialism, AITs only infrequently lead to new knowledge dynamics, let alone transformative changes in the function of knowledge. Managerialism needs to be replaced by genuine knowledge networking.

Challenge #3: A Knowledge Politics of Conceptualizing

The Distinctive Features of Organizational Knowledge Networking

If organizations really do resocial, the cyberfacture labor process will emerge as an important site for those who see in cyberspace a new social reproduction dynamic. It is now common to describe organizations as communities of practice and, therefore, for example, to see organizational learning (Argyris and Schön 1978) as something which depends upon community. Reducing problems of knowledge reproduction networking in organizations to "building community" will not further knowledge networking. While performances of community and the networking of knowledge have substantial similarities, the differences are important enough to keep them conceptually distinct.

Knowledge networking is also like individual identity formation, the process through which individuals often formulate competing, even mutually exclusive notions of their selves, frequently via discourse. Individuals also act out multiple self-concepts across their interactions with intimates and broader networks of others. However, while it is perfectly possible for individuals to perform multiple selves, knowledge networking requires a relatively stable set of approved practices, for example, regarding evaluating the quality of particular performances. As a collective activity, knowledge networking necessarily tolerates less internal evaluative conflict than identity formation, and certainly not incoherence.

This is also an important way in which knowledge networking is similar to education, where it is also difficult to proceed without a presumption of coherence. However, in both its disputatiousness and its extensive metadiscourse, actual knowledge networking differs from the "stable content" conception of knowledge presumed by standard "banking" approaches to education. The subtle complexity of the knowledge networking moment in social reproduction means it cannot be reduced to the community, identity formation, or education moments. As its own thing, knowledge networking requires its own politics.

Recognizing knowledge networking as a distinct moment in social reproduction and adopting a knowledge theory of value open space for numerous interventions into social formation reproduction. This space can be illustrated in some key academic arenas.

Anthropological Politics

Ethnography has become increasingly popular as a way to study knowledge and AITs. As an experiential way of knowing, ethnography has developed ways to hold both people's practices and their cultural constructions of them in the same frame (Hakken 1999a). By subordinating neither the moment of action nor that of cultural construction, ethnography constitutes an appropriate approach to the study of knowledge networking.

Ethnographic study of AITs in practice facilitates theorizations of knowledge that allow scholars and activists to carry out a metadiscourse about knowledge as sophisticated as the knowledges that they would collectively construct. However, ethnography's contribution to knowledge networking is limited by inadequacies in its own account of knowledge. Often hobbled by the same implicit presumption of transformation manifest in the broader cyberculture, contemporary anthropological study of cyberspace change in knowledge has emerged only slowly. Nonetheless, the work of the colleagues discussed in chapter 5 attempting to develop a new anthropology of knowledge justifies further basic research on knowledge networking itself.

Anthropology needs an ethnology of knowledge, a comparative study of knowledge networking, an empirical study of multiple knowledges focused especially on the social practices central to acknowledgment across different social formations. An ethnology of knowledge networking is, like nonmodern ethics and aesthetics, an important prerequisite to designing AITs that can really address the new knowledge problems of organization.

Concern about the general problem of how to conceptualize anthropological knowledge today has been given further impetus by changes in knowledge networking associated with automated information technologies. At the same time, efforts to answer the knowledge question in cyberspace can make a distinctive contribution to the general anthropology of knowledge and the clarification of anthro-knowledge in particular.

Informatics Politics

As an engineering discipline, informatics aims to produce systems that work. As in all "practical" practices, "working" is a relative concept, degrees of success often spreading across multiple dimensions of effectiveness and efficiency. Engineering fields are full of examples of systems considered to work despite shaky conceptual foundations.

To enact the special ethical onus that falls on those who must act in an anticipatory mode, informatics needs to foster a dual practice: on the one hand, making actual the potential of AITs to transform knowledge, while on the other, cultivating a critical stance that insists that knowledge technologies be "really useful." Such a dual practice can only be constructed on a good knowledge construct, so an informatics of knowledge depends upon stronger conceptual foundations than those currently dominant in the field.

Informatics, like the social sciences, has a complex history of knowledge-related notions. One unfortunate consequence of the initial public uses of computers primarily in scientific contexts is their vapid social construction as fancy machines for calculating. The Danish informatician Niels Ole Finnemann (1996) argues that such social constructions misrepresent computers' much broader potential as general symboling machines. While previous language innovation processes have held referent constant while altering code or visa versa, computers make simultaneous change possible. As such, they can support unprecedented cultural innovation.

Unfortunately, the field remains committed to an abstract, modernist knowledge construct that misshapes informatics' ability to answer the cyberspace knowledge question. This modernist discourse causes problems for both the creation of informatics' own knowledge and that of the many other fields which it now impacts. Informatics has largely structured how we think about knowledge in cyberspace; indeed, this field is responsible for some of the peculiar forms in which the cyberspace knowledge question has been placed on both intellectual and popular agendas. Moreover, there is a pattern of rhetorical slippage in this field, as in attempts to technologize everything from intellectual property to "medical informatics."

There is a viable alternative way to think about knowledge in the field. Notions like Participatory Design and Computer Supported Collaborative (or Cooperative) Work accept the sociality of work. Organizations like Computer Professionals for Social Responsibility and movements for open computing like free software/open source provide informaticians with important opportunities to practice an alternative disciplinary politics. The social informatics behind these notions and movements offer the necessary alternative to dominant technical informatics.

However, social informatics should not be used to mask a surreptitious colonization of the problem space of social science. Insufficient attention to cultural and social dimensions has meant a tendency to treat computing in different places as "the same thing," whereas appreciation of context is essential. Shifting attention from using computers for data and information to knowledge networking is an important step in a long overdue reconfiguring of basic disciplinary concepts.

Philosophical Politics

Chapter 4 presented a nonmodern, process approach to philosophical thinking about knowledge that grounded the complementary anthropology and informatics presented subsequently. This revitalization of epistemology depends on empirical study of existing knowledge practices. Fortunately, knowledge networking, the actual discourses and other practices that create, reappropriate, and spread knowledge, is highly accessible to empirical study. A first step in developing a philosophy of knowledge more cognizant of its social context was to separate what was useful in the core Western knowledge concept of "justified true belief" from the Cartesian, Modernist presumptions that normally and popularly encumber it. The etymology of the word "knowledge" was traced, as a way to give priority to the overt, even public process by which claims to truth come to be acknowledged.

Ethnography shows that, far from being either a unidirectional abstraction or concretization, knowledge emerges out of a networking dialectic that moves between various moments of abstraction and concretization. Based on a "turn to the social" and consideration of the complex relationships among data, information, and knowledge, a framework was created that incorporates both positivist and deconstructionist moments into a unified account of knowledge networking. The modernist notion of "raw" data is acknowledged to be a culturally constructed silencing too often "disappeared" in subsequent discourse. While silencing data's origin may on occasion be part of making scientific knowledge useful in the world, a general knowledge science would systematically resituate data in social relations and practices like labor processes. With its emphasis on context, the contemporary revival of pragmatism in philosophy offers much to those who would answer the cyberspace knowledge question.

The Politics of Knowledge Creation

Contemporary cultures privilege science, treating it as the most central form of knowledge networking. Ironically, this privileging of science undermines the argument of those who see AITs as transforming knowledge creation by deplacing it. This is because, long before computing, science had dispersed knowledge networking among universities, research centers, corporations, and homes often at considerable distance from each other.

That it is normally dispersed means that some performances of community/celebrations of unity—Festschriften, banquets, award presentations with large elements of collective self-congratulation—are important aspects of the disciplined creation of knowledge. Unfortunately, the term "community of practice" has become increasingly used to characterize the main instrumentality of knowledge production. Diana Crane's notion of "invisible college" invokes the same community image. Yet because they only evoke the symbols of what is shared, performances of community tend to construct boundaries.

Like community, knowledge creation networking does involve development of shared standards, by which we collectively distinguish mere information from what we together "know to be the case." Such symbols, connected to explicit metadiscourses about what constitutes knowledge, are ethnophilosophies with their own ontologies and epistemologies. Knowledge creation networking also definitely involves boundary work when participants assert what separates their accounts of knowledge from those of others.

However, knowledge creation networking has a dynamic in many ways very unlike community, one of divergence *within* rather than between networks. In contrast to community, knowledge creation networking *evokes* rather than suppresses differences of view. As network participants disagree about each other's assertions or dispute their right to make them, competing ideas emerge, in relation to both what standards should be shared and how they are to be applied in concrete cases. Thus, knowledge networking's equality is of a rougher sort than that of community. Its performances of difference are less like community and more like those of a social movement.

As Haraway argues, the "'knowers' of scientific knowledge claims," are neither individuals, nor 'no one at all' . . . "(1997, 36). We need to know who these collectivities are, who is in them, and the positions from which they construct knowledge. Haraway's characterization of them as "social communities" is misleading, however. Similarly, to describe a collaboratory as a "virtual community" is to emphasize sharing, that aspect of knowledge creation networking which humans find relatively easy to perform. The other, more difficult conflictual aspects have to be supported as well, so that they support pursuit of multiple, often contradictory knowledge projects.

Knowledge creation networking is better understood as a distinct kind of social activity, more like online searching than a celebration of nations. Knowledge is "really" shared only to the extent that those in a knowledge network have a common approach to the rules through which both epistemological and indexical status are established. Networkers embody such knowing to the extent to which they participate in the establishment of such rules. The "trust" involved in a knowledge network is conditional, not "trusting." It is based on social experiences—often of compromise, of negotiated optimality that corresponds directly to the extent of confidence that rules will be applied with similar senses of scope and coverage.

AIT systems to support knowledge creation networking must therefore enable extensive participation in often conflictual metadiscussion of the criteria by which to judge which kinds of data and arguments are telling. These systems must promote awareness of just how "known" something is considered to be as well as what it is claimed to be. Inevitable in this kind of networking is considerable non-, even *anti-*, community performance: conflict over ideas and competition, if not necessarily backstabbing and sabotage. Much more central to knowledge creation than the occasional performance of cohesion is the simultaneous promotion and management of conflict. Those promoting AITed interventions have given insufficient attention to such problems.

Educational Politics

To create a knowledge society, even if sustained and substantial in the creation aspect, change would also have to be equally great in knowledge reproduction and sharing. The ethnographic picture of actual computer-mediated learning in schools is similar to that produced by studies of AITs in organizations. In both domains,

- There are as yet few concrete examples of AITs that facilitate truly transformed learning and organization, certainly not enough to speak as if an "revolution" is underway.
- Attempts to mechanize learning and organization based on modernist conceptions of knowledge may be, on theoretical grounds, conceptually incompatible with transformative effects.
- Nonmodernist conceptions provide a better basis for imagining and perceiving transformative organizational and educational practices.
- In order for schools and enterprises to become "learning organizations," they must engage substantially in transformations that go far beyond the realm of technology.
- These broader processes of transformation must extend participation in organizational governance, even achieving consensus over what learning/organization is for.

Their potential for change may explain the interest in dispersed knowledge management and AITed knowledge sharing, but actual knowledge networking achievements do not explain why so many resources have been and continue to be expended, especially on "pure" (for example, asynchronous online) forms. In contrast, commodification is already a clearly important contemporary development in education. This takes place directly when schools try to appropriate employees' working knowledge via claiming copyright over online materials, but this is only its most overt, and probably not its most important, form. The significant danger is not my losing some potential personal income because a university has appropriated my "knowledge capital," but the more general transformation of knowledge networking going on behind my back. As teachers alienate our collective thinking through turning it into "products," we hobble our capacities as critics of ethnocentrism. Like indigenous peoples forced to turn their cultural resources into property, we risk undermining basic justifications for the relative social privileges (for example, academic freedom) that we enjoy.

How should new concerns for fair economic return be balanced against a continuing need to foster both previous and new forms of knowledge networking? The disappearance of most of the private organizations set up to make a profit from putting education "online" means a hiatus in creeping commodification, but this may only be temporary. Responding to it will require more than better empirical understanding of commodification. We need to theorize the strong absences of evidence for transformation with regard to AITed knowledge creation, knowledge adaptation, and knowledge sharing networking. The existence of so much ado about so little suggests a deeper political project, one that confronts political economies that foster commodification.

General Conclusion

What is it reasonable to expect from computers? What is the future of social life? Are work institutions entering a new stage, one of greater sociality, or will computers accomplish even greater control of workers? How can we relate ethically to the vast contradictions in social reproduction engendered by computing: that between, on one hand, the potential for renewal of sociality in organizations and, on other, how the profound flexibility of the computer as a medium to increase the danger of stricter surveillance and hyperabstraction; or between the hopes the computer engenders for controlling market-fostered creative destruction and the dangers of loss of social infrastructure?

From arenas as diverse as education to voting to publishing, for war and peace, from individual identity construction to group emancipation, for fourth world revolutionaries as well as geeks, cyberspace has become the repository for many human/cyborg dreams. With new technology viewed as a key means of liberation, cyberspace exercises hegemony over large portions of the cultural imaginaries throughout the world. Any effort to reform the condition of individuals, groups, or humans as a whole must now incorporate, or at least relate to, some version of cyberspace.

To answer the knowledge question in cyberspace, we surveyed *The Knowledge Landscapes of Cyberspace*. Perhaps the paramount necessity revealed by this excursion is the necessity of a difficult scholarly balancing act, to acknowledge in practice both the potential for transformation and the reality of continuity. The former has meant attending to practices in which the potential takes palpable form, to embody Steve Fuller's (2000) vision of a social science of knowledge that is truly liberatory, that both recognizes how science and society are connected and fosters a program for deliberately reconnecting them more effectively. The latter has required acknowledging the ways in which the dynamics of practice still overwhelmingly manifest familiar, at least structurelike qualities. Knowledge management and knowledge integration are recent examples of the continuing failure of other "computer as brain" approaches (see also, for example, artificial intelligence and expert systems).

The cyberspace knowledge question is part of the broader computerization or "computer revolution" issue. To address it, social science needs to be as clear about what its knowledge is for as what it is. Good accounts of the dynamics of past and present knowledge networking are not only central to understanding the general dimensions of knowledge and knowledging; they are also essential if we are interested in whether current changes in knowledge's production, reproduction, and/or sharing are of transformative significance, and thus if cyberspace is an appropriate locus for the cultural imaginary.

Notes

Chapter 1

1. To refer to the phenomenon normally seen as causing so much change, I prefer to use the phrase "automated information technology" rather than simpler terms like "the computer," "technology," or even "information technology." The term "computer" inappropriately features only one aspect of the multiple capabilities of electronic machines for manipulating the information representation capacities of symbols; I'd be happier were we to call them "symbolers." "Technology" alone is too general to get at the strategic role in recent social reproduction of one kind of technology, that which automates information. Use of "information technology" on its own tends to reinforce an ethnocentric blindness to the universality of information-related technologies, which include gesture and story as well as writing and printing. What is new is a class of technologies that make storage, retrieval, and manipulation of information automatic, raising numerous issues about the character of information-related phenomena, including what knowledge is. This all being said, for ease of exposition, I occasionally use "computing" as a gloss on "use of automated information technologies."

Chapter 2

1. I have here emphasized the "logic" of my argument to signal my intent to put forward my case more formally than is now characteristic of contemporary social science writing. I want readers to engage my points in a similar manner. I am quite aware of the arguments in favor of a less direct and linear style or poetics in contemporary ethnographic writing, but I find these less helpful in attaining the kind of goal toward which I aim, a collective understanding of the broad parameters of cyberspace. It is too easy for the less direct style to mimic the optative rhetorics that (often deliberately) confuse understanding of cyberspace. In this and related ways, the new rhetorics inhibit the knowledge networking necessary to an informed approach to contemporary issues of social policy, in areas like education and telecommunications. The detailed case for thinking seriously about social causation in cyberspace is made in chapter 2 of *Cyborgs@Cyberspace?*; that for a philosophy of knowledge specific to cyberspace in chapter 4 of the present book.
2. Such simplistic quantitative arguments have, unfortunately, strong roots in the scientific bibliometrics pioneered by Diana Crane. Her 1972 book on science, *Invisible Colleges*, builds on the work of Derek de Solla Price (e.g., 1979). The problems with this approach are addressed in detail in chapter 6.
3. In addition to the controversy over whether the new roles of science-based knowledge empower or deskill workers, science itself has become a site of particularly important self-conscious knowledge discourses. These are the so-called science wars (Ross 1996; Gross and Levitt 1994). At base, these contestations are epistemological, about how we know. What is the status of knowledge produced by scientific activity? Is it to be privileged above all others or only one among many? Gauntlet throwers like Paul Gross (in SCI-TECH-STUDIES 1996) argue that there is only one epistemology and that is practiced most fully by science, but others follow Foucault in positing multiple epistemic moments and often champion a postmodern sensibility (e.g., Lyotard 1984). Leigh Star (1995) frames current efforts to foster diversity in an epistemic approach as an ecological approach to scientific knowledge. These issues are examined further, especially in relation to anthropology, in chapter 4.

4. The primary failure of most organizational knowledge networking to date is its failure to come to terms adequately with the difference between information and knowledge, a failure based on a similar inattention to the information/knowledge transformation's sociality. This issue is explored more fully in chapter 3.
5. In an important sense wisdom, as knowledge-yet-to-be-articulated, also fits more easily into this regression. This is also true of what ethnographers call "local knowledge" or "counter discourses." Finding ways to think as systematically about them as STSers do about science is a central task of the new anthropology of knowledge discussed in chapter 5.

Chapter 3

1. In fairness, applicants' failure to maintain the information/knowledge distinction rigorously in their proposed activities reproduced the Science Foundation's (and Drucker's) own actions in this regard, described in chapter 2. It is interesting that one of the projects I reviewed embodied a strong sense of skepticism about what KNing could accomplish, proposing to investigate whether knowledge is ever really portable, separable from its initial context!

Chapter 4

1. Yet it remains the case that most of the professional philosophy that I have read in working out the ideas presented here continues to take the individual level as the proper level for the framing of inquiry into knowledging; see, for example, the entries in the Goldberg telepistemologies book (2001). An important impetus for the ethnography and ethnology of knowledge described in the next chapter is to help philosophers to theorize knowing as primarily a collective, rather than individual, phenomenon.
2. As philosophers take up the study of "really existing" epistemologies, they would do well to pay attention to moments of deliberate change in social formation reproduction—for example, the decisions of some Plains Indian tribes to change from a matri- to a patrilineal system of kin reckoning, or even of Swedes to change from driving on the left to the right side of the road. Such moments focus attention on cultural variation, a central issue in the study of complex social formations, and one to often ignored by ordinary language philosophy, as in Quine's simplistic appeals to linguistic communities (1990).
3. In his introduction to *The Robot in the Garden*, Ken Goldberg tries to place a different issue at the center of cyberspace empistemology, or "telepistemology." This is an alleged difficulty, when activity takes place at a distance, of ascertaining whether our perceptions are accurate. How do I differentiate between *actually* using a video link to observe a robot control to water plants in a garden in Linz, Austria, and only apparently doing so—for example, when a library of previously recorded video segments is presented as if the robot's actions are in "real time."

   Hubert Dryfus and Catherine Wilson seems to agree with Goldberg that this is the core problem of telepistemology, a situation which is therefore a sufficient basis for reviving the Cartesian framing of epistemology. Albert Borgman, Jeff Malpas, and Alvin Goldman take a different position, offering instead a range of different phenomena, like the increased number and time of representations to which I drew attention, as what is characteristic of cyberspace. Goldman's stress on the deplacing of learning, for example, addresses not truth value but what it means to know. He raises questions similar to Bowker et al. and shares their concern about whether knowledge is in any sense portable. These issues are not Cartesian.

Chapter 5

1. Malinowski did attempt to explain why fostering a strong emotional dependence of the fieldworker on the native would lead to better data. His advocacy in this regard can be seen as ultimately of some inspirational value to the nonmodern study of knowledge. (See my comments regarding Hastrup.)

2. As with community computing, the problem with Japanese CSCW may be trying to use computers to do things that humans do well enough on their own, rather than discovering new tasks which they can do better; see *Cyborgs@ . . .* , chapter 4.

Chapter 6

1. Although physics and mathematics share the same suffix, they do not fit under the label of engineering, but this is another artifact of, in this case, Greek history. In my experience there are two basic attitudes toward math. One is that it is real, part—indeed, perhaps the center—of existence. A second view is that math is a language, invented and developed by humans to express their ideas about all kinds of relationships, its proper usage more a matter of convention than of some reality "out there" which it inherently reflects. This second, artificial view of math is quite compatible with my sense of informatics; indeed, the best statement of the scientific object of computer science I ever got was "applied finite math."
2. Stonier is not the only informatician who has tried to theorize the conceptual terrain of knowledge (see also Devlin 2001). The thoroughness of his theorizing, however, is somewhat unique. As Whitehead indicated, most informaticians deploy without much examination the notions about knowledge that they find in popular culture. I choose to focus on Stonier because his thoroughness means his arguments are clearer than most. Also, although his theorizing may not be the most influential, its thoroughness has also given it greater influence than might otherwise be the case. In short, he combines something of a "good" with an "influential" case. By bringing out the contradictions evident in his development of a scientistic modernist informatics of knowledge, I hope to indicate the likely sterility of any attempt along these lines.
3. Claude Shannon's is arguably the most influential theory of information (1995). In essence, Shannon argued that the difference between a signal carrying information and one that didn't was that the former had meaning while the latter didn't—it was just "noise." The effectiveness of an information system could thus be measured in terms of things like the "signal to noise" ratio. Interestingly, in Shannon, once one had established that a signal had meaning, one could ignore that meaning, including measuring "how much" meaning the signal contained. This is yet another example of the centrality of the form/content dichotomy in informatics.
4. In some ways, their attitude toward experimentation is the most fascinating aspect of Diana Forsythe's description, in "Artificial Intelligence Invents Itself" (2001), of the 1980s-era knowledge engineers building AI science. To be doing something artificial, they had to be making something, but to build things that are actually used is to be a technician, not a scientist. Therefore, AI scientists described themselves as building "toy systems," ones not intended actually to do anything! Moreover, what was definitive of "a real AI scientist" was deployment in her toys of "certainty factors": "According to their creators, cf's [*sic*] are not probabilities and do not rest on any formal mathematical underpinning; rather, they constitute a technique developed through trial and error that was adopted because 'it worked'" (89). Here the "techno" in technoscience is suppressed, but not eliminated.
5. I also note in passing Stonier's switch to an "economics" register (e.g., "cost effective"), a rhetorical move so common in engineering as to suggest that it is really a branch of business/economics. More on this connection in chapter 10.
6. A list of the orientations of standard knowledge engineering might include the following:
   1. Positivism
   2. Rationalism
   3. Managerialism
   4. "ONE BEST WAY"ism
   5. Lack of a critical perspective
   6. Cartesian dualism (e.g., form versus content, rational versus irrational)
   7. Transformationalism
   8. Technological determinism
   9. Crackpot realism

   10. Mindless professionalism
   11. Methodological individualism
   12. Technologicalism
   13. Modernism
   14. Totalism
   15. Formalism
   16. Reductionism
7. At the time of writing, the issue of ontology has again been raised forcefully within informatics, most noticeably in Tim Berners-Lee's notion of a semantic web (1999). At the center of his project is the idea of a universal thesaurus of object types that could be used and reused as part of a renewed knowledge program. While a renewed focus on knowledge problems is to be welcomed (as well as expected), I am concerned that the new project will also fail because of its pursuit of foundationalist goals—this time, a universal map of semantic space.
8. These splits include that between Raymond (1999), for whom it is OK to make money off open source code as long as it is not effectively proprietized, and Richard Stallman, the Gnu pioneer who advocates always "free software" (Williams 2002). Both sides use the term "free," Raymond on the model of a "free press" or "free speech," Stallman on an overtly anticommodity "free of cost" model. Torvalds, inventor of Linux, stakes out a position somewhere in the middle, personally abjuring benefit from code but not objecting to others doing it, as long as the source code is "copyleft" protected, under the General Public License. On this license, anyone is free to do anything they want to the source code and publish it, as long as others are free in turn to do what they want to the modified code at no cost. I am not the only scholar who sees in the open computing movement that includes all these approaches the seeds of an interesting alternative political economy; see chapter 11.

Chapter 7

1. In *Cyborgs@Cyberspace?* I followed a similar analytic strategy. I argued that in order to decide if we were living through a technology-induced social transformation or "computer revolution," we needed to bring together analyses of computing's correlates at multiple levels, from individual identity to global social formation reproduction. If the dynamics of computing's social correlates were both substantial and similar at all these levels, it would suggest that a fundamental change was indeed afoot. Conversely, if they were either not substantial or were quite different, this would undermine the transformationalist case. My ultimate goal was to persuade intellectual cybernauts that, by locating their work in terms of these levels, they could locate their individual efforts within a broader analytic project.

   In this section of *The Knowledge Landscapes . . .* I pursue a similar strategy of decomposing an alleged general process, construction of a "knowledge society," into smaller components to foster more effective analysis. However, rather than being "horizontal" sections or "the levels of sociocultural integration," as in *Cyborgs@ . . .* , the components here are "vertical," parallel sections, alternate forms, aspects, or "submoments" of a single universal moment in social formation reproduction. An important goal of the research of my own described in this current chapter is to determine if this vertical sectioning provides a similarly serviceable analytic strategy for bringing together the work of those studying knowledge in cyberspace.
2. There is of course a danger in adopting any analytic scheme that the picture of dynamics that emerges from analysis is more an artifact of the scheme than it is reflective of what is actually going on in the world. The argument made here for adopting this particular analytic frame is institutional. It presumes that the institutional differentiations between research, business, and educational institutions reflect basic aspects of general knowledge networking. However, these institutional differentiations might be more a result of very different dynamics—for example, historical or hidden social formation reproductional—rather than the functional differentiations posited. As a 1960s educational activist, I argued that the *real* reason for schools was to reproduce class, not for students to learn.

Moreover, as described below, it became obvious in the "Knowledge Networking Knowledge Networking" project that the three aspects intermingle to a high degree. There is much knowledge sharing in the creation of new knowledge. Indeed, these aspects may be more accurately viewed as "submoments" of the general moment of knowledge networking, analytically distinguishable but of doubtful existential status.

Still, for analysis to proceed, some differentiation is necessary. As also described below, adopting an institutional stance aided development of the research process. As in action research in general, adoption of a provisional analytic frame is legitimate as long as the analyst remains sensitive to the possibility that the provisional frame may well need to be abandoned at some point in the future. The increased possibility of having something substantial to say about the generic design problem in knowledge networking support infrastructures is an additional reason for adopting this analytic strategy.

3. These opportunities include an ongoing interest in the automatizing of data collection and analysis in anthropology, sociology, computer science, business/organization studies, trade union studies, education, and other natural science, artificial science, human science, and humanities fields. In addition to anthropological field studies of systems developers, computer scientists, economic developers, trade union officials, worker educators, and social service and public policy specialists, I have spent a great deal of time in academic institutions. Not only have I taught anthropology, sociology, and computer science, but I have participated actively in the institutional life of a technologically oriented academic institution. I have also reviewed grants for and/or served on interdisciplinary panels of the U.S. National Science Foundation, the Social Science Research Council, and other research organizations. While much of my professional work has been in anthropology, I have been an active professional in science, technology, and society, as well as, to a lesser extent, in education, sociology, social work, and computer science. Those interested in my personal experience of academic knowledge creation may wish to review the professional curriculum vitae on my personal website: http://www1.sunyit.edu/~hakken.
4. The ethnographer faces a basically insoluble problem in presenting her analysis. It comes from considerable participant observation in phenomena of interest. No one has the patience to read an exhaustive account of what was learned in each aspect of this experience. If she merely summarizes it, perhaps supplemented by illustrative examples (the typical expository strategy), she leaves herself open to the charge of presenting only "anecdotal" evidence. Placing a particular project in a career context is hopefully an improvement on either of these unsatisfactory alternatives.
5. There is also a considerable ethnography of science, as well as a robust ethnography of education, reasonable alternative foci for the project. I did not choose the former at least in part because of the confusion over what ethnography is, coming from the European SSK (Knorr-Certina, Latour, Woolgar) appropriation of a more sociological than anthropological tradition in ethnography. Also, the focus on scientific as opposed to technoscientific knowledge of both sociology and anthropology of science tends to replicate some additional confusions. While often a source of difficulty, the clear split between anthropological and other approaches to ethnography in "high tech" organizations means that focusing on the former was a way of avoiding these problems.
6. However, current systems development ethnography with its object orientation and occasional "quick and dirty" approach (Crabtree 1998), is more closely informed by a sociological tradition of ethnography, tied to ethnomethodogy (Garfinkel 1984). Some (e.g., Hakken 1999a; Nyce and Loewgren 1995 ; Shapiro 1995) have questioned this development.
7. Also, KN^2 was overtly framed by several events indicative of renewed explicit scholarly interests in knowledge. These included the developments in computer science, discussed in chapter 6, that stressed both the need for and purported capability of going beyond mere information processing to generating real new knowledge, and that championed open source and open software as means to achieve more CS knowledge more quickly. Economics theorizings were positing knowledge as a distinct, possibly primary, factor of production, accompa-

nied by efforts to commodify knowledge as a distinct, measurable organizational resource. More explicit efforts at cross-culturally valid or ethnological approaches to the study of knowledge were developing within anthropology, while the neopragmatism discussed in chapter 5, informed in part by studies of science, technology, and society (STS), emerged in philosophy as an approach to questions of epistemology. Creating a dispersed collaboratory for ethnographers involved in organizational KNing projects, was intended to bring an interesting form of actual KNing practice into dialogue with these scholarly developments.

8. The preliminary experience with KN^2 was responsible for my decision to write *The Knowledge Landscapes of Cyberspace*. I began to see the knowledge management experience in the context of the strong knowledge engineering program of computing described in chapter 6, and thus how its problems were predictable. These problems looked closely connected to the difficulties inherent in any totalizing program to formalize knowledge, because they invariably involve separating the knowledge from the networking. Maybe separating form from content (or object from process?) is not a good idea; at least with regard to knowledge, software is inherently both. It may be that better understanding of the idea of functionality in software will provide a key to developing a more satisfactory approach to the problem of mediating knowledge networking via AIT. Perhaps there is hope in finding AIT-mediated forms of sociality which achieve some functional equivalent of existing face-to-face and dispersed forms of KNing. Indeed, as has periodically been the case—for example, with the notion of "communities of practice"—cultural perspectives may provide inspiration for more successful approaches to this issue.
9. This was to include an advanced, XML-capable Tomaino database, run on a version of Linux. Front ends for the prototypes were to be portals built with the Uportal tool kit.
10. I have long been involved in their promotion, as during the time I spent in the mid-1970s as a professional associate in the anthropology program at NSF. One task involved using advanced, computer-based multivariate analysis of several data sets, including bibliometric ones, relevant to the academic effectiveness of funding for dissertations and conferences versus senior research grants. Another involved designing and implementing a computerized database of potential reviewers.
11. It seems fair to me to hold de Solla Price and Crane responsible in part for the mindless quantitative orientation of so much of the popular discourse on knowledge—for example, President Clinton's simplistic treatment of the number of books published as an indicator of growth in knowledge. These scholars really do take the number of articles published in a research area as a good indicator of the amount of its knowledge, rate of growth in publications as indicating rate of growth in knowledge, and so on. Thus, Crane doesn't bother to define knowledge or to argue for how she measures its growth.

Chapter 8

1. Sometimes taken to stand in for "civil society," it is not always clear whether attention is being drawn to the influence on social reproduction of the NGO/nfp staff or of the activities of organizations' broader membership—for example, letter writing, political campaigns, demonstrations, and so on.
2. Paralleling a renewed (mostly theoretical) interest in community, this shift from form to include process is sometimes argued to be another example of how cyberspace issues are associated with various reshapings of the agenda of social science. As I will argue in chapter 10, however, it makes more sense to see most of what's going on here as shifts among forms of the employment relationship rather than as indicators that, as a social institution, employment is no longer primary. Even temporary workers still get their income from organizations, just as governments get their money from taxes on wages, and not-for-profits from wage checkoffs.
3. I have made a similar critique of attempts at community computing (e.g., Shuler 1996; Rheingold 1993), which in my view too quickly presumes that the goal of community AIT

networks is to reinforce existing place-based local or physical community (Hakken 1999a). The alternative (or rather, the complement) for both organizations and communities is to use technology to build new dimensions. For communities, these would be based on relationships different from those of place, whereas for organizations, these would be based on new kinds of knowledge networking.

4. This is one example of how the tacit/explicit dichotomy is still too simplistic to capture what is of importance with regard to knowledge in organization (Hansen 1991; pace Holtshouse 1998). Nonaka and Takeuchi locate the advantage of Japanese firms in this alleged ability to convert tacit to explicit knowledge. This is unlikely, however, because its very "tacitness" provides much of what makes tacit knowledge of value, and this is what is lost in formalizing it (Ehn 1988). Rather than converting it to explicit knowledge, the task would seem to be to find AITed ways to foster creation of more tacit knowledge and its application. This, of course, is another reason for thinking differently about how to use machines to symbol.
5. Computers' ability to automate the application of instruction sets or algorithms meant they were perceived as a threat to middle management. The 1990s downsizing of American organizations, taken up again after the collapse of the dot-com bubble economy, indeed did mean that many middle managers were dispensed with. While this was often justified as a consequence of computerization, the "leaner" organizations often had substantial, knowledge-related weaknesses. These managers, far from just making decisions that could be automated, often individually and collectively possessed knowledge essential to making the organization work. As real facilitators, not as separate decision makers, managers do reproduce knowledge often essential to organizations.

Chapter 9

1. My educational research has focused as much on informal—that is, noncredentialed—education as on the formal learning which is the focus of this chapter. Informal education includes not only worker education like that institutionalized in the Workers Educational Association, trade unions, and political parties; it also includes the potential for "self-education" which is a justification for the Internet frequently cited in, for example, Microsoft "Where Do You Want to Go Today?" commercials. The character of my college students' Internet research—which, despite some interesting flashes of potential, is largely anemic—does not make me sanguine about the actual educative accomplishment of net surfing. Still, study of the educative as opposed to mere information sharing implications of this latter phenomenon would be difficult but is potentially rewarding, as suggested by the debate over whether Internetting dampens or fosters sociality.

   It was my interest in informal education that led to my dissertation research on the important role of the largely informal institutions of workers' education in the reproduction of working-class culture in Sheffield, England (1978). In this work, I aimed to identify both the contribution of and the limits imposed by these institutions on extending class cultural reproduction. Unlike Castells and others committed to the "Third Way" (for example, Giddens 1998), I, like Bourdieu (1978), think class cultures remain important mediators of the reproduction of contemporary social formations, especially through informal institutions like those described in my dissertation. Diana Forsythe's research on "patient education" is suggestive of the continuing importance of the informal domain to the sharing submoment of knowledge networking (2001).

   The relationship between formal and informal educational institutions in the reproduction of culture is a central issue in the social study of education. However, in this chapter, I concentrate on the formal institutions, credentialed education, because it has been the focus of most education research as well as the prime locus of "knowledge society" educational rhetoric. Also, schools are of overwhelming significance to social formation reproduction, formal institutions being those where the dangers of both commodification and mythification are most evident. If these latter are to be avoided, schools are an important terrain on which they will be defeated.

2. The reason for the popularity of this image probably has something to do with the ease with which American heads are turned by technicist "machine" rhetoric (Leo Marx's *The Machine in the Garden* 1964). Indeed, hucksterism may be so present in educational technology because, as pointed out above, it ties together so well two American "hot buttons."
3. All too often in actual constructivist science education, while the learning is approached processually—to be accessed via a postmodernist, socialed pedagogy—the *content* of scientific knowledge remains framed in modernist, "thing," individuated terms. A typical result of failure to acknowledge the underlying inconsistency is that *participation in activities framed as contributing to* the development of scientific knowledge (for example, collecting data on weather) is *substituted for* real coconstructive learning. Related variations of this failure are frequent enough so that they are sometimes mistaken as characteristic of constructivism.
4. Over the last thirty years, a great deal of social policy has promoted computer-based educational technologies. Because this policy has been implemented in public and not-for-profit sectors, the people doing it have not been able to use the "bottom line" of the private sector to justify their efforts and thereby avoid hard thinking. Of course, there is an unfortunate tendency of publics/voluntary organizations to make policy in the deliberate absence of information about results, not to demand proper evaluation of what has been done, of whether goals were met and other consequences transpired. Nonetheless, several of these authors have refused to treat evaluation as merely just some silly administrative requirement to be complied with formally and ignored substantively. Evaluation is, but is not only, helpful for "midcourse correction." Rather, like any good research, evaluation is a prod to more theoretical work.
5. Andrew Feenberg has concluded after several years of research and practice regarding online education that the support of "learning community" is the crucial element of success. He has developed Textweaver, an open source utility to foster collaborative student-directed supplementary discussion of course materials. In a graduate course on ethnography is which Textweaver was beta tested, I found my students quite enthusiastic about it despite several elements of "kludgenes." As part-time, largely female adults students, often with families and full-time jobs, they wanted to extend class space. In my view, they had the advantage of highly similar work cultures, many dong jobs in social service.
6. Mark Slouka (1995) gives Turkle-type arguments a compputropian twist. He claims that the displacement of shared experience in a single "real" world by virtual experience in cyberspace undermines the capacity to act ethically. Among other reasons, this is because ethical behavior depends upon seeing the consequences of our actions on other people.
7. Given the breadth of Americanist anthropology (that is, the effort to include both basic and applied aspects of cultural anthropology, archaeology, physical anthropology, and anthropological linguistics in a single profession), it is difficult to get us to talk together about any particular thing. Indeed, some have taken the recent split in the Stanford department, between "scientific" and "cultural studies" anthropologists, to be indicative of the unraveling of the Americanist project. This breadth also means we have very little experience formulating the kinds of policies necessary to take advantage of the dispersed AITed knowledge networking opportunity. After years of adopting structures that made it ever harder to formulate policy, only in 1998 did the AAA regain an elected policy committee.
8. Work by the Institute for Research on Learning in Palo Alto, California justifies conceptualizing learning organizations as a strongly coconstructivist. In its interpretation, knowledge is locatable first in organizational or other social relationships rather than in members themselves, and only secondarily in artifacts like books or databases. In the social coreconstruction that is learning, chances increase that the knowledge itself, and the learners, are "transformed" (B. Jordan 1998).

   Such perspectives on learning have profound implications for both the enterprises discussed in the preceding chapter and the way we think and teach. For example, the slogan "Be a Learning Organization," which has recently had considerable cachet in literature on directed organizational change and among organizational development gurus, takes on much more significance when one thinks of knowledge as normally inhering in social relationships rather than in books or minds.

Chapter 10

1. Chapter 2 of *Cyborgs@Cyberspace?* addressed the need for a theory of social formation reproduction if one is to address successfully the "computer revolution" hypothesis, while chapter 6 of that book put forward the case for developing a macrostructural discourse on cyberspace. The current chapter takes off from this discussion, briefly reiterating the reasons for such discourse and then critiquing the particular analytic categories in which it is carried out with regard to knowledge. Readers who find this discussion confusing or who question the argument's presumptions might prefer to begin with this prior book.
2. At least, he usually does. In line with the title he gives his recent (2000b) *British Journal of Sociology* article, one could read his intervention as more tentative: "Materials for an *exploratory* theory of the network society" (emphasis added). There is thus some ambiguity regarding Castells's theoretical project. However, for some twenty years he has been making statements like the following: "The network society is the social structure characteristic of the Information Age...It permeates most societies in the world . . . as the industrial society characterized the social structure of both capitalism and stateism for most of the twentieth century" (2000b, 5). Similarly, he characterizes his recently republished (2000 millennial edition!) three-volume *The Information Age* as making the empirical case for this analysis.
3. See chapter 2 of *Cyborgs@Cberspace?* for an extended development of this point. For Castells, it was interestingly not misleading enough to require renaming (for example, from *The Information Age* to *The Network Age*) his three-volume magnum opus! Below I sketch out similar problems with the "network society" label, more accurately describable as the "automated information technology network-driven society."
4. I suspect a rather more complex picture shown be drawn of the considerable AIT efforts of the Soviet era in Eastern Europe; this at least was my suspicion when I began in 1987 to develop a project on computing in Bulgaria. Subsequent events wiped out much of the indigenous AIT infrastructure, which is perhaps now, as in places like Gujarat in India, reemerging as part of the open source movement.
5. While these scholars are right to frame their work, as does Toffler (1983), in terms of evolution, the specific sequence of the technological forms they offer is questionable. Like most sociologists, they foreshorten human history prior to the "industrial revolution" into one long, effectively atechnological, "traditional" period. A metaphysical leap in dialectics brings them into an antithetical, "modern, technological" era, the synthesis being the "third wave."

   Chapter 2 of *Cyborgs@Cyberspace?* outlined a more varied set of evolutionary options for cyberspace: as a new, cyborgified, species; a new mode of production or social formation; a new, "fourth" form within the labor/commodity mode of production/social; or merely another, perhaps more concentrated manifestation of the existing machinofacture stage of the labor process. Equally important was the notion that cyberspace might just as logically "devolve" to a prior form. This framing provides much more space to capture the many possible nuances of change than the Castells or Davidson/Toffler options. More nuanced structuralistics enhances our capacity to identify which account best describes the actual, empirically observable relationship between AIT-based actor networks and broader cyberspace-related social changes. Are these highly correlated? If so, what are the implications of their most likely causal links?

   Our mid-'80s research convinced Barbara Andrews and me that the cyberspace-related patterns of Sheffield culture were similar to pre-Fordist social patterns of unemployment and class degradation, more compatible actually with a devolution to a previous form of the labor social formation than with some new stage or a nonlabor form. At the same time, some interesting interventions and people's general willingness to appropriate AIT discourses in new identity work seemed to be indicators of potentially new social arrangements. Perhaps the most typical correlate of AIT, however, was to shift the terrain of class power. New skills and jobs rarely carry the same gender, trade union, class cultural, and/or workplace-based political implications as the ones they replace.

6. Advocates for Schumpeterian perspectives also tend to foster other unwarranted assumptions, such as the notion that new technologies necessarily produce more value than the ones they replace.
7. Some may object that the phrasing I have chosen here hypostatizes capital—it attributes agency to an abstraction and, in the process, marginalizes human agency. However, more convoluted phrasing, such as "The character of social formation reproduction appears to suggest that the reproduction of capital is the strongest influence on its course," just makes agency more ambiguous. The problem with any phrasing that attributes central influence to conscious human action—as in, for example, the idea of a class of humans centering social reproduction on the reproduction of capital in order to reproduce their social privilege—is that this kind of account seems to falter in the absence of a general conspiracy.

   The value of the phrasing chosen is that it communicates an important Marxian insight, the extent to which things really are in a certain kind of control. That is, the capitalist who fails to maximize his capital by demanding the highest possible return on his investment really does find his capital shrinking. It is in this sense that it is humans who serve capital, even though capital is just a fetishized social relationship.
8. I don't think this was Bourdieu's intent, but it is, as manifest in the frequency of citation of his work in organization studies, a consequence for which he had some intellectual responsibility.

Chapter 11

1. A number of radical political economists have followed Ernest Mandel (1978) in referring to the current era as "late capitalism." This terminology is intended to suggest that contradictions like the above are so overwhelming that capitalism's state is one of senility at best. I do not choose this terminology because the last decade has surely demonstrated capitalism's resiliency, in both symbolic and political economic domains. Its long-term fate remains doubtful, but its demise is not imminent.
2. By defining "innovation" in purely neoclassical terms, as "the creative process through which additional economic value is extracted from the stock of knowledge," the conference organizers did themselves no favor in the search for alternatives, however.
3. Such a contention is quite debatable. Saskia Sassen (2000) has argued that the globalization of finance reprivileges a small number of core cities that effectively facilitate the face-to-face interaction that is paradoxically essential to the high level of trust required. Similar arguments have been made in regard to the small number of "hot house" loci (the Boston Route 128 corridor; Silicon Valley, Fen, and Glen; the Grenoble region of France; Kista in Sweden; and so on) of apparently central importance to the "new economy," whatever it turns out to be.

Chapter 12

1. My initial activism on issues of professional ethics within the American Anthropological Association is described in Hakken 1991. Over the years, working with Dr. Carolyn Fleuhr-Lobban has greatly stimulated my thinking on the ethics of fieldwork (Hakken 1991, 2003). A conference at the New York Academy of Sciences (Hakken 2000) provided an opportunity to reconsider previous thoughts (1995b) on cyberspace ethnography ethics.
2. The "Scandinavian" school emerged as an alternative "use" approach to information systems development in the 1970s, one which stressed "computing in social context" rather than a more narrowly technical perspective (Ehn 1988; Dahlbom and Mathiessen 1993). In insightful comments on changes in the Scandinavian approach, some of its practitioners have argued, in effect, that it failed to "get right" the relationship between systems design and poli-

tics. Gro Bjerknes and Tone Bratteteig (1995), for example, argue that early conceptions of systems development as overtly political practice, as in coalition with workers' organizations, were displaced by conceptions of professional practice as political in its own right. Bo Dahlbom (1990) argued that these new directions burdened systems developers with impossible demands. More recently, Hans Glimmel (personal communication) has warned that the most recent aesthetic turn in Scandinavian systems development approaches practically obviate the possibility of any political practice.

These are by no means the only difficulties encountered by those who would try to state professional ethics and aesthetics for swimming in the convergences of systems development, use, and politics (Allwood and Hakken 2001). Still, the logic of this book's analysis of knowledge in cyberspace suggests that these issues cannot be avoided. Thus, it is appropriate to articulate the kinds of design considerations that make sense; at a minimum, making them explicit opens them to the critical moment characteristic of knowledge creation networking.

3. To be more concrete about the kind of cyberspace analysis that I found persuasive, I offered a "rethinking" of actor network theory in *Cyborgs@* . . . My reconstruction of ANT along realist (RANT) as opposed to textualist (TANT) lines built on its strengths, like its refocusing of technology away from artifacts alone and toward the complexes of relationships among human and nonhuman—artifactual, organizational, and other life-formal—entities (for example, Callon 1986). The construction of a particular technological network or "system" is an active process. Methodologically, ANT developed out of the attempt of STS scholars to free themselves of presentist or "Whiggish" readings of the history of technoscience. Another important ANT corollary is the potential for nonhuman agency.

   Some of the failure of ANT to lead to cumulative knowledge of cyberspace can be traced to forms of its practice overly affected by postmodernist presumptions, especially textualism. The critique of "political constructivists" is that radical interpretivism marginalizes study of power and exploitation.

   An alternative, realist ANT embraces the existence of an external reality beyond human discourse, one substantially open to empirical study. Unlike empiricism, it recognizes the role of human action, including discourse, in the construction of reality. It avoids reductionism by distinguishing among various levels of reality. In "reproductionist" Marxist realism (Hakken 1987), such levels are understood to be related dialectically in that, for example, certain developments at the level of the actual can have a cumulative impact on the underlying real generative processes.

   The benefits of RANTing can be seen in relation to the issue of nonhuman agency. The material consequences of their history, not just human interpretations of them, also explain why some artifacts become the focus of a complex network, actors in a TAN manifest at the actual level. Sufficiently stable, actual-level TANs may even impact the generative "real." This is the way technologies have "real" impacts—mediated by, but not reducible to, the discourses through which humans apprehend, and influence, TAN reproduction. On RANT, it makes sense to talk of the agency of nonhuman entities in TANs, that such networks can be transformed—not just translated—by nonhuman actors. RANT is also an advantage when confronting an issue that is of particular concern to political constructivists: whether technologies have politics. On RANT, the relative degree of autonomy of any particular TAN can be analyzed in terms of its manifestations in the various moments of reality (for example, empirical, actual, and/or generatively real) in which it is implicated.

   Bryan Pfaffenberger (1992) urges us to see the technological systems embedded in artifacts as the outcome of previous political struggles, but to see these struggles themselves as framed in symbolic, textual discourses. In this reading, technologies have material, determinant qualities precisely because they embody the momentum of previous human activity, a momentum particularly difficult to change in the short run when an actual TAN—for example, a system of national highways—integrates widely dispersed practices. This view of technological momentum is not incompatible with a perspective which recognizes how current practices can lead to new apprehension of old technologies, the malleability of TANs strongly emphasized in the Nordic scholarship of Berg and Lie (1993). Still, as argued by Donna Haraway (1991), it is the limitations on human action constituted by such momentums that justify the attribution of "agency" to the nonhuman components of technology actor networks.

4. On the critique of the "site-boundedness" of older ethnographic research and the need for "multisited" research, see Marcus (1998), or "non-site-bound" research (Hakken 1999a). On problem-oriented versus people-oriented ethnography, see Powdermaker (1966).
5. This phenomenon refers to social phenomena thought to be characteristic of the social practices being studied but which are actually epiphenomenal to the process of study. It is named after famous early studies in a factory in Hawthorne, Illinois (Roethlisberger and Dickson 1934).
6. These were encouraged as providing, for example, a means for the reader to identify how such values might unconsciously affect ethnographic practice. Although too often used disingenuously as a cheap basis for demonstrating "ethnographer bias," this articulation of our values, an act of individual self-reflection, was still a good practice to institutionalize.
7. On their way to such a foundational rethinking, cyberspace anthro-ethicists encounter already elaborated ethical machineries, like the Institutional Review Boards in the United States. A common theme among my cocontributors to the 1991 Fleuhr Lobban ethics volume was the inappropriateness of the hospital-, laboratory-based IRB process to the ethical challenges facing the ethnographer. If operationalized as written agreement prior to research, the "informed consent" that is the touchstone of IRB ethics is incompatible with many ethnographic forms of practice. Consequently, in the past I personally thought of ethical research primarily via avoiding misrepresenting myself, explaining my intent at the earliest possible moment, and so on, activities to which the IRB process was largely irrelevant.

   As more ethnography has been done in cyberspace, the shortcomings in such personal "don't ask, don't tell" versions of research ethics have become more evident. Under the slogan "Informed Consent: We are not exempt," Carolyn Fleuhr-Lobban now argues for more proactive measures which accomplish goals consistent with the substance of informed consent.
8. Such a position is, of course, also an ethical contention; although not intended in this way, it could even be given a Kantian, categorical imperative reading.

Glossary of Abbreviations

AAA American Anthropological Association, the main professional association of anthropologists.

AI Artificial intelligence, a research program in informatics that aimed to create intelligent machines.

AIT Automated information technology, a general term to refer to all those artifact systems in which programmable machines for storing and manipulating data (computers) are important.

CKO Chief knowledge officer, a new executive position proposed for organizations at the height of the knowledge management boom of the late 1990s, modeled on the chief executive officer (CEO) and chief information officer (CIO).

CMC Computer-mediated communication.

CMI Computer-mediated instruction.

CSB Community Services Board, the local advisory group set up by New York State to facilitate deinstitutionalization of previously hospitalized recipients of mental health and metal retardation services.

CSCW Computer Supported Collaborative (or Cooperative) Work, a suite of information technology systems that aim to support groups rather than individuals.

CVS Competing values framework, Alan Belasen's typology for the conflicting roles of contemporary management.

IRL "In real life," versus cyberspace.

KD/I Knowledge/Distributed Intelligence, a late-1990s grant program of the U.S. National Science Foundation.

KM Knowledge management, a late-1990s craze in organizations that used AITs to adapt knowledge.

KNing Knowledge networking, the process by which groups create knowledge or "knowledge."

KN^2 Knowledge Networking Knowledge Networking, a USNSF-funded project to support knowledge networking among ethnographers studying knowledge networking activities like knowledge management in organizations.

KNOC Knowledge Networking Oneida County, a project to implement "leading edge" knowledging practices in public and not-for-profit organizations in a region of upstate New York.

LLC Limited liability corporation, a cultural innovation of the nineteenth century to foster economic projects of high risk; now the dominant business organization type.

NGO Nongovernmental organization, the term used more generally outside of the United States to refer to organizations neither part of the state (government) nor intended to make a profit; see nfp.

nfp Not-for-profit, the term used in the United States to refer to organizations neither part of the state (government) nor intended to make a profit.

NSF The United States National Science Foundation, the main funder of basic research in the natural and social sciences.

OCDMH Oneida County Department of Mental Health.

OECD The Organization for Economic Cooperation and Development, an action organization of the leading industrial nations, which also has a significant research function.

OED *Oxford English Dictionary.*

OS Organization studies, an interdisciplinary research field with strong roots in the concerns of management of for-profit corporations.

SLN SUNY Learning Network, the primary modality for delivery of asynchronous online courses in the State University of New York.

SM Scientistic modernism, an epistemological approach that presumes the existence of transcendent, universal truths whose most direct expression is to be found in science.

SSK The sociology of scientific knowledge, a research area that focuses on developing social explanations for the character of knowledge claims in science.

STS Science, technology, and society, or more recently science and technology studies, a partially disciplinized interdisciplinary research field that aims to provide integrated accounts for the character of technosciences and technoscience in general.

SUNY State University of New York.

TANF Temporary Aid to Families in Need, the popular name for the U.S. welfare reform of 1996.

References Cited

Abbott, Andrew. 2001. *Chaos of Disciplines.* Chicago: University of Chicago Press.

Agre, Phil. 1996. *Computation and Human Experience.* Cambridge: Cambridge University Press.

Allwood, Carl Martin, and David Hakken. 2001. "Deconstructing 'Use': Diverse Discourses on 'Users' and 'Usability' in Information System Development and Reconstructing a Viable Use Discourse." *AI & Society* 15: 169–199.

Althusser, L., and Etienne Balibar. 1970. *Reading capital.* London: New Left Books.

Anderson, Benedict. 1983. *Imagined Communities: Reflections on the Origin and Spread of Nationalism.* London: Verso.

Andrews, Barbara, and David Hakken. 1977 "Educational Technology: A Theoretical Approach." *College English* 39 (1): 68–108.

Appadurai, Arjun. 1996. *Modernity at Large: The Cultural Dimensions of Globalization.* Minneapolis: University of Minnesota Press.

Argyris, Chris, and Don Schön. 1978 *Organizational Learning: A Theory of Action Perspective.* Reading, Mass.: Addison-Wesley.

Aronowitz, Stanley, and William DiFazio. 1995. *The Jobless Future: Sci-tech and the Dogma of Work.* Minneapolis: University of Minnesota Press.

Attewell, Paul. 1994. "Information Technology and the Productivity Paradox." In *Organizational Linkages: Understanding the Productivity Paradox,* ed. Douglas Harris, 13–53. Washington, D.C.: National Academy Press.

Audi, Robert. 1999. *The Cambridge Dictionary of Philosophy.* (2d ed.) Cambridge: Cambridge University Press.

Ayer, A. J. 1968 *The Origins of Pragmatism: Studies in the Philosophy of Charles Sanders Peirce and William James.* San Francisco: Freeman, Cooper.

Baba, Marietta. 1999. "Dangerous Liaisons: Trust, Distrust, and Information Technology in American Work Organizations. *Human Organization* 58 (3): 331–346.

———. 2002. "Comment" to Panel on Technology and Communication, Annual Meeting, American Anthropological Association, New Orleans.

Bacon, Sir Francis. 1944 *Advancement of Learning and Novum Organum.* New York: Willey Book Co.

Bailey, James. 1996. *After Thought: The Computer Challenge to Human Intelligence.* New York: Basic Books.

Bainbridge, William Sims. 2001. Sociology on the World Wide Web. Available at: http://www.sdsc.edu/CC/positions/William_Bainbridge.html

Barley, Stephen, and Julian Orr, eds. 1997. *Between Craft and Science: Technical Work in the United States.* Ithaca, N.Y.: Cornell University Press.

Barth, Frederick. 1974 "On Responsibility and Humanity: Calling a Colleague to Account." *Current Anthropology* 15: 99–102.

Barwise, Jon, and Jerry Seligman. 1997. *Information Flow: The Logic of Distributed Systems.* Cambridge: Cambridge University Press.

Baudrillard, Jean. 1995. *The Gulf War Did Not Take Place.* Bloomington: Indiana University Press.

Belasen, Alan. 2000. *Leading the Learning Organization: Communication and Competencies for Managing Change.* Albany: SUNY Press.

Bell, Daniel. 1973 *The Coming of Post-industrial Society.* New York: Basic Books.

Berg, Anne-Jurunn, and Marete Lie. 1993. "Do Artifacts Have a Gender? Feminism and the Domestication of Technical Artifacts." Paper presented to the CRICT Conference, Brunel University.

Berger, John. 1991. *Ways of Seeing.* Harmondsworth, England: Viking Penguin.

Berger, Peter, and Thomas Luckmann. 1972. *Social Construction of Reality: A Treatise in the Sociology of Knowledge.* New York: Doubleday.

Bernal, Martin. 1989. *Black Athena: The Afroasiatic roots of classical civilization.* Vol. 1. New Brunswick, N.J.: Rutgers University Press.
Bernard, Russell. 2001. *Research Methods in Anthropology.* Walnut Creek, Calif.: Alta Mira Press.
Berners-Lee, Tim. 1999. *Weaving the Web: The Past, Present, and Future of the World Wide Web by its inventor.* London: Orion Business Books.
Bird-David, Nurit. 1999. "'Animism' Revisited: Personhood, Environment, and Relational Epistemology." *Current Anthropology* 40 (Supplement): S67–S91.
Bjerknes, Gro, and Tone Bratteteig. 1995. "User Participation and Democracy: A discussion of Scandinavian Research on System Development." *Scandinavian Journal of Information Systems* 7 (1): 73–98.
Bjerknes, Gro, and Bo Dahlbom., eds. 1990. *Organizational Competence in System Development: A Scandinavian Contribution.* Lund, Sweden: Studentlitteratur.
Blauner, Robert. 1964 *Alienation and Freedom.* Chicago: University of Chicago Press.
Blim, Michael. 1996. "Cultures and the Problems of Capitalisms." *Critique of Anthropology* 16 (1): 79–93.
Bloor, David. 1976 *Knowledge and Social Imagery.* (2d ed.) Chicago: University of Chicago Press.
Blomberg, Jeanette. 1998. "Knowledge Discourses and Document Practices: Neogotiating Meaning in Organizational Settings." Paper presented to the Annual Meeting, American Anthropological Association, Philadelphia.
Bourdieu, Pierre. 1978. *Outline of a Theory of Practice.* Cambridge: Cambridge University Press.
———. 1990. *Distinction: A Social Critique of the Judgment of Taste.* Cambridge: Harvard University Press.
———. 1993. *The Field of Cultural Production.* New York: Columbia University Press.
Bowker, Geoff, Susan Leigh Star, William Turner, and Les Gasser. 1997. "Introduction," In *Social Science, Technical Systems, and Cooperative Work: Beyond the Great Divide*, ed. Geoff Bowker et al., x–xxiii. Mahwah, N.J.: Lawrence Earlbaum Associates.
Brachman, Ronald, and Hector Levesque, eds. 1985. *Readings in Knowledge Representation.* San Francisco: Morgan Kaufmann Publishers.
Braverman, Harry. 1974 *Labor and Monopoly Capital.* New York: Monthly Review Press.
Brown, John Seeley, and Paul Duguid. 1998. "Organizing Knowledge." *California Management Review* 40 (3): 90–111.
———. 2000. *The Social Life of Information.* Boston: Harvard Business School Press.
Bruckman, Amy, and Carlos Jensen. 2002. "The Mystery of the Death of MediaMOO: Seven Years of Evolution of an Online Community." In *Building Virtual Communities: Learning and Change in Cyberspace*, ed. K. Ann Renninger and Wesley Shumar, 34–59. Cambridge: Cambridge University Press.
Burrows, Roger, and Sarah Nettleton. 2002. "Reflexive Modernization and the Emergence of Wired Self-Help." In *Building Virtual Communities: Learning and Change in Cyberspace*, ed. K. Ann Renninger and Wesley Shumar, 249–268. Cambridge: Cambridge University Press.
Calhoun, Craig. 1995. *Critical Social Theory.* Cambridge, Mass.: Blackwell.
Callon, Michel. 1986. "The Sociology of an Actor Network: The Case of the Electric Vehicle." In *Mapping the Dynamics of Science and Technology,* ed. M. Callon, 19–34. London: Macmillan.
Castells, Manuel. 2000a. *The Rise of the Network Society.* Cambridge: Blackwell Publishers.
———. 2000b. "Materials for an Exploratory Theory of the Network Society." *British Journal of Sociology* 51 (1): 5–24.
Chandler, Alfred. 1990. *Strategy and Structure: Chapters in the History of the Industrial Enterprise.* Cambridge: MIT Press.
Clark, Paul. 1995. *Managing Innovation and Change: People, Technology, and Strategy.* London: Sage Publications.
Clegg, Stewart, Cynthia Hardy, and Walter R. Nord, eds. 1996. *Handbook of Organizational Studies.* London: Sage Publications.
Clifford, James, and George Marcus, eds. 1986. *Writing Culture: The Poetics and Politics of Ethnography.* Berkeley: University of California Press.
Coff, Russell. 1999. "How Buyers Cope with Uncertainty When Acquiring Firms in Knowledge-Intensive iIndustries: *Caveat Emptor." Organization Science* 10 (2): 144–161.
Cognition and Technology Group. 1995. *Review of Computer-Mediated Instruction.* Nashville: Vanderbilt University Learning Technology Center.

Cohen, Anthony. 1985. *The Symbolic Construction of Community*. Chichester: E. Horwood Publications.
Collins, Harry. 1990. *Artificial Experts: Social Knowledge and Intelligent Machines*. Cambridge: MIT Press.
Cook, Scott, and John Seely Brown. 1999. "Bridging Epistemologies: The Generative Dance between Organizational Knowledge and Organizational Knowing." *Organization Science* 10 (4): 381–400.
Coombe, Rosemary J. 1998. *The Cultural Life of Intellectual Properties: Authorship, Appropriation, and the Law*. Durham, N.C.: Duke University Press.
Counts, George. 1969. *Dare the Schools Build a New Social Order*? New York: Arno Press.
Crabtree, Andrew. 1998. "Ethnography in Participatory Design." Paper presented at the Participatory Design Conference, Seattle.
Crane, Diana. 1972. *Invisible Colleges: Diffusion of Knowledge in Scientific Communities*. Chicago: University of Chicago Press.
Cuthbert, Alex, Douglas Clark, and Marcia Linn. 2002. "WISE Learning Communities: Design Considerations. In *Building Virtual Communities: Learning and Change in Cyberspace*, ed. K. Ann Renninger and Wesley Shumar, 215–248. Cambridge: Cambridge University Press.
Dahlbom, Bo. 1990. "Using Technology to Understand Organizations." In *Organizational Competence in System Development*, ed. Gro Bjernes and Bo Dahlbom, 127–47. Lund: Studentliteratur.
Bahlbom, Bo, and Lars Mathiessen. 1993. *Computers in Context: The Philosophy and Practice of Systems Design*. Cambridge: NCC/Blackwell.
Daniel, John. 1999. "Virtually All You'll Need to Know." *Guardian Weekly*, July 8–14, 24.
Darrah, Charles. 1998. "Knowledge Management in the Big Company Town." Paper presented to the Annual Meeting, American Anthropological Association, Philadelphia.
Davenport, Thomas, and Lawrence Prusak. 1997. *Working Knowledge: How Organizations Manage What They Know*. Boston: Harvard Business School Press.
David, Ken. 1998. "Distance Working: Cultural and Power Issues in Dispersed Projects." Paper presented to the Annual Meeting, American Anthropological Association, Philadelphia.
Davidson, Ann, and Janet Shofield. 2002. "Female Voices in Virtual Reality: Drawing Young Girls into an Online World." In *Building Virtual Communities: Learning and Change in Cyberspace*, ed. K. Ann Renninger and Wesley Shumar, eds., 34–59. Cambridge: Cambridge University Press.
Davidson, Carl, Ivan Handler, and Jerrry Harris. 1994. "The Promise and the Peril of the Third Wave." *cy.Rev* 1: 28–39.
Dennett, Daniel. 1991. *Consciousness Explained*. Boston: Little, Brown.
Derkson, Tom, ed. 1998. *The Promise of Evolutionary Epistemology*. Tilburg, Netherlands: Tilburg University Press.
Dery, Mark, ed. 1994. *Flame Wars: The Discourse of Cyberculture*. Durham, N.C.: Duke University Press.
Desai, Meghnad. 2002. *Marx's Revenge: The Resurgence of Capitalism and the Death of Statist Socialism*. London: Verso.
De Sanctis, Gerardine, and Marshall Poole. 1994. "Capturing the Complexity in Advanced Technology Use: Adaptive Structuration Theory." *Organization Science* 5 (2): 121–147.
De Solla Price, Derek. 1979. "The Citation Cycle." Reprinted in *Essays of an Information Scientist*. Vol 4, 621–633.
Devlin, Keith. 2001. *Infosense: Turning Information into Knowledge*. New York: W. H. Freeman.
Dewey, John. 1997. *Experience and Education*. New York: Macmillan.
Downey, Gary. 1998. *The Machine in Me: An Anthropologist Sits among Computer Engieneers*. New York: Routledge.
Drucker, Peter. 1970 *Technology, Management, & Society*. New York: Harper & Row Publishers.
Dubinskas, Frank, ed. 1988. *Making Time: Ethnographies of High-Technology Organizations*. Philadelphia: Temple University Press
Durham, William. 2001. Presentation on Electronic Publishing at the Annual Meeting, American Anthropological Association, San Francisco.
Dyer-Whitheford, Nick. 1999. *Cyber-Marx: Cycles and Circuits of Struggle in High-Technology Capitalism*. Urbana: University of Illinois Press.
Ehn, Pelle. 1988 *Work-Oriented Design of Computer Artifacts*. Stockholm: Almqvist & Wiksell.
Ehn, Pelle, *et. al.* 1995. "What Kind of Car Is This Sales Support System? On Styles, Artifacts and Quality-in-Use." Paper presented to the Third Decennial Computers in Context Conference, Aarhus, Denmark.

Engelbart, Doug. 1963 "A Conceptual Framework for the Augmentation of Man's Intellect." In *Vistas in Information Handling*, ed, P. Howerton, vol. 1, 1–29. Washington, D.C.: Spartan.
English-Lueck, Jan. 2002. *Cultures@Silicon Valley*. Palo Alto, Calif.: Stanford University Press.
Erickson, Ken. 1998. "It's All in the Cards: Knowledge Management at Cinderella Greetings, Incorporated." Paper presented to the Annual Meeting, American Anthropological Association, Philadelphia.
Ess, Charles, ed. 1996. *Philosophical Perspectives on Computer-Mediated Communication*. Albany: SUNY Press.
Etzkowitz, Henry, and Loet Leydesdorff. 1998. "A Triple Helix of University-Industry-Government Relations: Introduction." *Industry & Higher Education* 12 (4): 197–258.
Feenberg, Andrew. 1999. "Whither Educational Technology?" *Peer Review* 1 (4).
Fenstermacher, Gary, and Mathew Sanger. 1997. "What Is the Significance of John Dewey's Approach to the Problem of Knowledge?" *The Elementary School Journal* 98 (5): 467–479.
Finnemann, Niels Ole. 1996. *Thought, Sign, and Machine—The Idea of the Computer Reconsidered*. Aarhus, Denmark: University of Aarhus.
Forsythe, Diana. 2001. *Studying Those Who Study Us: An Anthropologist in the World of Artificial Intelligence*. Ed. David Hess. Palo Alto, Calif.: Stanford University Press.
Fosnot, Catherine. 1995. *Constructivism: Theory, Perspectives, and Practice*. New York: Teachers College Press.
Fox, Richard. 2000. "Ethics in Anthropology." In *Ethics and Anthropology: Facing Future Issues in Human Biology, Globalism, and Cultural Property*, Anne-Marie Cantwell, Eva Friedlander, and Madeleine L. Tramm, 112–135. New York: New York Academy of Sciences.
Frake, Charles. 1980. *Language and Cultural Description*. Palo Alto,, Calif.: Stanford University Press.
Fraser, Nancy. 1989. *Unruly Practices: Power, Discourse, and Gender in Contemporary Social Theory*. Mineapolis: University of Minesota Press.
Frazer, James A. 1997. *The Golden Bough: A Study in Magic and Religion*. Harmondsworth, England: Penguin Classics.
Freire, Paulo. 2000. *Pedagogy of the Oppressed*. New York: Continuum International Publishing Group.
Fujimura, Joan. 1996. *Crafting Science: A Sociohistory of the Quest for the Genetics of Cancer*. Cambridge: Harvard University Press.
———. 1999. "Authorizing Knowledge in Science and Anthropology." *American Anthropologist* 100 (2): 347–360.
Fuller, Steve. 1992. "Talking Metaphysical Turkey about Epistemological Chicken, and the Poop on Pidgens." Paper presented to the Annual Conference, Society for Social Studies of Science, Gothenburg, Sweden.
———. 2000. *Thomas Kuhn: A Philosophical History for Our Times*. Chicago: University of Chicago Press.
Furlong, Chris. 1998. "Islamic Science." Paper presented to the Annual CASTC Workshop, New York.
Fölsing, Albrecht. 1997. *Albert Einstein*. Harmondsworth, England: Penguin.
Gamst, Fred, ed. 1995. *Meanings of Work: Considerations for the Twenty-First Century*. Albany: SUNY Press.
Garfinkel, Harold. 1984. *Studies in Ethnomethodology*. Cambridge: Polity Press.
Garsten, Christina. 1994. *Apple World*. Stockholm: Stockholm University Studies in Social Anthropology.
Geertz, Clifford. 2000. *Available Light: Anthropological Reflections on Philosophical Topics*. Princeton: Princeton University Press.
———. 2001. "We Are All Anthropologists Now." *Guardian*, July 12.
Giddens, Anthony. 1987. *Social Theory and Modern Sociology* Stanford, Calif.: Stanford University Press. 1991.
———. *Modernity and Self-Identity*. Palo Alto, Calif.: Stanford University Press.
———. 1998. *The Third Way: The Renewal of Social Democracy*. Cambridge: Polity Press.
Giddens, Anthony, and Will Hutton, eds. 2000. *Global Capitalism*. New York: The New Press.
Glaser, Barney, and Anselm Strauss. 1967 *The Discovery of Grounded Theory: Strategies for Qualitative Research*. Berlin: Aldine de Gruyter.
Gluesing, Julia. 1998. "Team Interactions and Quality Performance in the Product Development Process." Paper presented to the Annual Meeting, American Anthropological Association, Philadelphia.

———. 2000. "Global Teams are Different." Paper presented to the Annual Meeting, American Anthropological Association, Washington, D.C.

Goffman, Erving. 1959 *The Presentation of Self in Everyday Life*. Garden City, N.Y.: Doubleday.

Goldberg, Ken. 2001. *The Robot in the Garden: Telerobotics and Telepistemology in the Age of the Internet.* Cambridge: MIT Press.

Goodman, Robert, and Walter Fisher. 1995. *Rethinking Knowledge: Reflections across the Disciplines.* Albany: SUNY Press.

Gough, Kathleen. 1968. "Anthropology: Child of Imperialism." *Monthly Review* 19 (11): 12–27.

Gould, Stephen Jay. 2002. *The Structure of Evolutionary Theory*. Cambridge: Harvard University Press.

Greenbaum, Joan. 1995. *Windows on the Workplace: Computers, Jobs, and the Organization of Office Work in the Late Twentieth Century*. New York: Monthly Review Press.

Greeno, James. 1988 *Situations, Mental Models, and Generative Knowledge*. Palo Alto, Calif.: IRL.

Gross, Paul. 1996. posting to SCI-TECH-STUDIES listserv: see http://helix.ucsd.edu/~bsimon/sts/lists.htm

Gross, Paul, and Normal Levitt. 1994. *Higher Superstition*. Baltimore: John Hopkins University Press.

Habermas, Jürgen. 1987. *Theory of Communicative Action: Lifeworld and System: A Critique of Functionalist Reason*. Vol. 2. Boston: Beacon.

———. 1990. *Legitimation Crisis*. Boston: Beacon.

Hakken, David. 1978 *Workers' Education: The Reproduction of Working Class Culture in Sheffield, England, and "Really Useful Knowledge."* Ann Arbor, Mich.: University Microfilms.

———. 1987. "Reproduction and Culture in Complex Social Formations." *Dialectical Anthropology* 12 (2): 193–204.

———. 1990. "Has There Been a Computer Revolution? An Anthropological Approach." *The Journal of Computing and Society* 1 (1): 11–28.

———. 1991. "Anthropological Ethics in the 1990s: A Positive Approach," in *Ethics and the Profession of Anthropology*, ed. C. Fleuhr-Lobban, 72–91. Philadelphia: University of Pennsylvania Press.

———. 1992. "Computing and Social Change: New Technology and Workplace Transformation, 1980 90." *Annual Review of Anthropology* 22: 107–32.

———. 1995a. "The Cultural Reconstruction of Science: A Response to Labinger." *Social Studies of Science* 25 (2): 317–320.

———. 1995b. "Cybertalk and the Study of Computing and Social Change." Presented to the 93d Annual Meeting, American Anthropological Association, Washington, D.C.

———. 1999a. *Cyborgs@Cyberspace?: An Ethnographer Looks to the Future*. New York: Routledge.

———. 1999b "A Manifesto for Cyberspace Anthropology" (in Catalan). *Revista d'Ethnologia de Catalunya.*

———. 2000. "Ethical Issues in the Ethnography of Cyberspace." In *Ethics and Anthropology: Facing Future Issues in Human Biology, Globalism, and Cultural Property*, ed. Anne-Marie Cantwell, Eva Friedlander, and Madeleine L. Tramm, 170–186. New York: New York Academy of Sciences.

———. 2001. "Knowledge, Cyberspace, and Anthropology." Presented at the annual meeting, American Anthropological Association, Washington, D.C.

———. 2002. "Afterword." In *Building Virtual Communities: Learning and Change in Cyberspace,* ed. Ann Renninger and Wesley Shumar, 355–367 Cambridge: Cambridge University Press.

———. 2003. "An Ethics for an Anthropology in and of Cyberspace." In *Ethics and the Profession of Anthropology: A Dialogue for Professional Conduct* (2d ed.) Ed. Carolyn Fleurh-Lobban, Walnut Creek, Calif.: Alta Mira Press.

Hakken, David, with Barbara Andrews. 1993. *Computing Myths, Class Realities: An Ethnography of Technology and Working People in Sheffield, England.* Boulder, Colo.: Westview Press.

Hakken, David, Jan DeAmicis, Mary Ann Ewen, Arthur Sanders, and James White. 1991. *Oneida/Herkimer Human Needs Assessment Research Report.* Utica, N.Y.: Greater Utica United Way.

Hakken, David, and Johanna Lessinger, eds. 1987. *Perspectives on U.S. Marxist Anthropology.* Boulder, Colo.: Westview Press.

Hakken, David, with Nuria Soeharto. 2004. "Introduction" to Special Issue on "Questions of Identity on the Internet: Research 'Software' Towards a New Indonesia." *Antropologi Indonesia.*

Hamada, Tamoko. 1998. "The Anthropology of Business Organization: Introduction." *Anthropology of Work Review* 18 (2–3):1–6.

Hammer, Michael, and James Champy. 1994. *Reengineering the corporation.* New York: Harper Business.

Hannerz, Ulf. 1996. *Transnational Connections: Culture, People, Places*. London: Routledge.

Hansen, Morten. 1991. "The Search-Transfer Problem: The role of Weak Ties in Sharing Knowledge across Organization Subunits." *Administrative Science Quarterly* 441): 82–84.

Hanseth, Ole. 2000. "IT Systems Development and the Fashion Economy." Presented to the Annual Conference, Society for Social Studies of Science, Vienna, Austria.

Haraway, Donna. 1991. *Simians, Cyborgs, and Women: The Reinvention of Nature*. London: Free Association Books.

———. 1997 *Modest_Witness@Second_mllenium.FemaleMan©_Meets_Oncomouse*™. New York: Routledge.

Harding, Sandra. 1991. *Whose Science? Whose Knowledge?: Thinking from Women's Lives*. Ithaca, N.Y.: Cornell University Press.

Harris, Marvin. 1968. *Rise of Anthropological Theory: A History of Theories of Culture*. New York: HarperCollins.

Hartsock, Nancy. 1983. *Money, Sex, and Power: Toward a Feminist Historical Materialism*. Boston: Northeastern University Press.

Harvey, David. 1989. *The Condition of Postmodernity*. Oxford: Basil Blackwell.

Harvey, Penny. 1999. "Why 'Smart Objects' Need Narratives: Re-orienting Objects in a New Museum Database." Manchester: Department of Social Anthropology, University of Manchester.

Hastrup, Kirsten. 1994. "Anthropological Knowledge Incorporated," In *Social Experience and Anthropological Knowledge*, ed. Kirsten Hastrup and Peter Hervik, 224–240. London: Routledge.

Heath, Deborah. 1994. "Recombinant Fieldsites: Chimeric Ethnography and the Co-production of Technoscientific Research." Paper presented to the annual meeting, American Anthropological Association, Atlanta.

Heathcote, Dorothy, Cecily O'Neil, and Liz Johnson, eds. 1991. *Collected Writings on Education and Drama*. Evanston, Ill.: Northwestern University Press.

Heaton, Lorna. 1999. "Preserving Communication Context: Virtual Workspace and Interpersonal Space in Japanese CSCW." In Proceedings: *Cultural Attitudes Towards Communication and Technology '98*, ed. Charles Ess and F. Sudweeks. University of Sydney, Austrialia.

Helmreich, Stefan. 1999. *Silicon Second Nature: Culturing Artificial Life in a Digital World*. Berkeley: University of California Press.

Hess, David. 1992. "Introduction: The New Ethnography and the Anthropology of Science and Technology." In *Knowledge and Society: The Anthropology of Science and Technology*, ed. David Hess and Linda Layne, 1–26. Greenwich, Conn.: JAI Press.

———. 2001a. *Selecting Technology, Science, and Medicine: Alteernative Pathways in Globalization*. Vol. 1. Nyskayuna, N.Y.: Author.

———. 2001b. "Boundary Tensions and the Conditions of Knowledge Production." Presented at the Annual Meeting, American Anthropological Association, Washington, D.C.

Hiltz, Starr Roxanne, and Murray Turoff. 1993. *The Network Nation: Human Communication via Computer*. Cambridge: MIT Press.

Hoadley, Christopher, and Roy Pea. 2002. "Finding the Ties That Bind: Tools in Support of a Knowledge-Building Community." In *Building Virtual Communities: Learning and Change in Cyberspace*, ed. K. Ann Renninger and Wesley Shumar, 321–354. Cambridge: Cambridge University Press.

Holtshouse, Dan. 1998. "Knowledge Research Issues." *California Management Review* 40i3: 277–80.

Hunter, Beverly. 2002. "Learning in the Virtual Community Depends upon Changes in Local Communities." In *Building Virtual Communities: Learning and Change in Cyberspace*, ed. K. Ann Renninger and Wesley Shumar, 96–128. Cambridge: Cambridge University Press.

Information Strategy. 1998. "Special Report: The Facts about Knowledge." *Information Strategy Knowledge Management Survey*: http://www.info-strategy.com/knwosur1/.

Ingold, Tim. 1995. "'People Like Us': The Concept of the Anatomically Modern Human." *Cultural Dynamics* 7 (2): 187–214.

Ito, Mizuko. No date. "Networked localities." http://www.itofisher.com/PEOPLE/mito/.

Jacobsen, David. 1999. "Impression Formation in Cyberspace: Online Expectations and Offline Experiences in Text-Based Virtual Communities." *Journal of Computer-mediated Communication*: www.ascusc.org/jcmc/vol5/issue1/jacobson.html.

Jarvie, I. C. 1964. *The Revolution in Anthropology*. Chicago: Henry Regnery.

Jordan, Ann. 1998. "The Complexity of Becoming in Self-Directed Teams of Knowledge Workers." Paper presented to the Annual Meeting, American Anthropological Association, Philadelphia.

Jordan, Birgitte. 1998. "Authoritative Knowledge in Corporate Settings." Paper presented to the Annual Meeting, American Anthropological Association, Philadelphia.

Katz, Jon. 2000. *Geeks: How Two Lost Boys Rode the Internet out of Idaho.* New York: Villard (Random House).

Kidder, Tracy. 1981. *The Soul of a New Machine.* New York: Avon Books.

Kiel-Slawik, Reinhardt. 1995. "From Mechanization of the Brain towards an Ecology of the Mind: A Personal Perspective on Contextualizing Design in Informatics." Paper presented to the Third Decennial Conference on Computers in Context, Aarhus, Denmark.

Kim, Jaegwon. 1994. "What is 'Naturalized Epistemology.'?" In *Naturalizing Epistemology,* ed. Hilary Kornblith, 33–57. Cambridge: MIT Press

Knorr-Certina, Karin. 1999. *Epistemic Cultures: How the Sciences Make Knowledge.* Cambridge: Harvard University Press.

Kogut, B., and U. Zander. 1992. "Knowledge of the Firm, Combinative Capabilities, and the Replication of Technology." *Organization Science* 3: 383–397.

Kolb, David. 1996. "Discourse across Links." In *Philosophical Perspectives on Computer-Mediated Communication,* ed. Charles Ess, 15–26. Albany: SUNY Press.

Kornblith, Hilary. 1994. "Introduction: What Is Naturalistic Epistemology?" In *Naturalizing Epistemology,* ed. Hilary Kornblith, 1–14. Cambridge: MIT Press.

Kotter, Wade. 2001. "Researching Anthropology in the Digital Library: Problems and Prospects." Presented to the Annual Meeting, American Anthropological Association, Washington, D.C.

Kozol, Jonathan. 1992. *Savage Inequalities: Children in America's Schools.* New York: HarperCollins.

Kuhn, Thomas. 1970. *The Structure of Scientific Revolutions.* Chicago: University of Chicago Press.

Kunda, Gideon. 1992. *Engineering Culture: Control and Commitment in a High-Tech Corporation.* Philadelphia: Temple University Press.

Kurzweil, Ray. 1999. *The Age of Spiritual Machines: When Computers Exceed Human Intelligence.* New York: Penguin Group (USA).

Kusterer, Ken. 1978 *Know-How on the Job.* Boulder, Colo.: Westview Press.

Labinger, Jay. 1995. "Science as Culture: A View from the Petri Dish." *Social Studies of Science* 25 (2): 285–306.

Lahsen, Myana. 1995. "Climate Scientists: The Shapes and Limits of Globalization." Paper presented to the Annual Conference, Society for Social Studies of Science, Charlottesville, Va.

Lakoff, George. 1990. *Women, Fire, and Dangerous Things.* Chicago: University of Chicago Press.

Lamphere, Louise. 1979. "Fighting the Piece Rate System: New dimensions of an Old Struggle in the Apparel Industry." In *Case Studies on the Labor Process,* ed. Andrew Zimbalist, 257–276. New York: Monthly Review Press.

Langefors, Börje. 1993. *Essays in Infology.* Göteborg, Sweden: Göteborg University.

Latour, Bruno. 1987. *Science in Action.* Cambridge: Harvard University Press.

———. 1993. *We Have Never Been Modern.* Cambridge: Harvard University Press.

———. 1999. *Pandora's Hope: Essays on the Reality of Science Studies.* Cambridge: Harvard University Press.

Latour, Bruno and Steve Woolgar. 1979. *Laboratory Life: The Social Construction of Scientific Knowledge.* Beverly Hills, Calif.: Sage.

Lave, Jean, and Etienne Wenger. 1994. *Situated Learning: Legitimate Peripheral Participation.* Cambridge: Cambridge University Press.

Lee, B., and W. Shu. 1997. "Where Has Information Technology Productivity Been Buried? A Macroeconomic Analysis of Information Technology Investment with Fast Structural Changes and Slow Productivity Growth." In *Proceedings of the Fourth International Decision Science Institute Meeting,* Sydney, Australia.

Lee, Richard. 1979. *The !Kung San: Men, Women, and Work in a Foraging Society.* Cambridge: Cambridge University Press.

———. 1998. "Non-capitalist Work: Baseline for an Anthropology of Work or Romantic Delusion?" *Anthropology of Work Review* XVIII (4): 9–13.

———. 1998. Paper in *Anthropology of Work Review.*

Levi-Strauss, Claude. 1966 *The Savage Mind.* Chicago: University of Chicago Press.

Levin, James, and Raoul Cervantes. 2002. "Understanding the Life Cybles of Network-Based Learning

Communities." In *Building Virtual Communities: Learning and Change in Cyberspace*, ed. K. Ann Renninger, and Wesley Shumar, 269–292. Cambridge: Cambridge University Press.
Leydesdorff, Loet, and Henry Etzkowitz. 1998. "The Triple Helix as a Model for Innovation Studies." (Conference Report.) *Science & Public Policy* 25 (3): 195–203
Lincoln, Yvonna. 1992. "Virtual Community and Invisible Colleges: Alternation in Faculty Scholarly Networks and Professional Self-image." Paper presented at the Annual Meeting, American Anthropological Association, San Francisco.
Lundsgaarde, Henry. 1992. "Knowledge Engineering and Ethnography." Paper presented to the Annual Meeting, American Anthropological Association, San Francisco.
Lynd, Robert, and Helen Merrill. 1925 (1989.) *Middletown: A Study in Modern American Culture*. New York: Harcourt.
Lyotard, François. 1984. *The Postmodern Condition*. Minneapolis: University of Minnesota Press.
Malpas, Jeff. 2001. "Acting at a Distance and Knowing from Afar: Agency and Knowledge on the Internet." In *The Robot in the Garden*, ed. Goldberg, Ken, 108–125. Cambridge: MIT Press.
Mandel, Earnest. 1978 *Late Capitalism*. New York: Knopf Publishing Group.
Mankin, Donald, Susan Cohen, and Tora Bikson. 1996. *Teams and Technology: Fulfilling the Promise of the New Organization*. Cambridge: Harvard University Press.
Marcus, George. 1998. *Ethnography through Thick & Thin*. Princeton: Princeton University Press.
Marcus, George, and Michael Fisher. 1986. *Anthropology as Cultural Critique: An Experimental Moment in the Human Sciences*. Chicago: University of Chicago Press.
Marglin, Steve. 1974 "What Do bosses Do?" *Socialist Revolution*.
Margolis, Joseph. 1993. *Pragmatism without Foundations: Reconciling Realism and Relativism*. Oxford: Blackwell Publishers.
Marquardt, Michael, and Greg Kearsley, eds. 1998. *Technology-Based Learning: Maximizing Human Performance and Corporate Success*. Boca Raton, Fla.: St. Lucie Press.
Marx, Karl. 1871 (1967) *Capital*. New York: International Publishers.
Marx, Leo. 1964. *The Machine in the Garden*. London: Oxford University Press.
Mathews, Michael, ed. 1998. *Constructivism in Science Education: A Philosophical Examination*. Dordrecht, Netherlands: Kluwer Publishers.
Maturana, Humberto, and Francisco Varela. 1998. *The Tree of Knowledge: The Biological Roots of Human Understanding*. (Rev. ed.). Boston: Shambhala.
Mead, George Herbert. 1962 *Mind, Self, and Society*. Chicago: University of Chicago Press.
Menands, Louis. 1997. "The Return of Pragmatism. "*American Heritage* 48 (6): 48–57.
Merton, Robert K. and Piotr Sztompka (editor). 1996. *On Social Structure and Science*. Chicago: University of Chicago Press.
Meyer, Eric, and Rob Kling. 2000. "Technology & Unequal Participation: Access to Electronic Working Paper Repositories and Scholarly Participation in Elite Scientific Communities." Presented to the American Sociological Association Conference, Washington, D.C.
Mills, C. Wright. 1959 *The Sociological Imagination*. New York: Oxford University Press.
Mitchell, J. Clyde. 1969 *Social Networks in Urban Situations*. Manchester: Manchester University Press.
Mol, Annemarie, and John Law. 1994. "Regions, Networks and Fluids: Anemia and Social Topology." *Social Studies of Science* 24: 641–671.
Moody, Glyn. 2001. *Rebel Code: Inside Linux and the Open Source Revolution*. Cambridge, Mass.: Perseus Publishing.
Moore, Henrietta, ed. 1999. *Anthropological Theory Today*. Cambridge, England: Polity Press.
Nader, Laura. 1996. *Naked Science: Anthropological Inquiry into Boundaries, Power, and Knowledge*. New York: Routledge.
Nahapiet, Janine, and Sumantra Ghosal. 1998. "Social Capita, Intellectual Capital, and the Organizational Advantage." *Academy of Management Review* 23 (2): 242–266.
Nardi, Bonnie. 1993. *A Small Matter of Programming*. Cambridge: MIT Press.
National Science Foundation (NSF—United States). 1999. Knowledge and Distributed Intellegence Home Page. http://www.her.nsf.gov/kdi.
Noble,. 2001. *Digital Diploma Mills: The Automation of Higher Education*. New York: Monthly Review Press.
Nolan, D. Jason, and Joel Weiss. 2002. "Learning in Cyberspace: An Educational View of Virtual Community." In *Building Virtual Communities: Learning and Change in Cyberspace*, ed. K. Ann Renniner and Wesley Shumar, 293–320. Cambridge: Cambridge University Press.

Nonaka, Ikujiro, and Hirotaka Takeuchi. 1995. *The Knowledge-Creating Company: How Japanese Companies Create the Dynamics of Innovation*. New York: Oxford University Press.

Norman, Donald. 1999. "The Challenges and the Problems." Presented to the IBM Workshop on Advancing Distributed Learning, Cambridge, Mass.

Nowotny, Helga, Peter Scott, and Michael Gibbons. 2001. *Rethinking Science*. Oxford: Blackwell Publishers.

Nyce, James, and Paul Kahn. 1991. *From Memex to Hypertext: Vannevar Bush and the Mind's Machine*. San Diego: Academic Press, Inc.

Nyce, James, and Jonas Loewgren. 1995. "Toward Foundational Analysis in Human-Computer Interaction" In *The Social and Interactional Dimensions of Human-Computer Interfaces*, ed. P. J. Thomas, 37–46. Cambridge: Cambridge University Press

Oettinger, Anthony, and Sema Marks. 1969 *Run Computer, Run: The Mythology of Educational Innovation: An Essay*. Cambridge: Harvard University Press.

Oldenburg, Ray. 2000. *Celebrating the Third Place: Inspiring Stories about the "Great Good Places" at the Heart of Our Communities*. Emeryville, Calif.: Avalon Publishing Group.

Organization for Economic Cooperation and Development. 1994. *The OECD Jobs Study: Facts, Analysis, Strategy*. Paris: OECD.

Orlikowski, Wanda. 2000. "Using Technology and Constituting Structures: A Practice Lens for Studying Technology in Organizations." *Organization Science* 11 (4): 404–428.

Orr, Julian. 1996. *Talking about Machines: An Ethnography of a Modern Job*. Ithaca, N.Y.: Cornell University Press.

Ortner, Sherry. 1984. "Theory of Anthropology since the Sixties." *Comparative Studies in Society and History* 126 (1): 126–66.

Oxford English Dictionary. 1968. "Knowledge." Oxford: Oxford University Press.

Pfaffenberger, Bryan. No date. "The Second Self in the Third World." Unpublished MS, University of Virginia.

———. 1988. The Social Meaning of the Personal Computer: Or, Why the Personal Computer Revolution Was No Revolution. *Anthropological Quarterly* 61(1): 39–47.

———. 1990. *Democratizing Information: Online Data-Bases and the Rise of End-User Searching*. Boston: G. K. Hall.

———. 1992. "The Social Anthropology of Technology." *Annual Review of Anthropology* 21: 491–516.

———. 1999. "Open Source Software and Software Patents: A Constitutional Perspective." *Knowledge, Technology, and Policy* 12 (3): 94–112.

Pitt, Martyn. 1998. "Strategic Intervention: Statements of the Art or in Search of a Chimera?" *Human Relations* 51(4): 547–564.

Plotkin, Henry. 1994. *Darwin, Machines, and the Nature of Knowledge*. Cambridge: Harvard University Press.

Poitou, Jean-Pierre. 1996. "Building a Collective Knowledge Management System: Knowledge-Editing versus Knowledge-Eliciting Techniques." In *Proceedings of the 1996. ACM Conference on Computer Supported Cooperative Work*, ed. Mark Ackerman, 235–256. New York: ACM Press.

Polanyi, Michael. 1983. *Tacit Dimension*. Glouster, Mass.: Peter Smith Publisher.

Powdermaker, Hortense. 1966 *Stranger and Friend: The Way of an Anthropologist*. New York: W. W. Norton.

Powell, Walter. 1990. "Neither Market nor Hierarchy: Network Forms of Organization." *Research in Organizational Behavior* 12: 295–336.

Putnam, Robert, and Lewis Feldstein. 2001. *Bowling Alone: The Collapse and Revival of American Community*. New York: Simon & Schuster.

Quine, W. V. O. 1994. "Epistemology Naturalized." In *Naturalizing Epistemology*, ed. Hilary Kornblith, 15–32. Cambridge: MIT Press.

Quinn, James Bryan. 1992. *Intelligent Enterprise: A Knowledge- and Service-Based Paradigm for Industry*. New York: The Free Press.

Rabinow, Paul, ed. 1984. *The Foucault Reader*. New York: Pantheon.

———. 1992. "Serving the Ties: Fragmentation and Dignity in Late Modernity." In *Knowledge and Society: The Anthropology of Science and Technology*, ed. David Hess and Linda Layne, 9: 169–187. Greenwich, Conn.: JAI Press.

Raymond, Eric. 1999. "The Cathedral and the Bazaar." *Knowledge, Technology, & Policy* 12 (3): 23–49.

Read, Dwight. 1991. "Making Visible the Invisible: Computer Modeling of Hidden Structure." Paper presented to the Annual Meeting, American Anthropological Association, Chicago.

Renninger, K. Ann, and Wesley Shumar, eds. 2002a. *Building Virtual Communities: Learning and Change in Cyberspace*. Cambridge: Cambridge University Press.

———. 2002b. "Community Building with and for Teachers at the Math Forum." In *Building Virtual Communities: Learning and Change in Cyberspace*, ed. K. Ann Renninger and Wesley Shumar, 60–95. Cambridge: Cambridge University Press.

Rheingold, Howard. 1993. *The Virtual Community*. New York: Harper Perennial.

Rifkin, Jeremy. 1995. *The End of Work: The Decline of the Global Labor Force and the Dawn of the Post-market Era*. New York: G. P. Putnam's Sons.

Roberts, Joanne. 2000. "From Know-How to Show-How? Questioning the Role of Information and Communication Technologies in Knowledge Transfer." *Technology Analysis & Strategic Management* 12 (4): 429–443.

Roethlisberger, F., and Dickson, W. 1934 *Management and the Worker*. Cambridge: Harvard University Press.

Ross, Andrew, ed. 1996. *Science Wars*. Durham, N.C.: Duke University Press.

Rothschild, Joan, ed. 1983. *Machina ex Dea: Feminist Perspectives on Technology*. New York: Pergamon Press.

Rothstein, Francis A., and Michael Blim, eds. 1992. *Anthropology and the Global Factory*. New York: Bergin and Garvey.

Rybczynski, Witold. 2000. *A Clearing in the Distance: Frederick Law Olmsted and America in the Nineteenth Century*. New York: Simon & Schuster.

Sachs, Patricia. 1994. "Thinking through Technology: The Relationship of Technology and Knowledge at Work." *Information Technology and People*.

Sacks, Karen. 1982. "Caring by the Hour: Women, Work and Organizing at Duke Medical Center." In *My Troubles are Going to Have Trouble with Me*, ed. Karen Sacks and Dorothy Remy, 172–92. New Brunswick, NJ: Rutgers University Press.

Sahlins, Marshall. 1972. *Stone Age Economics*. Chicago: Aldine-Atherton.

———. 1976a. *Culture and Practical Reason*. Chicago: University of Chicago Press.

———. 1976b. *The Use and Abuse of Biology*. Ann Arbor: University of Michigan Press.

———. 1996. *How "Natives" Think: About Captian Cook, for Example*. Chicago: University of Chicago Press.

———. 2000. *Culture in Practice: Selected Essays*. New York: Zone Books.

Sahlins, Marshall, and E. R. Service, eds. 1960. *Evolution and Culture*. Ann Arbor: University of Michigan Press.

Sapir, Edward. 1990. *Selected Writings of Edward Sapir on Language, Culture, and Personality*. Ed. David G. Mandelbaum. Berkeley: University of California Press.

Sassen, Saskia. 2000. *Global City: New York, London, Tokyo*. Princeton: Princeton University Press.

Schlager, Mark, Judith Fusco, and Patricia Schank. 2002. "Evolution of an Online Education Community of Practice." In *Building Virtual Communities: Learning and Change in Cyberspace*, ed., K. Ann Renninger and Wesley Shumar, 129–158. Cambridge: Cambridge University Press.

Scholte, Bob. 1986. "The Charmed Circle of Geertz's Hermeneutics: A Neo-Marxist Critique." *Critique of Anthropology* 6 (1): 5–15.

Schön, Donald. 1990. *The Reflective Practitioner: How Professionals Think in Action*. New York: Basic Books.

Schumpeter, Joseph. 1976 *Capitalism, Socialism*. New York: HarperCollins.

Sejersted, Francis. 1997. "Technological Development and the Right to Work." In *OECD Proceedings: Creativity, Innovation, and Job Creation*, ed. anon. Paris: Organization for Economic Cooperation and Development.

Sennett, Richard. 1999. *The Corrosion of Character: The Personal Consequences of Work in the New Capitalism*. New York: W. W. Norton.

Service, E. R. 1971 *Cultural Evolutionism: Theory in Practice*. Melborne, Fla.: Krieger.

Shank, Gary and Donald Cuningham. 1996. "Mediated Philosopher Dots: Toward a Post-Cartesian Model of CMC via the Semiotic Superhighway." In *Philosophical Perspectives on Computer-mediated Communication*, ed. Charles Ess, 27–43. Albany: SUNY Press.

Shannon, Claude Elwood. 1995. *Collected Papers*. New York: IEEE.

Shapiro, Dan. 1995. "The Limits of Ethnography: Combining Social Sciences for CSCW." In Association for Computing Machinery, *CSCW '94 Proceedings.*

Shields, Rob. 2000. *A Critical Analysis of Knowledge Management Initiatives in the Canadian Federal Public Service.* Ottawa: Carlton University Innovation Management Research Unit.

Shuler, Douglas. 1996. *New Community Networks.* New York: Addison-Wesley.

Simon, Herbert. 1996. *The Sciences of the Artificial.* (3d ed.) Cambridge: MIT Press.

Sinding-Larsen, Henrik. 1987. "Information Technology and Management of Knowledge." *AI & Society* 1: 93–101.

———. 1991. "Computers, Musical Notation, and the Externalization of Knowledge." in *Understanding the Artificial,* ed. M. Negrotti, 101–125. London: Springer.

Slouka, Mark. 1995. *War of the Worlds: Cyberspace and the High-tech Assault on Reality.* New York: Basic Books.

Smith, Adam. 1776 (1991). *Wealth of Nations.* Amherst, N.Y.: Prometheus Books.

Smith, Brian Cantwell. 1996. *On the Origin of Objects.* Cambridge: MIT Press.

Smith, Dorothy. 1989. *The Everyday World as Problematic: A Feminist Sociology.* Boston: Northeastern University Press.

Smith, Keith. 1996. "Creativity, Innovation, and Job Creation." In *OECD Proceedings: Creativity, Innovation, and Job Creation,* ed. anon. Paris: Organization for Economic Cooperation and Development.

Solow, Robert. 1987. "We'd Better Watch Out." *New York Times Book Review,* July 12.

Spitulnik, Debra. 1993. "Anthropology and the Mass Media." *Annual Review of Anthropology* 22: 293–315.

Sproull, Lee, and Sara Kiesler. 1995. *Connections: New Ways of Working in the Networked Organization.* Cambridge: MIT Press.

Star, Susan Leigh. 1995. *The Cultures of Computing.* Oxford: Blackwell

Stewart, Thomas A. 2002. *The Wealth of Knowledge: Intellectual Capital and the 21st Century.* New York: Doubleday.

Stocking, George. 1990. *Observers Observed: Essays on Ethnographic Fieldwork.* Madison: University of Wisconsin Press.

Stoll, Clifford. 1996. *Silicon Snake Oil: Second Thoughts on the Information Highway.* New York: Doubleday.

Stonier, Tom. 1992. *Beyond Information: The Natural History of Intelligence.* New York: Springer Verlag.

Strathern, Marilyn, ed. 1995. "Foreword," and "The Nice Thing about Culture Is That Everyone Has It." In *Shifting Contexts: Transformations in Anthropological Knowledge (The Uses of Knowledge),* 1–11 and 153–176. London: Routledge.

———. 1996. "Cutting the Network." *Journal of the Royal Anthropological Institute* (New Series) 2: 517–535.

Suchman, Lucy. 1987. *Plans and Situated Actions.* Cambridge: Cambridge University Press.

———. 2000. "Organizing Alignment: A Case of Bridge-Building." *Organization* 7 (2): 311–327.

Taussig, Michael. 1993. *Mimesis and Alterity: A Particular History of the Senses.* New York: Routledge.

Taylor, Frederick Winslow. 1998. *The Principles of Scientific Management.* Mineola, N.Y.: Dover Publications.

Tierney, Patrick. 2000. *Darkness in El Dorado: How Scientists and Journalists Devastated the Amazon.* New York: W. W. Norton.

Toffler, A. 1983. *Previews and Promises.* New York: Bantam Press.

Torvalds, Linus, and David Diamond. 2001. *Just for Fun: The Story of an Accidental Revolutionary.* New York: Harper Business.

Toulmin, Stephen. 1992. *Cosmopolis: The Hidden Agenda of Modernity.* Chicago: University of Chicago Press.

———. 1995. "Foreword," In *Rethinking Knowledge: Reflections across the Disciplines,* ed. Robert Goodman and Walter Fisher, ix–xv. Albany: SUNY Press.

———. 2001. *Return to Reason.* Cambridge: Harvard University Press.

Traweek, S. 1988 *Beamtimes and Lifetimes: The World of High Energy Physicists.* Cambridge: Harvard University Press

Turkle, Sherry. 1984. *The Second Self: Computers and the Human Spirit.* New York: Simon & Schuster.

———. 1995. *Life on the Screen: Identity in the Age of the Internet.* New York: Simon & Schuster.

Turnbull, Colin. 1972. *The Forest People*. New York: Simon & Schuster.
United States Chamber of Commerce/Center for International Private Enterprise. 1996, January 4 Hypertext extension to "Generating Jobs, Raising Wages" (WWW document). http://www.cipe.org/e17/laborC_3_95.html
Van Maanen, John. 1983. "The Fact of Fiction in Organizational Ethnography." In *Qualitative Methodology*, ed. John Van Maanen, Newbury Park, Calif.: Sage.
———. 1988 *Tales of the Field: On Writing Ethnography*. Chicago: University of Chicago Press.
Verran, Helen. 2001. *Science and an African Logic*. Chicago: University of Chicago Press.
Vygotsky, Lev. 1986. *Thought and Language*. Cambridge: MIT Press.
Wagner, Ina. 1989. "Restructuring Hospital Work: Negotiable Issues for Nurses and Radiology Assistants." *Office, Technology, and People.*
Weinberg, G.M. 1971. *The Psychology of Computer Programming*. New York: Van Nostrand.
Weizenbaum, Joseph. 1976. *Computer Power and Human Reason: From Judgement to Calculation*. New York: W. H. Freeman.
Wellman, Barry, and Caroline Haythornthwaite, eds. 2003. *The Internet in Everyday Life* (Information Age Series). Oxford: Blackwell.
Wenger, Etienne, William M. Snyder, and Richard McDermott. 2002. *Cultivating Communities of Practice: A Guide to Managing Knowledge*. Boston: Harvard Business School Publishing.
Werle, Raymond. 1998. Call for Proposals, IVth Conference of the European Sociological Association: Will Europe Work? SCI-TECH-STUDIES listserv 12/8/98.
Whitehead, Alfred North. 1925. *Science and the Modern World*. New York: New American Library.
———. 1933. *Adventures of Ideas: A Brilliant History of Mankind's Great Thoughts*. New York: New American Library.
Whitehead, Alfred North, and Bertrand Russell. 1927. *Principia Mathematica*. Cambridge: Cambridge University Press.
Whyte, William Foote. 1991. *Participatory Action Research*. Newbury Park, Calif.: Sage Publications.
Williams, Bernard. 1985. *Ethics and the Limits of Philosophy*. Cambridge: Harvard University Press.
Williams, Michael. 1999. *Groundless Belief: An Essay on the Possibility of Epistemology*. 2d ed. Princeton: Princeton University Press.
Williams, Sam. 2002. *Free as in freedom: Richard Stallman's Crusade for Free Software*. Sebastapol, Calif.: O'Reilly.
Winner, Langdon. 1977 *Autonomous Technology*. Cambridge: MIT Press.
———. 1980. "Do Artifacts have Politics?" *Daedalus* 109: 121–33.
———. 1984. "Mythinformation in the High Tech Era." *IEEE Spectrum* 21 (6): 90–96.
———. 1994. "How to Criticize Technology without Becoming a Luddite." Paper presented at the Annual Conference, Society for Social Studies of Science, New Orleans.
Winograd, Terry, and F. Flores. 1986. *Understanding Computers and Cognition: A New Foundation for Design*. Norwood, N.J.: Ablex.
Wood, Stephen, ed. 1982. *The Degradation of Work?* London: Hutchinson.
Woolley, Benjamin. 1992. *Virtual Worlds: A Journey in Hyped Hyperreality*. Oxford: Blackwell.
Worsley, Peter. 1997. *Knowledges: Culture, Counterculture, Subculture*. New York: The New Press.
Young, Iris. 2000. *Inclusion and Democracy*. Oxford: Oxford University Press.
Young, Michael. 1994. *The Rise of the Meritocracy*. Somerset, N.J.: Transaction Publishers.
Zimbalist, Andrew, ed. 1979. *Case Studies on the Labor Process*. New York: Monthly Review Press.

Index